Materials Science and Nanotechnology

Materials Science and Nanotechnology

Edited by **Andrew Green**

WILLFORD PRESS

New York

Published by Willford Press,
118-35 Queens Blvd., Suite 400,
Forest Hills, NY 11375, USA
www.willfordpress.com

Materials Science and Nanotechnology
Edited by Andrew Green

International Standard Book Number: 978-1-68285-118-0 (Hardback)

Printed in the United States of America.

Contents

Preface

This profound book on materials science and nanotechnology focuses upon the design and discovery of various nanomaterials and nanocomposites. It consists of studies provided by internationally renowned experts from the fields of nanotechnology and materials science. The topics elucidated in this book include various processes and techniques for formation of different nanomaterials and composites, their structures, properties and applications. It brings forth new concepts and applied aspects for further research and discussion. The chapters included herein, with their detailed analyses and data, will prove immensely beneficial to professionals and students involved in these fields at various levels.

The information shared in this book is based on empirical researches made by veterans in this field of study. The elaborative information provided in this book will help the readers further their scope of knowledge leading to advancements in this field.

Finally, I would like to thank my fellow researchers who gave constructive feedback and my family members who supported me at every step of my research.

Editor

One-Step Formation of WO_3-Loaded TiO_2 Nanotubes Composite Film for High Photocatalytic Performance

Wai Hong Lee [†], Chin Wei Lai * and Sharifah Bee Abd Hamid [†]

Nanotechnology & Catalysis Research Centre (NANOCAT), Institute of Postgraduate Studies (IPS), University of Malaya, Kuala Lumpur 50603, Malaysia;
E-Mails: leewaihong@siswa.um.edu.my (W.H.L.); sharifahbee@um.edu.my (S.B.A.H.)

[†] These authors contributed equally to this work.

* Author to whom correspondence should be addressed; E-Mail: cwlai@um.edu.my;

Academic Editor: Klara Hernadi

Abstract: High aspect ratio of WO_3-loaded TiO_2 nanotube arrays have been successfully synthesized using the electrochemical anodization method in an ethylene glycol electrolyte containing 0.5 wt% ammonium fluoride in a range of applied voltage of 10–40 V for 30 min. The novelty of this research works in the one-step formation of WO_3-loaded TiO_2 nanotube arrays composite film by using tungsten as the cathode material instead of the conventionally used platinum electrode. As compared with platinum, tungsten metal has lower stability, forming dissolved ions (W^{6+}) in the electrolyte. The W^{6+} ions then move towards the titanium foil and form a coherent deposit on titanium foil. By controlling the oxidation rate and chemical dissolution rate of TiO_2 during the electrochemical anodization, the nanotubular structure of TiO_2 film could be achieved. In the present study, nanotube arrays were characterized using FESEM, EDAX, XRD, as well as Raman spectroscopy. Based on the results obtained, nanotube arrays with average pore diameter of up to 74 nm and length of 1.6 μm were produced. EDAX confirmed the presence of tungsten element within the nanotube arrays which varied in content from 1.06 at% to 3.29 at%. The photocatalytic activity of the nanotube arrays was then investigated using methyl orange degradation under TUV 96W UV-B Germicidal light irradiation. The nanotube with the highest aspect ratio, geometric surface area factor and at% of tungsten exhibited the highest photocatalytic activity due to more photo-induced electron-hole pairs generated by the larger surface area and because WO_3 improves charge separation, reduces charge carrier recombination and increases charge carrier lifetime via accumulation of electrons and holes in the two different metal oxide semiconductor components.

Keywords: WO_3-loaded TiO_2 nanotubes; electrochemical anodization; anodization voltage; photocatalytic degradation; active surface area

---

1. Introduction

The process of dyeing in the textile industry has resulted in the production of large amounts of wastewater with intense coloration which has to be eliminated before release into natural water streams. If left untreated, such dyes will remain in the environment for an extended period of time and cause serious environmental and health problems. Therefore, such compounds must be completely removed from aquatic system [1,2].

Titanium dioxide (TiO_2) is one of the most widely studied transition metal oxide semiconductor and has been widely applied in solar cells, hydrogen generation, gas sensing, and photocatalysis applications [3–5]. One of the most widely researched and an important application of TiO_2 photocatalyst is pollution treatment. The effectiveness of TiO_2 in these applications is further complimented by its unique properties of non-toxicity, cost effective, long-term stability, widespread availability, corrosion stability, and high photocatalytic ability. However, researchers have shown that TiO_2 nanotubes are only able to utilize around 2%–3% solar light that reaches the earth due to a large band gap of 3.20 eV [6]. Therefore, the doping of TiO_2 nanostructures with transition metals to enable the TiO_2 nanostructures to react to a much larger visible region is currently widely researched.

In this present study, tungsten trioxide (WO_3) was selected as potential dopant to decorate the pure TiO_2 nanotubes. The reason is mainly attributed to the WO_3 with a smaller band gap of 2.3–2.8 eV (440 to 540 nm), which is advantageous for visible-light-driven photocatalysis [7]. Furthermore, the upper edge of the valence band and the lower edge of the conduction band are lower for WO_3 than for TiO_2. These differences in band edge positions create a potential gradient at the composite interface. This facilitates better charge separation and inhibits charge carrier recombination. Also, the properties and performance of the nanotubes are highly dependent on the dimensions of the nanotubes [8–10]. To optimize the properties and performance of the nanotubes, anodizing conditions such as applied potential, anodization time, and electrolyte composition can be tailored to control the dimensions of the nanotubes such as length, pore diameter and wall thickness [11–15]. The morphology and structure of the nanotubes layer are affected strongly by the electrochemical conditions, especially the anodization voltage, as it is the key factor controlling the tube diameter. Generally, nanotubes growth occurs proportional to the applied potential up to a voltage where dielectric breakdown of the oxide occurs [16]. Tube diameter is affected by voltage but it is not affected by time. Instead, the time of anodization influences the thickness of the nanotube layer [17]. Thus, the influence of anodization voltage on the formation of WO_3-loaded TiO_2 nanotube arrays was investigated in this study with the aim to fabricate nanotubes with optimum length, wall thickness, and pore diameter for better photocatalysis application. The nanotube diameter is expected to increase with increasing voltage due to higher field assisted oxidation rate of Ti metal to form TiO_2 layer and field assisted dissolution rate of Ti metal ions in the electrolyte [18].

2. Results and Discussion

2.1. Transient and Steady State Current Density Analysis

Figure 1 shows the current density curve for a WO_3-loaded TiO_2 nanotubes sample produced at 40 V and anodized for 30 min in electrolyte containing 0.5 g NH_4F. About 5 min after application of the voltage, the measured current density reduced from about 49 mA/cm^2 to around 15 mA/cm^2, point P2 on the plot. The reduced current density resulted from the field-assisted oxidation of the Ti metal surface, which forms a compact oxide layer [18,19]. The reaction occurred is represented by the equation below:

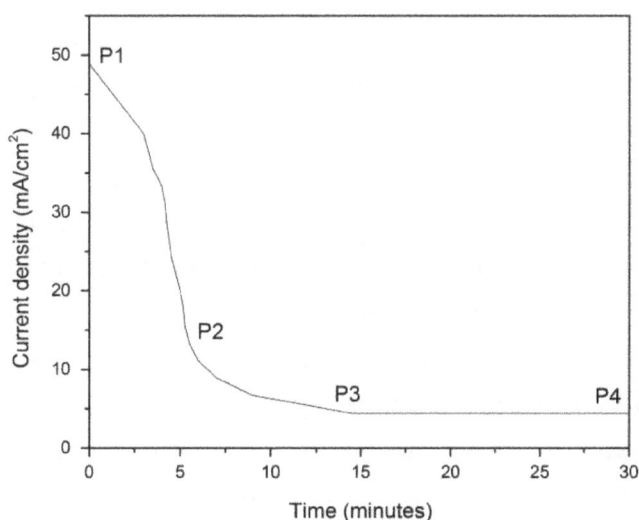

$$Ti^{2+} + H_2O_2 \rightarrow TiO_2 + 2H^+ \tag{1}$$

Figure 1. Anodization current behavior of tungsten trioxide (WO_3)-loaded titanium dioxide (TiO_2).

Region P2 to P3 represents the field-assisted dissolution of the oxide layer caused by high electric field across the thin layer. The current gradually drops with a corresponding increase in porous structure depth. Fine pits or cracks form on the oxide surfaces which arise from chemical and field-assisted dissolution of the oxide at local points of high energy. The reduced oxide layer thickness at these points decreases the current density [18,20,21]. The equation below represents the reaction that occurred:

$$TiO_2 + 4H^+ + 6F^- \rightarrow [TiF_6]^{2-} + 2H_2O \tag{2}$$

Point P3 shows the transition between the porous and nanotube structures. Nanotube array length increases to point P4 after which the current is cut off and the reaction is ended.

The current density curves for experiments conducted at 10 V, 20 V and 30 V are not plotted since they follow the same path as the 40 V experiment (within reasonable error). The formation of WO_3-loaded TiO_2 nanotubes is illustrated in Figure 2.

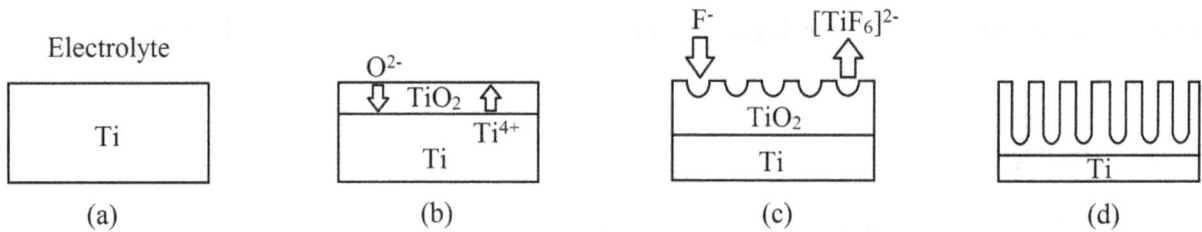

Figure 2. Formation of WO₃-loaded TiO₂ nanotubes: (**a**) Ti foil; (**b**) oxide layer formation; (**c**) chemical dissolution of oxide layer and (**d**) titania nanotubes.

The equation below represents the formation of WO₃ species for the synthesis of anodic WO₃-loaded TiO₂ nanostructure:

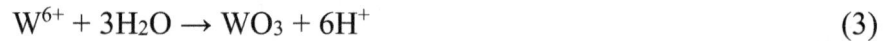

$$W^{6+} + 3H_2O \rightarrow WO_3 + 6H^+ \tag{3}$$

2.2. Morphological Studies and Elemental Analysis

The effect of anodization voltage on the morphology of anodic WO₃-loaded TiO₂ nanostructure was investigated. Figure 3 showed the surface morphologies of anodic WO₃-loaded TiO₂ layer of different anodization voltage from 10 V to 40 V. All samples were anodized for 30 min in electrolyte composed of EG, NH₄F and H₂O₂. Anodization voltage of 10 V produced nanotube arrays with smallest average pore's diameter of 47 nm and shortest length of approximately 0.9 µm. At anodization voltage of 20 V, nanotube arrays with average pore's diameter of 56 nm and length of approximately 1.2 µm were produced. As anodization voltage is increased to 30 V, the average pore's diameter and length of the nanotube arrays also increased to 65 nm and 1.4 µm respectively. Anodization voltage of 40 V produced nanotube arrays with the longest tube length of approximately 1.6 µm and largest average pore's size of 74 nm. The average diameter, length, wall thickness, aspect ratio (*AR*), and geometric surface area factor (*G*) of the nanotubes anodized at different applied voltage are summarized in Table 1. The aspect ratio and geometric area factor were calculated as follows:

$$AR = L/(D + 2w) \tag{4}$$

$$G = [4\pi L\,(D + w)]/[\sqrt{3}\,(D + 2w)^2] + 1 \tag{5}$$

where L = nanotube length in nm; D = pore size; w = wall thickness.

The diameter and length of nanotubes were found to increase with anodization voltage up to 40 V because of the high electric field dissolution at the barrier layer of nanotubes [18]. At low potential (10 V), the low field assisted oxidation rate and field-assisted dissolution rate during the anodization process resulted in small diameter of pores. Thus, short and small nanotubular structures were formed. However, at higher potential, these small nanotubular structures were then etched into larger pores due to the higher field assisted oxidation and dissolution rate. Higher voltage will provide higher driving force for ionic species (H⁺, F⁻, and O²⁻) to move through the barrier layer at the bottom of the nanotube, which allows for faster movement of the Ti/TiO₂ interface into the Ti metal [18,22]. Nanotube arrays with longer length will be produced from this improved pore deepening process.

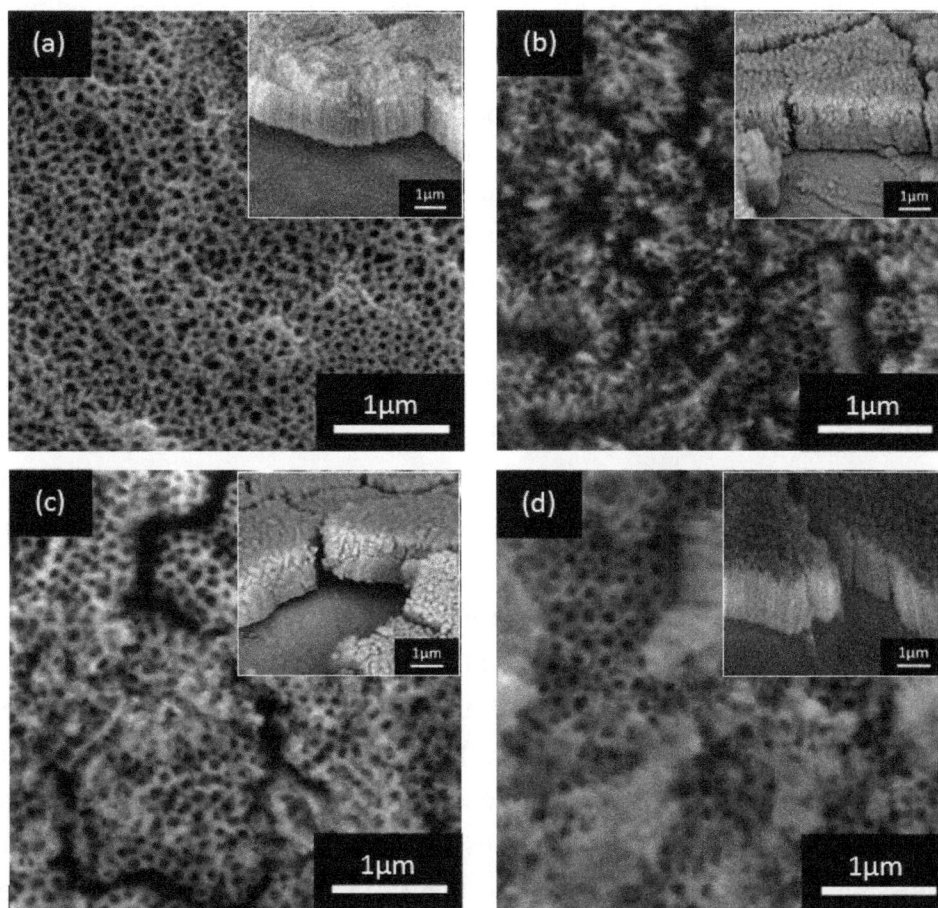

Figure 3. Field emission scanning electron microscopy (FESEM) images of WO$_3$-loaded TiO$_2$ nanotubes obtained for different anodization voltage at: (**a**) 10 V; (**b**) 20 V; (**c**) 30 V and (**d**) 40 V. Insets are the side views of the samples.

Table 1. Pore's diameter, length, wall thickness, aspect ratio, and geometric surface area factor of WO$_3$-loaded TiO$_2$ nanotubes formed with varying anodizing voltage.

Voltage (V)	Diameter (nm)	Length (μm)	Wall Thickness (nm)	AR	G
10	47	0.9	13	12.33	74.52
20	56	1.2	15	13.95	84.58
30	65	1.4	17	14.14	85.98
40	74	1.6	18	14.55	89.26

The quantitative elemental analysis of WO$_3$-loaded TiO$_2$ nanotubes was carried out by FESEM-EDAX and the average elemental compositions (at%) were obtained by taking eight spots in EDAX analysis. The percentage of each element is shown in Table 2. The WO$_3$-loaded TiO$_2$ nanotubes show the presence of Ti, O, W and C elements. Sample anodized at 40 V shows the highest at% of W which is 3.29 at%. The samples anodized at 30 V and 20 V showed 2.01 at% and 1.36 at% of W, respectively. The sample anodized at 10 V showed the lowest at% of W which is 1.06 at%. The presence of W within the nanotube arrays was found to increase with anodization voltage. This is because increasing voltage will increase the strength of electric field in the electrolyte solution, thereby increasing the mobility and rate of migration of W^{6+} ions towards the titanium foil [23]. Therefore, at higher anodization voltage, more W will be incorporated into the TiO$_2$ nanotubes.

Table 2. Energy-dispersive X-ray elemental analysis of WO_3-loaded TiO_2 nanotubes.

Samples	Atomic %			
	Ti	O	W	C
10 V	43.81	51.79	1.06	3.34
20 V	48.16	46.95	1.36	3.53
30 V	59.22	31.83	2.01	6.94
40 V	47.38	44.11	3.29	5.22

2.3. Phase Structure Analysis

Figure 4 is an XRD profile of the WO_3-loaded TiO_2 nanotubes after annealing at 400 °C in air atmosphere for 4 h. The result shows the presence of TiO_2 with anatase phase [JCPDS No. 21-1272]. The diffraction peaks at 25.37°, 38.67°, 48.21°, and 54.10° are corresponding to (101), (112), (200), and (105) crystal planes for the anatase phase, respectively. Additionally, for the sample synthesized at 40 V, small additional peaks at 23.62° and 29.16° corresponds with the (020) and (120) crystal planes of the monoclinic WO_3 phase. The intensity of the (101) peak at 25.37° increased with increasing anodization voltage, indicating the increased crystallinity of anatase phase. This increase in anatase intensity is due to more growth of TiO_2 nanotubes as voltage is further increased. Furthermore, TiO_2 layer formed at higher voltages are thicker and denser, resulting in higher anatase intensity [24]. However, the XRD patterns did not show any obvious WO_3 phase for samples synthesized at 10 V, 20 V and 30 V. A possible explanation would be that the XRD analysis was not sensitive enough to detect very low WO_3 content (<3 at% from EDX analysis) within the TiO_2 lattice due to the nearly similar ionic radius of W^{6+} and Ti^{4+} [25,26].

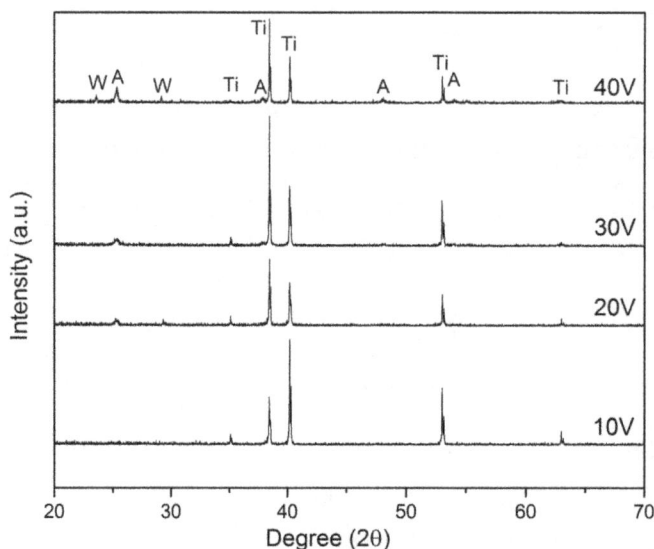

Figure 4. X-ray diffraction patterns of WO_3-loaded TiO_2 nanotubes produced at different anodization voltage.

2.4. Raman Analysis

Raman analysis was conducted to detect the presence of WO_3 and to confirm the XRD inferences of WO_3-loaded TiO_2 nanotubes. Figure 5 is the Raman spectrums obtained which shows five characteristic modes at 146, 198, 396, 516, and 640 cm^{-1}. The mode at 146 cm^{-1} is strong and assigned as the E_g phonon of the anatase structure and B_{1g} phonon of the rutile structure. The latter four modes are assigned as E_g, B_{1g}, B_{1g}, and E_g modes of the anatase phase, respectively. The positions and intensities of the five Raman active modes correspond well with the anatase phase of TiO2 [27–29]. The Raman spectrums show increasing intensity of peaks from 10 V to 40 V. Higher intensity of peaks corresponds to higher crystallinity [29]. The increase in anatase intensity from 10 V to 40 V is due to more growth of TiO2 nanotubes as voltage is increased. Furthermore, TiO2 layer formed at higher voltages are thicker and denser, resulting in higher anatase intensity [19]. However, Raman bands for WO3 was not detected because typical characteristic modes for WO3 are similar to those for anatase (e.g., 327, 714, and 804 cm^{-1}) and were overlapped by bands for the anatase phase [30].

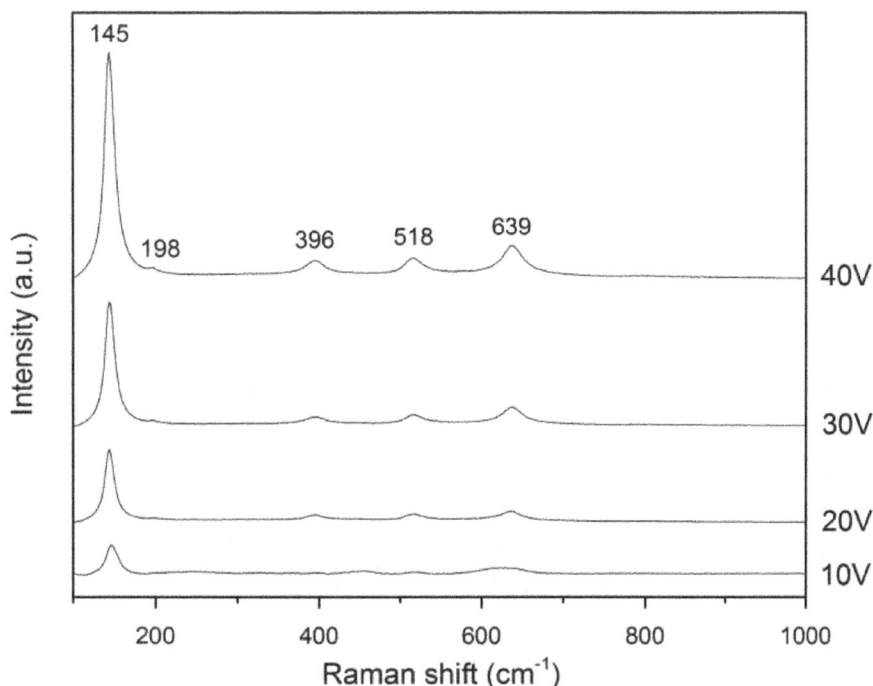

Figure 5. Raman spectrum of WO3-loaded TiO2 produced at different anodization voltage.

2.5. Photocatalytic Activity

The photocatalytic ability of the WO3-loaded TiO2 nanotube arrays was tested through the MO degradation under UV light irradiation. As shown in Figure 6, the nanotube arrays produced at 40 V presented the highest degradation percentages for the decomposition of MO where only 12% of initial MO concentration remained after 4 h of UV irradiation. Nanotube arrays produced at 30 V and 20 V showed lower efficiency of MO decomposition, where the MO concentration was reduced to 20% and 25% respectively. The sample anodized at 10 V showed the lowest activity, where the MO concentration was only reduced to 31% of initial MO concentration. WO3-loaded TiO2 nanotube arrays anodized at 40 V having the highest aspect ratio and geometric surface area factor exhibited the highest photocatalytic activity among the samples due to the larger active surface area to generate more photo-induced electron-hole pairs. The photoresponse of the WO3-loaded TiO2 nanotubes is affected by the nanotubes' length where longer tubes provide higher total light absorption. Also, with the larger surface area, more reactants can be adsorbed onto the inner and outer TiO2 nanotube surfaces and thus result in higher photocatalytic activity [3,11]. In order to compare the photocatalytic activity of WO₃-loaded TiO₂ nanotubes with pure TiO₂ nanotubes, pure TiO₂ nanotubes were produced using the same parameters as the WO₃-loaded TiO₂ nanotubes anodized at 40 V except replacing the tungsten cathode with a platinum cathode. As compared to WO₃-loaded TiO₂ nanotubes, pure TiO₂ nanotube arrays showed a lower efficiency of MO decomposition, where the MO concentration was reduced to 28% of initial MO concentration after 4 h. This shows that the coupling of WO₃ and TiO₂ gives significant improvement in the photocatalytic activity of the nanotube arrays due to suppression of the recombination of the photogenerated carriers and increased charge separation of TiO₂ [31–33].

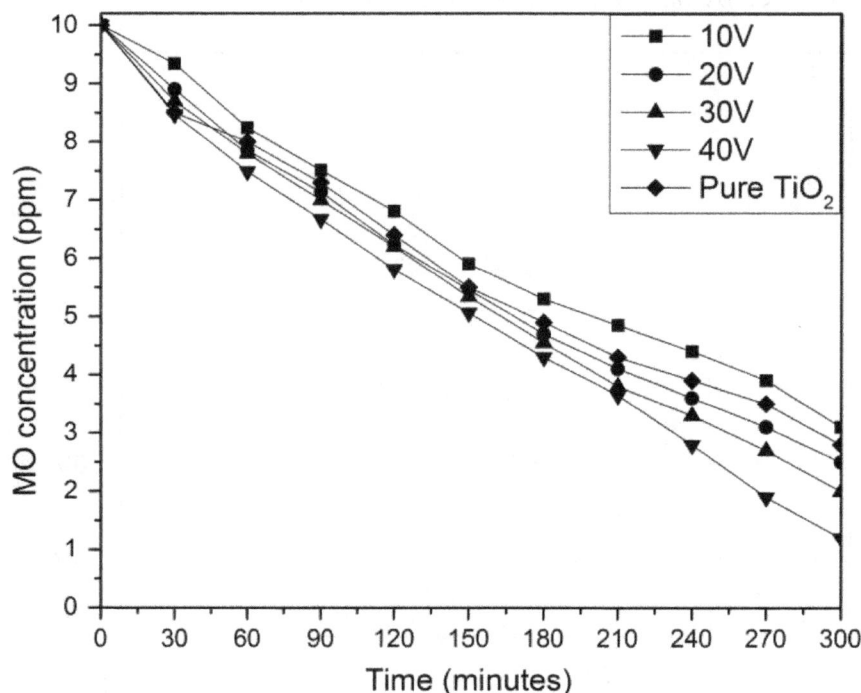

Figure 6. Photodegradation of methyl orange (MO) solution by WO₃-loaded TiO₂ nanotubes anodized at different voltage.

The kinetics analysis of MO degradation is illustrated in Figure 7. The linear curves suggests that the photocatalytic degradation of MO can be described by the first order kinetic model, $\ln(C_0/C) = kt$, where C_0 is the initial concentration and C is the concentration at time t. The plots of the concentration data gave a straight line. The results of fitting experimental data to pseudo-first-order kinetics are given in Table 3. The rate constant increases with increasing anodization voltage. This shows that the WO₃-loaded TiO₂ nanotubes anodized at 40 V demonstrated the best photocatalytic activity for the degradation of MO among the samples produced.

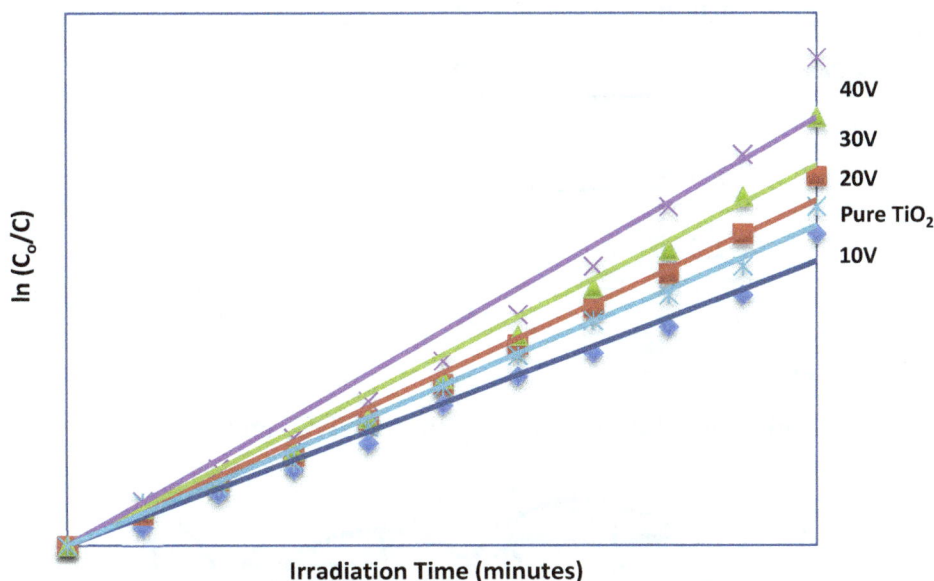

Figure 7. Pseudo-first-order kinetics for methyl orange photodegradation using WO$_3$-loaded TiO$_2$ nanotubes anodized at different voltage and pure TiO$_2$ nanotubes.

Table 3. Rate constants for catalytic photodegradation of MO.

Samples	Rate Constant (k)	R^2
10 V	0.0036	0.9874
20 V	0.0043	0.9920
30 V	0.0048	0.9759
40 V	0.0054	0.9715
Pure TiO$_2$	0.0040	0.9927

Figure 8 shows the energy band diagram of TiO$_2$ and WO$_3$. As shown in Figure 9, UV light radiation excites electrons from the valence band to the conduction band which results in electrons and holes separation. When the electrons and holes reach the semiconductor-environment interface, they will react with appropriate redox species (H$_2$O and O$_2$) to form reactive intermediates (OH• and O$_2$•). These radicals and photogenerated holes are extremely strong oxidants which are able to oxidize all organic materials to CO$_2$ and H$_2$O, leading to the degradation of MO solution [3]. The coupling of TiO$_2$ and WO$_3$ can lead to electron and hole transfer from one semiconductor particle to another upon light excitation [31]. The valence and conduction band potentials of TiO$_2$ are more cathodic than that of WO$_3$. Thus, photogenerated electrons can transfer from the conduction band of TiO$_2$ down to the conduction band of WO$_3$. This suppresses the recombination of the photogenerated carriers, leading to increased photo-oxidation efficiency [32]. The lower band gap of WO$_3$ also increases the charge separation of TiO$_2$ and extends the energy range of photoexcitation of the system. If a photon with not enough energy to excite TiO$_2$ but is of enough energy to excite WO$_3$ is incident, the hole that is created in the WO$_3$ valence band is excited to the conduction band of TiO$_2$, while the electron is transferred to the conduction band of TiO$_2$. It is this electron transfer that increases the charge separation and increases the efficiency of the photocatalytic process [33].

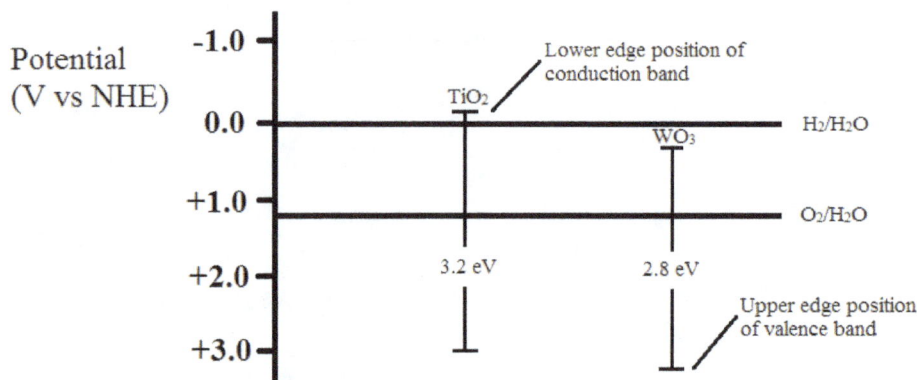

Figure 8. Energy band diagram of TiO_2 and WO_3.

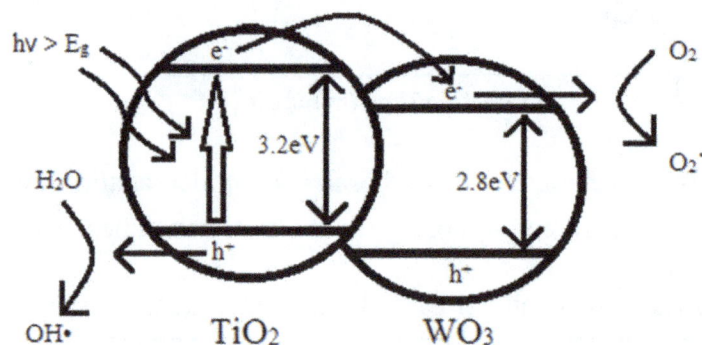

Figure 9. Photocatalytic mechanism of WO_3-loaded TiO_2 under UV light irradiation.

2.6. Optical Properties Analysis

The determination of the energy band gap of the WO_3-loaded TiO_2 nanotubes is a key point for application purpose. To investigate the optical properties of the WO_3-loaded TiO_2 nanotubes, we have performed photoluminescence (PL) analysis on the sample that showed the best performance in the photocatalytic activity test. The PL emission spectrum is a useful characterization tool which can be used to test the opticl properties of the nanocomposite. The band gap energy (E_{bg}) of the sample is calculated as follows: $E_{bg} = hc/\lambda$, where E_{bg} is the band gap energy, h is Planck's constant (4.135667×10^{-15} eVs), c is the velocity of light (2.997924×10^8 m/s), and λ is the wavelength (nm) of PL emission. In the photoluminescence spectra, the wavelength corresponding to the highest PL emission intensity is the light wavelength at which the sample is most active. By taking this wavelength value as λ, the energy band gap of the sample can be estimated. Figure 10 shows the photoluminescence spectra for WO_3-loaded TiO_2 nanotubes anodized at 40 V. From this photoluminescence spectrum, the sample shows the highest PL emission intensity at wavelength of 580 nm. By taking this wavelength into account, we estimate the energy band gap of the sample to be 2.14 eV. This band gap value is much lower than that of WO_3 alone (2.8 eV), attributed to the presence of carbon species within the TiO_2 nanotubes. Previous studies have also shown that carbon can be doped onto TiO_2 nanotubes from organic electrolyte such as ethylene glycol during anodization [34–36]. The presence of carbon significantly enhanced the visible light responsiveness of the WO_3-loaded TiO_2 nanotubes because the mixing of the delocalized p state of the carbon dopants with O 2p orbital in valence band of TiO_2 will shift the valence band edge of TiO_2 upwards, thus narrowing down the band gap energy of TiO_2.

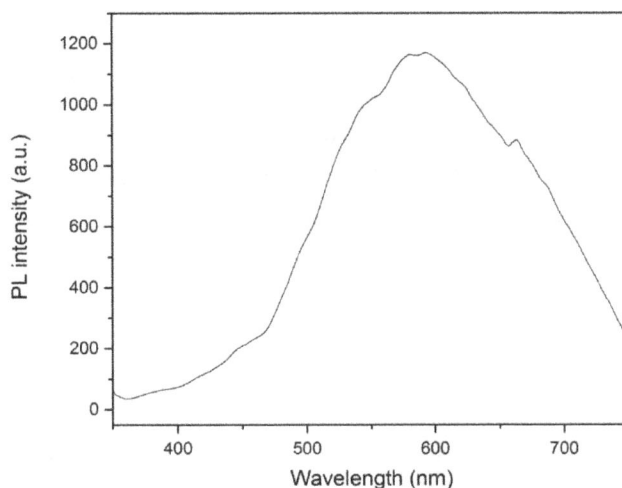

Figure 10. Smooth photoluminescence (PL) curve for WO$_3$-loaded TiO$_2$ nanotubes anodized at 40 V.

3. Experimental Section

The experiments were carried out in a two electrodes electrochemical cell as shown in Figure 11, where the two electrodes were placed 2 cm apart. Titanium (Ti) foil (0.127 mm, purity 99.6%, Sigma Aldrich, St. Louis, MO, USA) (5 cm × 1 cm dimension) over which WO$_3$-loaded TiO$_2$ nanotubes were grown was used as anode while tungsten foil (0.127 mm, purity 99.9%, Sigma Aldrich, St. Louis, MO, USA) was the counter electrode. The electrolytes were 0.5 wt% ammonium fluoride (NH$_4$F, Merck, Kenilworth, NJ, USA) dissolved in anhydrous ethylene glycol (EG, Friendemann Schmidt, Germany) and hydrogen peroxide (H$_2$O$_2$, Friendemann Schmidt, Germany). The function of H$_2$O$_2$ is to replace H$_2$O as oxygen provider to increase the oxidation rate for synthesizing highly ordered and smooth TiO$_2$ nanotubes at a rapid rate [37]. Anodization was carried out in a range of anodization voltage of 10–40 V. The anodization period was restricted to only 30 min, which was a typical time observed for growth of 1 μm long TiO$_2$ nanotubes [3,4]. As-anodized anodic WO$_3$-loaded TiO$_2$ samples were cleaned using deionized water followed by sonication in acetone (Friendemann Schmidt, Germany) to remove the remaining occluded ions from the anodized solutions or barrier oxide layer. The samples were then subjected to calcination at 400 °C for 4 h in air atmosphere.

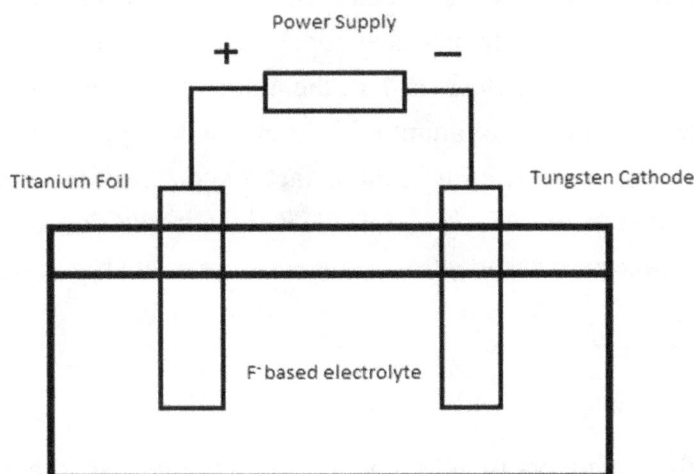

Figure 11. Schematic drawing of an electrochemical cell in which the Ti electrode is anodized.

The morphologies of anodic WO₃-loaded TiO₂ nanostructures were observed by field emission scanning electron microscopy (FESEM, FEI Quanta 200F Environmental SEM with EDAX, Hillsboro, OR, USA) microanalysis at 5 kV. The structural variations measurement and phase determination were done using X-ray diffraction (XRD, Bruker D8 Advance diffractometer, Billerica, MA, USA) analysis conducted from 10 to 80 with Cu Kα radiation ($\alpha = 1.5406$ Å). The phase composition was determined using Raman Spectroscopy (Renishaw inVia, Renishaw plc, Gloucestershire, UK) with a 514.5 nm Ar$^+$ laser as an excitation source.

Photocatalytic degradation studies were performed by dipping sintered sample in 100 mL of 10 ppm methyl orange (MO) solution in a photoreactor consisting of quartz glass, as shown in Figure 12. After leaving the samples in the reactor for 30 min in dark environment for dark adsorption, the samples were photoirradiated at room temperature by using TUV 96W UV-B Germicidal light. To monitor the degradation of methyl orange (MO) after UV irradiation, 5 mL solution was withdrawn from quartz tubes for every 30 min. A UV spectrometer was used to measure the concentration of the degraded MO solution.

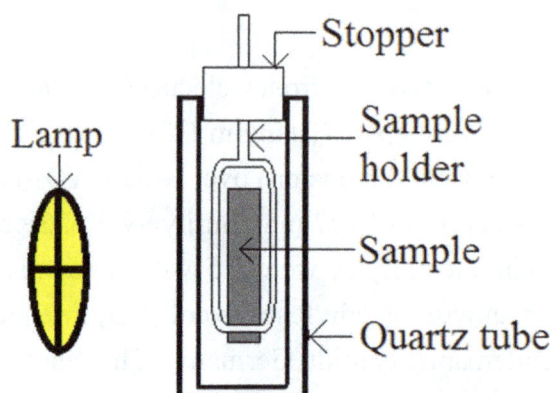

Figure 12. Schematic diagram of photocatalytic reactor in which photocatalytic degradation was performed.

4. Conclusions

In this study, the effect of anodization voltage on the formation of WO₃-loaded TiO₂ nanotube arrays using single step anodization was performed. WO₃-loaded TiO₂ nanotube arrays were successfully produced at 10 V, 20 V, 30 V and 40 V. The nanotube arrays anodized at 40 V produced the largest pore's size (74 nm) and longest tube length (1.6 μm). Besides that, the amount of tungsten in the nanotube arrays increased with anodization voltage up to maximum of 3.29 at%. Clearly, WO₃-loaded TiO₂ nanotube arrays with the highest aspect ratio, geometric surface area factor and at% of tungsten exhibited the more favorable photocatalytic degradation of MO dye under UV light irradiation due to the larger active surface area to generate more photo-induced electron-hole pairs, better charge separation and less charge carrier recombination.

Acknowledgments

The authors would like to thank University of Malaya for funding this research work under University of Malaya Research Grant (UMRG), (RP022-2012D) and Fundamental Research Grant Scheme (FRGS), (FP055-2013B).

Author Contributions

Wai Hong Lee and Chin Wei Lai designed the experiments. Wai Hong Lee carried out the anodization and sample preparations. Wai Hong Lee, Chin Wei Lai and Sharifah Bee Abd Hamid carried out analysis of FESEM-EDAX, XRD, Raman, photoluminescence and UV spectrometer data. Wai Hong Lee, Chin Wei Lai and Sharifah Bee Abd Hamid prepared the manuscript.

Conflicts of Interest

The authors declare no conflict of interest.

References

1. Olukanni, O.D.; Osuntoki, A.A.; Gbenle, G.O. Textile effluent biodegradation potentials of textile effluent-adapted and non-adapted bacteria. *Afr. J. Biotechnol.* **2006**, *5*, 1980–1984.
2. Palamthodi, S.; Patil, D.; Patil, Y. Microbial degradation of textile industrial effluents. *Afr. J. Biotechnol.* **2013**, *10*, 12657–12661.
3. Roy, P.; Berger, S.; Schmuki, P. TiO2 nanotubes: Synthesis and applications. *Angew. Chem. Int. Ed.* **2011**, *50*, 2904–2939.
4. Mohapatra, S.; Misra, M.; Mahajan, V.K.; Raja, K.S. A novel method for the synthesis of titania nanotubes using sonoelectrochemical method and its application for photoelectrochemical splitting of water. *J. Catal.* **2007**, *246*, 362–369.
5. Liao, J.Y.; Lei, B.X.; Wang, Y.F.; Liu, J.M.; Su, C.Y.; Kuang, D.B. Hydrothermal fabrication of quasi-one-dimensional single-crystalline anatase TiO2 nanostructures on FTO glass and their applications in dye-sensitized solar cells. *Chem.-A Eur. J.* **2011**, *17*, 1352–1357.
6. Kitano, M.; Matsuoka, M.; Ueshima, M.; Anpo, M. Recent developments in titanium oxide-based photocatalysts. *Appl. Catal. A: Gen.* **2007**, *325*, 1–14.
7. Higashimoto, S.; Ushiroda, Y.; Azuma, M. Electrochemically assisted photocatalysis of hybrid WO3/TiO2 films: Effect of the WO3 structures on charge separation behavior. *Top. Catal.* **2008**, *47*, 148–154.
8. Liu, Z.; Zhang, X.; Nishimoto, S.; Jin, M.; Tryk, D.A.; Murakami, T.; Fujishima, A. Highly ordered TiO2 nanotube arrays with controllable length for photoelectrocatalytic degradation of phenol. *J. Phys. Chem. C* **2008**, *112*, 253–259.
9. Lai, Y.; Sun, L.; Chen, Y.; Zhuang, H.; Lin, C.; Chin, J.W. Effects of the structure of TiO2 nanotube array on Ti substrate on its photocatalytic activity. *J. Electrochem. Soc.* **2006**, *153*, D123–D127.
10. Mohamed, A.E.R.; Rohani, S. Synthesis of Titania nanotube arrays by anodization. *AIDIC Conf. Ser.* **2009**, *9*, 121–129.
11. Chen, X.; Mao, S.S. Titanium dioxide nanomaterials: Synthesis, properties, modifications, and applications. *Chem. Rev.* **2007**, *107*, 2891–2959.
12. Baker, D.R.; Kamat, P.V. Disassembly, reassembly, and photoelectrochemistry of etched TiO2 nanotubes. *J. Phys. Chem. C* **2009**, *113*, 17967–17972.
13. Wang, J.; Lin, Z. Anodic formation of ordered TiO2 nanotube arrays: Effects of electrolyte temperature and anodization potential. *J. Phys. Chem. C* **2009**, *113*, 4026–4030.

14. Kang, S.H.; Kim, J.Y.; Kim, H.S.; Sung, Y.E. Formation and mechanistic study of self-ordered TiO_2 nanotubes on Ti substrate. *J. Ind. Eng. Chem.* **2008**, *14*, 52–59.

15. Raja, K.S.; Gandhi, T.; Misra, M. Effect of water content of ethylene glycol as electrolyte for synthesis of ordered titania nanotubes. *Electrochem. Commun.* **2007**, *9*, 1069–1076.

16. Macak, J.M.; Tsuchiya, H.; Ghicov, A.; Yasuda, K.; Hahn, R.; Bauer, S.; Schmuki, P. TiO_2 nanotubes: Self-Organized electrochemical formation, properties and applications. *Solid State Mater. Sci.* **2007**, *11*, 3–18.

17. Minagar, S.; Berndt, C.C.; Wang, J.; Ivanova, E.; Wen, C. A review of the application of anodization for the fabrication of nanotubes on metal implant surfaces. *Acta Biomater.* **2012**, *8*, 2875–2888.

18. Lai, C.W.; Sreekantan, S. Effect of applied potential on the formation of self-organized TiO_2 nanotube arrays and its photoelectrochemical response. *J. Nanomater.* **2011**, *2011*, doi:10.1155/2011/142463.

19. Song, Y.Y.; Gao, Z.D.; Wang, J.H.; Xia, X.H.; Lynch, R. Multistage coloring electrochromic device based on TiO_2 nanotube arrays modified with WO_3 nanoparticles. *Adv. Funct. Mater.* **2011**, *21*, 1941–1946.

20. Grimes, C.A. Synthesis and application of highly ordered arrays of TiO_2 nanotubes. *J. Mater. Chem.* **2007**, *17*, 1451–1457.

21. Paulose, M.; Shankar, K.; Yoriya, S.; Prakasam, H.E.; Varghese, O.K.; Mor, G.K.; Latempa, T.A.; Fitzgerald, A.; Grimes, C.A. Anodic growth of highly ordered TiO_2 nanotube arrays to 134 µm in length. *J. Phys. Chem. B* **2006**, *110*, 16179–16184.

22. Crawford, G.; Chawla, N. Porous hierarchical TiO_2 nanostructures: Processing and microstructure relationships. *Acta Mater.* **2009**, *57*, 854–867.

23. Issaq, H.J.; Atamna, I.Z.; Muschik, G.M.; Janini, G.M. The effect of electric field strength, buffer type and concentration on separation parameters in capillary zone electrophoresis. *Chromatographia* **1991**, *32*, 155–161.

24. Park, I.S.; Woo, T.G.; Lee, M.H.; Ahn, S.G.; Park, M.S.; Bae, T.S.; Seol, K.W. Effects of anodizing voltage on the anodized and hydrothermally treated titanium surface. *Metals Mater. Int.* **2006**, *12*, 505–511.

25. Lai, C.W.; Sreekantan, S.; Krengvirat, W.; Pei San, E. Preparation and photoelectrochemical characterization of WO_3-loaded TiO_2 nanotube arrays via radio frequency sputtering. *Electrochem. Acta* **2012**, *77*, 128–136.

26. Leghari, S.A.K.; Sajjad, S.; Chen, F.; Zhang, J. WO_3/TiO_2 composite with morphology change via hydrothermal template-free route as an efficient visible light photocatalyst. *Chem. Eng. J.* **2011**. *166*, 906–915.

27. Park, S.E.; Joo, H.; Kang, J.W. Effect of impurities in TiO_2 thin films on trichloroethylene conversion. *Solar Energy Mater. Solar Cells* **2004**, *83*, 39–53.

28. Chafik, T.; Efstathiou, A.M.; Verykios, X.E. Effects of W^{6+} doping of TiO_2 on the reactivity of supported Rh toward NO: Transient FTIR and mass spectroscopy studies. *J. Phys. Chem. B* **1997**, *101*, 7968–7977.

29. Song, H.Y.; Jiang, H.F.; Liu, X.Q.; Jiang, Y.Z.; Meng, G.Y. Preparation of WO_x-TiO_2 and the Photocatalytic Activity under Visible Irradiation. *Key Eng. Mater.* **2007**, *336*, 1979–1982.

30. Lai, C.W.; Sreekantan, S. Effect of heat treatment on WO_3-loaded TiO_2 nanotubes for hydrogen generation via enhanced water splitting. *Mater. Sci. Semicond. Process.* **2013**, *16*, 947–954.

31. MansoobáKhan, M.; OmaisháAnsari, M.; HungáHan, D.; HwanáCho, M. Band gap engineered TiO$_2$ nanoparticles for visible light induced photoelectrochemical and photocatalytic studies. *J. Mater. Chem. A* **2014**, *2*, 637–644.

32. Serpone, N.; Maruthamuthu, P.; Pichat, P.; Pelizzetti, E.; Hidaka, H. Exploiting the interparticle electron transfer process in the photocatalysed oxidation of phenol, 2-chlorophenol and pentachlorophenol: Chemical evidence for electron and hole transfer between coupled semiconductors. *J. Photochem. Photobiol. A: Chem.* **1995**, *85*, 247–255.

33. Yu, C.; Jimmy, C.Y.; Zhou, W.; Yang, K. WO$_3$ coupled P-TiO$_2$ photocatalysts with mesoporous structure. *Catal. Lett.* **2010**, *140*, 172–183.

34. Lai, C.W.; Sreekantan, S. Single step formation of C-TiO$_2$ nanotubes: Influence of applied voltage and their photocatalytic activity under solar illumination. *Int. J. Photoenergy* **2013**, *2013*, doi:10.1155/2013/276504.

35. Lai, C.W.; Sreekantan, S. Optimized sputtering power to incorporate WO$_3$ into C-TiO$_2$ nanotubes for highly visible photoresponse performance. *Nano* **2012**, *7*, doi:10.1142/S1793292012500518.

36. Lai, C.W.; Sreekantan, S. Study of WO$_3$ incorporated C-TiO$_2$ nanotubes for efficient visible light driven water splitting performance. *J. Alloys Compounds* **2013**, *547*, 43–50.

37. Sreekantan, S.; Lai, C.W.; Lockman, Z. Extremely fast growth rate of TiO$_2$ nanotube arrays in electrochemical bath containing H$_2$O$_2$. *J. Electrochem. Soc.* **2011**, *158*, C397–C402.

Surface Characterization and Photoluminescence Properties of Ce^{3+},Eu Co-Doped SrF$_2$ Nanophosphor

Mubarak Y. A. Yagoub [1,2], Hendrik C. Swart [1,*], Luyanda L. Noto [1], Peber Bergman [3] and Elizabeth Coetsee [1,*]

[1] Department of Physics, University of the Free State, PO Box 339, Bloemfontein, ZA 9300, South Africa; E-Mails: yagoubm@ufs.ac.za (M.Y.); luyanda.noto@gmail.com (L.N.)

[2] Department of Physics, Sudan University of Science and Technology, Khartoum 11113, Sudan

[3] Department of Physics, Chemistry and Biology, Linköping University, Linköping S-581 83, Sweden; E-Mail: peber@ifm.liu.se

* Authors to whom correspondence should be addressed; E-Mails: swarthc@ufs.ac.za (H.C.S.);

Academic Editor: Dirk Poelman

Abstract: SrF$_2$:Eu,Ce^{3+} nanophosphors were successfully synthesized by the hydrothermal method during down-shifting investigations for solar cell applications. The phosphors were characterized by X-ray diffraction (XRD), scanning Auger nanoprobe, time of flight-secondary ion mass spectrometry (TOF-SIMS), X-ray photoelectron spectroscopy (XPS) and photoluminescence (PL) spectroscopy. XRD showed that the crystallite size calculated with Scherrer's equation was in the nanometre scale. XPS confirmed the formation of the matrix and the presence of the dopants in the SrF$_2$ host. The PL of the nanophosphor samples were studied using different excitation sources. The phenomenon of energy transfer from Ce^{3+} to Eu^{2+} has been demonstrated.

Keywords: SrF$_2$; cerium; TOF-SIMS; XPS; shake-down; energy transfer

1. Introduction

Strontium fluoride (SrF$_2$) is one of the most widely used optical materials because of its interesting luminescent, optical, and physical properties. It has a wide band gap, low phonon energy, low refraction index, high radiation resistance, and good mechanical strength [1,2]. The photoluminescence properties of SrF_2 doped by Ln^{3+} ions have been extensively investigated in which charge compensation is required when Ln^{3+} ions substitute Sr^{2+} cation. This gives rise to a rich multisite structure. It has therefore been considered as a good phosphor host material that can be doped by a number of lanthanide ions for various luminescent applications [1–4]. SrF_2 host material doped with Ce^{3+} lanthanide ions is an example of a phosphor material that is extensively being investigated specifically for light amplification [5,6]. Some of these light amplification studies proposed that the SrF_2:Ce^{3+} phosphor material could be a promising scintillator [5]. Shendrik *et al.* [5] reported efficient scintillation light output of SrF_2:Ce^{3+} with high temperature stability suggesting that this material can be applied in well-logging scintillation detectors. They have also reported that the optimal Ce^{3+} doping level for maximum luminescence was 0.3 mol% if prepared by the Stockbarger method. Ce^{3+} ions in SrF_2 showed a fully allowed broad band 4f–5d transition [5] and this transition strongly absorbs UV radiation that results in a high absorption coefficient.

In the other hand, several previous studies have described the luminescence of Eu^{3+} doped materials as a good downshifting ion [7–10]. Gao *et al.* [7] reported luminescence due to transitions from the $^5D_0$ excited level to the ^7F_J levels, where spectral conversion of 325–550 nm light to 570–710 nm light has been demonstrated. In our previous investigation of SrF_2:Eu we reported the emissions from both the Eu oxidation states (Eu^{3+} and Eu^{2+}) where emission from 400 to 710 was observed [10]. X-ray photoelectron spectroscopy (XPS) results confirmed that the samples contained both Eu^{2+} and Eu^{3+} ions. The Eu^{3+} ion doped materials emits narrow emission peaks in the range of the orange-red emission with large Stokes shifts (>150 nm) that originates from the 4f–4f weak absorption transitions [11,12], whereas the 4f–5d absorption transition of the Eu^{3+} ion in SrF_2 is situated at the far ultraviolet region, which can be less accessible. In some applications, high or suitable absorption cross-section is needed and this requires a sensitizer with a high absorption coefficient [2,9,13]. Therefore, the presence of the Eu^{2+} and Eu^{3+} ions in the SrF_2 host greatly enhanced the emission intensity of Eu^{3+} at high concentrations [10]. In this work, Ce^{3+} singly and co-doped Eu in SrF_2 was prepared by using the hydrothermal method. The surface and photoluminescence properties are discussed.

2. Results and Discussion

2.1. Structure Analysis

2.1.1. X-Ray Diffraction (XRD)

Figure 1 shows the XRD patterns of un-doped and doped SrF_2 as well as the standard data for SrF_2 from card 00-086-2418. Doping with Ce- or Eu ions as well as the co-doped systems result in a small shift to higher angles with comparison to the un-doped sample and the standard data. This can be attributed to the radius difference between Eu (Eu^{2+} is 0.125 nm, Eu^{3+} is 0.107 nm), Ce^{3+} (0.114 nm) and Sr^{2+} (0.126 nm) ions, which confirms that Eu- and Ce ions are successfully incorporated into the SrF_2 lattice. It should be mentioned that doping with Eu- and Ce ions (up to 10 mol%) does not change the structure of the SrF_2 host in this study. The calculated SrF_2 lattice parameter is found to be (5.785 ± 0.005) Å and this agreed well with the reported value of (5.7996 ± 0.0001) Å [14].

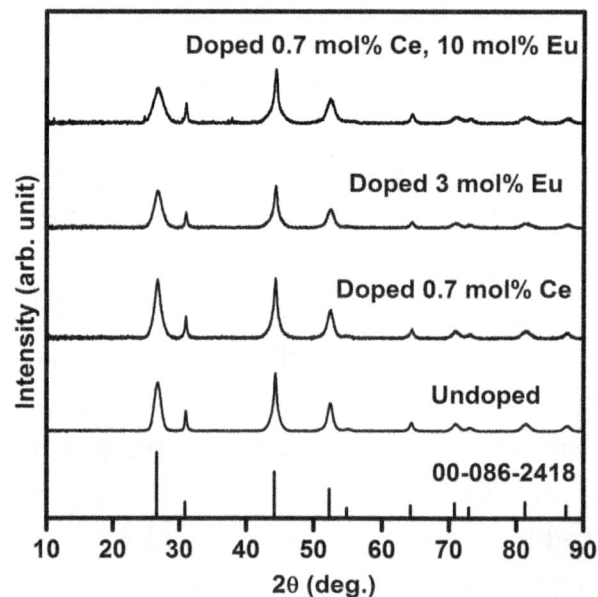

Figure 1. XRD patterns of pure and doped SrF_2.

The estimated average crystallite size (S) for pure and doped SrF_2 is calculated by using the diffraction peaks and Scherrer's equation [15], $S = 0.9\lambda/\beta\cos\theta$. S is the average crystallite size of the SrF_2 particles, λ is the wavelength of the X-rays (0.154 nm) and β is the full-width at half maximum of the X-ray peak at the Bragg angle θ. The average crystallite size of the pure SrF_2 was found to be 7.6 nm. The XRD peaks broaden with increasing the dopants ions (see Figure 1). The broadening of the XRD peaks were also observed by other groups [16,17]. H.A.A. Seed Ahmed et al. [16] attributed the XRD peak broadening to impurity broadening. Whereas, F. Wang et al. [17] assigned the XRD peak broadening to reduction in the nanoparticle size of the matrix. In our previous investigation of Eu doped SrF_2 samples, we assigned the XRD broadening as a result of a decrease in particle size of the matrix, which agreed well with F. Wang et al. [10]. Therefore, in the current study we can also assign these peaks' broadening to reduction in particle size of the matrix. The particle size reduced up to 3.9 nm for the SrF_2 sample that was doped with 0.7 mol% Ce^{3+} and 10 mol% Eu.

2.1.2. Auger and TOF SIMS analysis

An Auger profile of Ce and Eu co-doped SrF_2 was done to identify the sample's composition. The Auger spectrum of the $SrF_2:Ce^{3+}$,Eu is presented in Figure 2. The Auger peaks at 71, 1515, 1644 and 1713 eV are assigned to Sr while the F peak is situated at 656 eV [18]. The Auger spectrum not only confirmed the formation of the host matrix, but also showed the presence of the dopants. The Eu peaks were at 111, 142, 853 and 985 eV, while the peak at 89 eV corresponds to Ce. In addition C and O were also observed. The C contamination is attributed to adventitious hydrocarbons and the O is considered to be a common impurity in a fluoride compound [19,20]. The presence of the O in the sample did not change the structure of the sample (see Figure 1). Therefore, the O contamination was due to adventitious impurity species in the surface rather than oxygen impurity in the SrF_2 matrix.

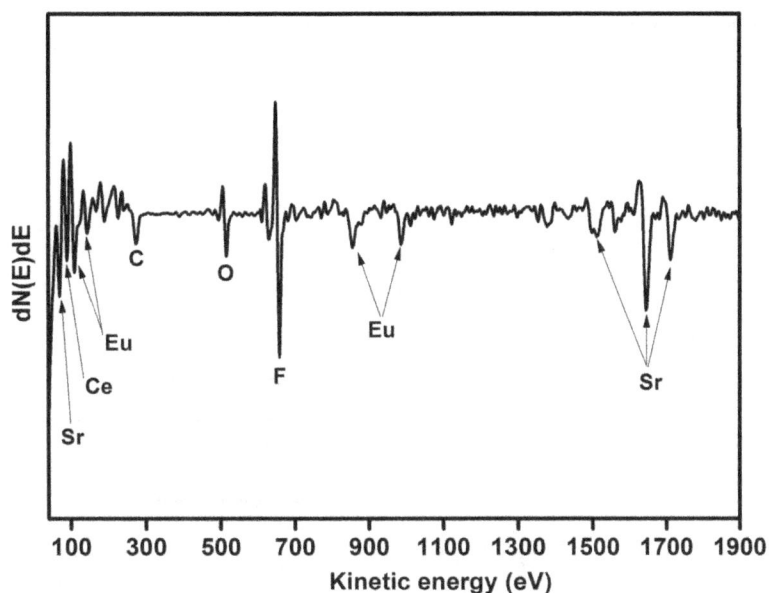

Figure 2. Auger spectrum of Ce and Eu co-doped SrF$_2$.

It could clearly be seen, not shown, that both the Ce and Eu ions were distributed quite homogeneously over the entire surface area of the Ce and Eu co-doped SrF$_2$. That indicated that the dopants were uniformly distributed in the SrF$_2$ matrix during the hydrothermal synthesis method.

2.1.3. X-ray Photoelectron Spectroscopy (XPS)

XPS measurements have been done in order to investigate the chemical composition and bonding state of the SrF$_2$:Ce,Eu phosphor powders. A higher dopant concentration (5 mol% for both Eu and Ce) was used in order to obtain a reasonable signal from the dopants. Figure 3 shows the peak fits for the (a) Sr 3d, (b) F 1s, (c) Eu 3d and (d) Ce 3d high resolution XPS peaks. The results also confirmed the presence of the host matrix elements (Sr and F) as well as the dopants (Eu and Ce) to their corresponding binding energies. During the peaks fit procedure, the C 1s peak at 284.8 eV was taken as a reference for all charge shift corrections. This is done because the C 1s peak resulted from hydrocarbon contamination and its binding energy generally remains constant, irrespective to the chemical state of the sample. In addition to that, all the Gaussian percentages were assumed to have a combined Gaussian-Lorentzian shape. The high resolution XPS peak for the Sr 3d showed two individual peaks. These two peaks are assigned to Sr 3d in SrF$_2$ that originate from the spin-orbit splitting 3$d_{5/2}$ (133.5 eV) and 3$d_{3/2}$ (135.3 eV), while the F 1s peak is situated at 684.7 eV. The spin-orbit splitting of Sr 3d is about 1.78 eV, it is in a good agreement with reported value of 1.75 eV [21].

The peak deconvolution for the Eu 3d high resolution XPS peaks are shown in Figure 3c. The 3d level of Eu ion is composed of four peaks. These four peaks can be attributed to Eu^{3+} and Eu^{2+} spin-orbit splitting 3$d_{5/2}$ and 3$d_{3/2}$ core level, respectively [21–24]. The spin-orbit splitting for both oxidation states Eu^{3+} and Eu^{2+} is about 29.96 eV. The Eu 3d results showed good agreement with our previous XPS investigation of SrF$_2$:Eu phosphors powder where Eu composed of its two oxidation states (Eu^{2+} and Eu^{3+}) [10].

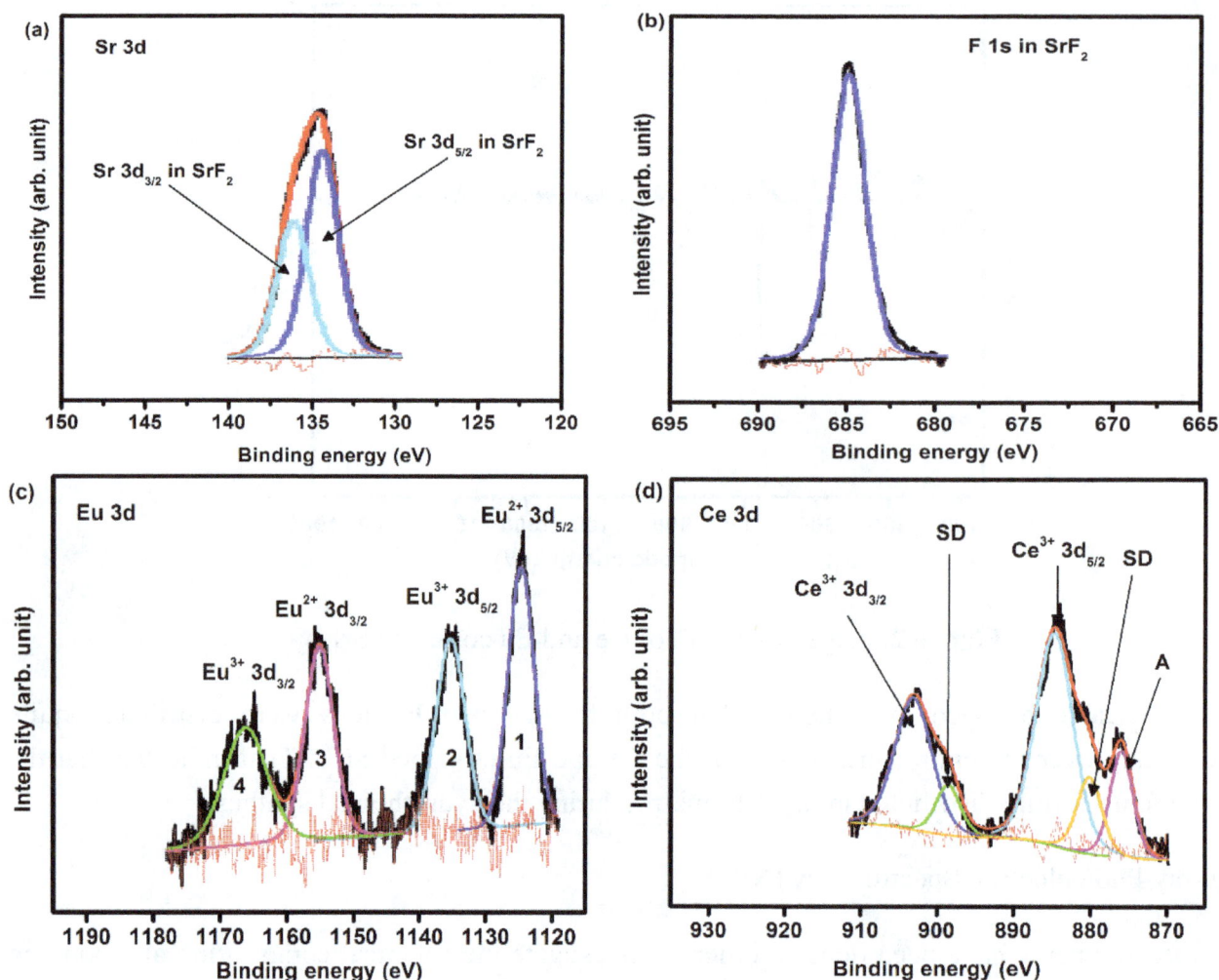

Figure 3. High resolution XPS peaks of (**a**) Sr 3d; (**b**) F 1s; (**c**) Eu 3d; and (**d**) Ce 3d for SrF$_2$:Ce,Eu phosphors powder.

The Ce $3d$ high resolution peak is shown in Figure 3d. The strong peaks correspond to the photoemission from the Ce^{3+} $3d$ state. Due to the spin-orbit interaction, the Ce^{3+} $3d$ photoemission peak consisted of two peaks that are assigned to the $3d_{3/2}$ and $3d_{5/2}$ peaks with 4f^1 final states, with an intensity ratio I($3d_{5/2}$)/I($3d_{3/2}$) = 3/2 [22,25,26]. The spin-orbit splitting value ($\approx$18.15 eV) is in good agreement with the estimated value ($\approx$18.10 eV). The energy peaks labelled SD are due to the strong Coulomb interaction between photoemission in the $3d$ level and electrons located near the Fermi level. These peaks originate from the screening of the $3d$ level by valence band electrons to the $4f$ states [22]. This is possible due to hybridization of the Ce $4f$ level with the conduction band states [26]. In the photoemission nomenclature, these peaks are a result from what is called, shake-down process [22]. The $3d$ shake-down peaks behave the same as the 3d spin-orbit splitting peaks but they are a result from the 3d^{9}f^2 final state. Therefore, the SD peaks can be assigned to the $3d_{3/2}$ and $3d_{5/2}$ XPS peaks with 4f^2 final states and this is in accordance with previous work done in Ce [25,26]. The shoulder peaks marked as A is related to the F KLL Auger electron peak. The XPS peak positions, area distributions and chemical bonding for all the peaks in as-prepared SrF$_2$:Ce,Eu are tabulated in Table 1.

Table 1. XPS peak position, area distribution and chemical bonding of as-prepared SrF$_2$:Ce,Eu phosphor powder.

Element	B.E (±0.1 eV)	Area distribution	Interpretation
F1s	684.7	2688	F in SrF$_2$
Sr3d	133.5	1986	Sr 3d$_{5/2}$ in SrF$_2$
	135.3	1311	Sr 3d$_{3/2}$ in SrF$_2$
Eu3d	1123.3	1613	Eu^{2+} 3d$_{5/2}$ in fluoride
	1133.05	1372	Eu^{3+} 3d$_{5/2}$ in fluoride
	1153.2	1064	Eu^{2+} 3d$_{3/2}$ in fluoride
	1163.0	905	Eu^{3+} 3d$_{3/2}$ in fluoride
Ce3d	880.3	1296	Shake-down satellite
	884.8	5141	Ce^{3+} 3d$_{5/2}$ in fluoride
	898.5	855	Shake-down satellite
	903.0	3393	Ce^{3+} 3d$_{3/2}$ in fluoride
	876.1	1592	F KL$_1$L$_1$ Auger electron peak

2.2. Photoluminescence Spectroscopy

2.2.1. SrF$_2$:Ce^{3+}

The emission and excitation spectra of the Ce^{3+} singly doped SrF$_2$ nanophosphor are shown in Figure 4. The excitation spectrum consists of a prominent peak that is centred at 295 nm. This peak has been previously assigned to Ce^{3+}:4f–5d excitation transition in SrF$_2$ [27]. By exciting the samples by 295 nm, a broad band emission peak is observed, which is attributed to the inter-configuration 5d^1–4f^1 allowed transition of Ce^{3+} ions. The inset graph in Figure 4 shows the emission intensity variation as a function of the Ce^{3+} concentration. The maximum luminescence intensity occurred for the sample doped with 0.7 mol% and a further increase in concentration resulted in a decrease in Ce^{3+} emission intensity. A previous study done by R. Shendrik *et al.* [5] on the SrF$_2$:Ce^{3+} sample reported that Ce^{3+} has a broad emission band that consist of two emission peaks (Ce^{3+} 5d to 4f ground state (^{2}F$_{7/2}$ and ^{2}F$_{5/2}$)) and the maximum intensity was observed at a Ce^{3+} dopant concentration of 0.3 mol%. In this study, the peaks were broadened and they fully overlapped, which might be the reason that only one broad peak was observed.

2.2.2. SrF$_2$:Ce,Eu

Figure 5a shows the PL emission spectra of SrF$_2$:Eu obtained by using the He-Cd laser PL system with a 325 nm excitation wavelength. The spectra clearly consist of a broad emission band that is centred at 416 nm with narrow bands in the range of 550–710 nm. The broad emission band is assigned to the inter-configuration 4f^{6}5d^1–4f^7 allowed transition of Eu^{2+} [11,12] and the narrow emission bands to the Eu^{3+} emission originating from the 4f–4f transition [28]. The Eu^{3+} emission consists of orange–red emission bands that is attributed to the ^{5}D$_0$→^{7}F$_J$ transitions (J = 1, 2, 3, 4). This implies that the SrF$_2$:Eu samples consist of both Eu oxidation states (Eu^{2+} and Eu^{3+}), with their emission ranging from 400 to 710 nm [10]. The Eu^{3+} emission bands increased with an increase in the Eu dopant concentration in the SrF$_2$ matrix. This can also be seen in Figure 5b, where the emission of Eu^{3+}

excited by 394 nm is portrayed. The PL emission intensity increased slightly at the lower concentrations but then increased dramatically at 10 mol%. The presence of both Eu oxidation states therefore strongly enhanced the emission intensity of the Eu^{3+} ions. Detailed investigations on the luminescence phenomenon of Eu^{3+} and Eu^{2+} have previously been studied by various workers [10,29–31].

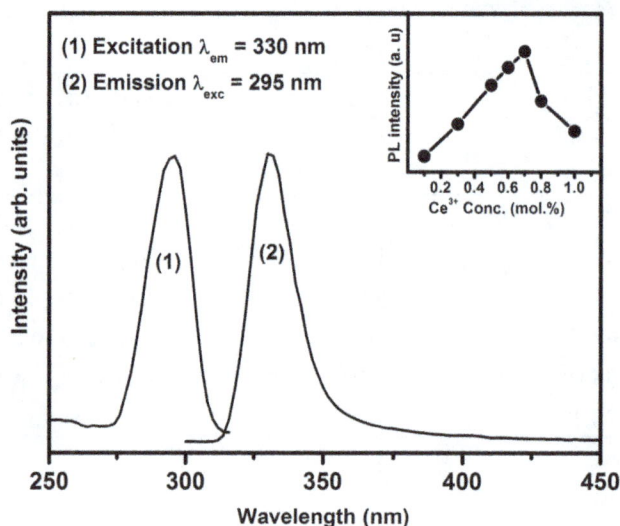

Figure 4. Excitation and emission spectra of the $SrF_2:Ce^{3+}$ (0.7 mol%) nanophosphor. The inset shows the 5d–4f transition's emission intensity as a function of Ce^{3+} concentration.

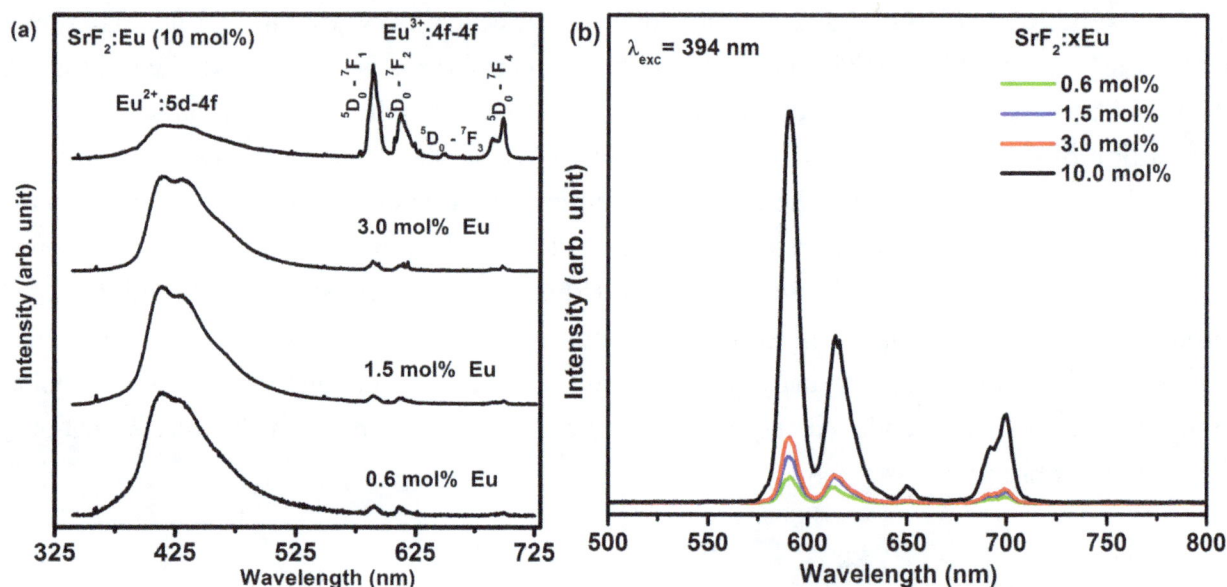

Figure 5. Photoluminescence spectra of $SrF_2:xEu$ excited by (a) using the He-Cd laser system with 325 nm excitation wavelength and (b) the Cary Eclipse with a wavelength of 394 nm.

Figure 6a depicts the PL emission of Ce^{3+} (0.7%) co-doped $SrF_2:xEu$ (where x = 0.2%, 0.6%, 5% and 10%) excited with the He-Cd laser system with a 325 nm wavelength. The spectra also consisted of both the Eu^{2+} and Eu^{3+} emissions. A shoulder peak (marked with a dollar sign ($)) at a lower wavelength only appeared for the smaller dopant concentrations' (0.2 and 0.6 mol%). This shoulder ($) is assigned to the 4f–5d emission of Ce^{3+}, which is completely quenched at the higher Eu concentration. With an increasing concentration of the Eu ions the relative PL emission intensity of the Eu^{2+} gradually decreased and the Eu^{3+} emission intensity increased. The emission intensity of the Eu^{3+} has dramatically increased at the high Eu doping concentration. This can clearly be seen in Figure 6b where the Eu^{3+} emission intensity plotted as function of Eu concentration for the Eu co-doped Ce^{3+} system. It can be noticed that Ce^{3+} co-doped SrF_2:Eu greatly enhanced the Eu^{3+} ions emission intensity at high Eu concentration. The increase of the Eu^{3+} emission intensity with an increase in the Eu concentration can be attributed to an increase in the Eu^{3+}/Eu^{2+} ratio in the presence of the Ce^{3+} ions. In the SrF_2 crystal, the Sr^{2+} ion is located at the body centre of a cube of eight F^- ions. The trivalent Ln^{3+} ions normally replace the Sr^{2+} cation. The extra charge of the Ln^{3+} ions is compensated by F^- anion charges situated elsewhere in an interstitial site. With increasing Ln^{3+} concentration, some kind of structural deformation occurs, the Ln^{3+}-F dipoles couple to dimers, trimers and higher aggregates. The interstitial F- ions and vacancies on the normal F^- site compose cuboctahedral clusters [32]. However, at low Eu concentration (less than Ce^{3+} concentration), the clusters are not completely formed. Besides, compare with the size of the Eu^{3+} (0.107 nm), the size of the Eu^{2+} (0.125 nm) is much closer to the size of the Sr^{2+} (0.126 nm), and hence the reduction of Eu^{3+} to Eu^{2+} ions is favored because it could reduce the lattice distortion of the doped SrF_2 crystal [33]. At high Eu concentration (bigger than the Ce^{3+} concentration), the dimensions of the Eu^{3+} ions cluster increased and hence the ratio of Eu^{3+}/Eu^{2+} increased. The increase of the Eu^{3+} ions therefore increased the Eu^{3+} emission intensity.

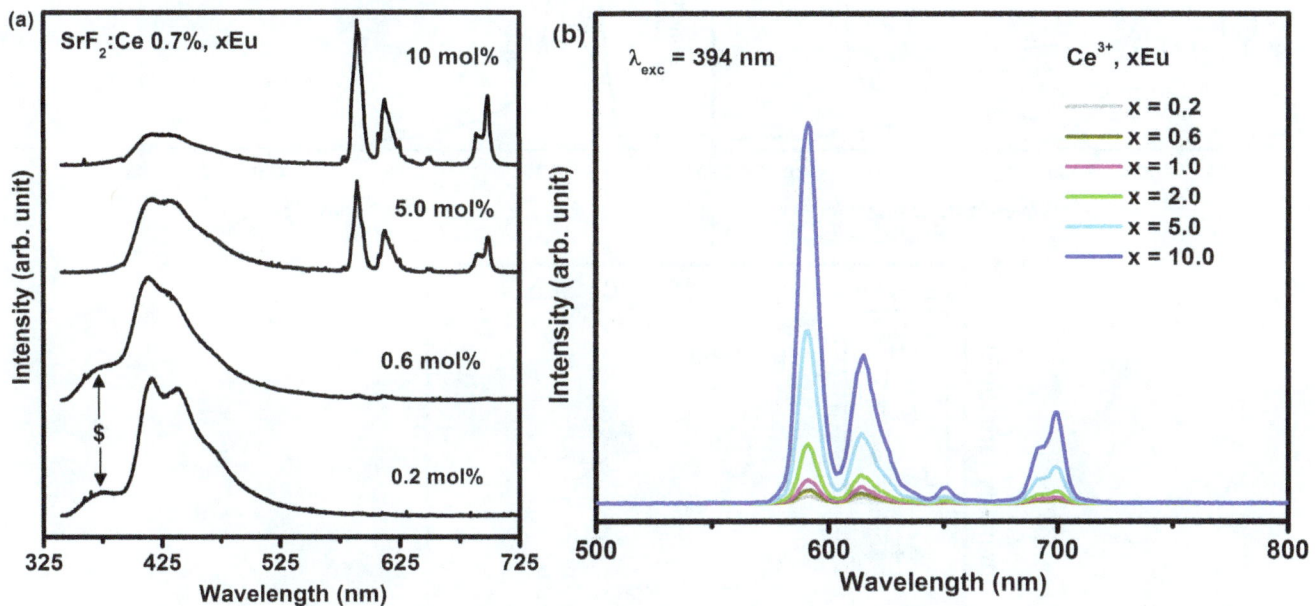

Figure 6. (a) PL spectra of SrF_2:Ce^{3+} (0.7 mol%), xEu excited with the laser system with a 325 nm excitation wavelength and (b) 394 nm using the xenon lamp.

The PL emission spectra of the SrF_2:Ce^{3+},Eu nanophosphor excited by the 295 nm excitation wavelength are plotted in Figure 7a. The broad emission band that is centered at a wavelength of 330 nm is a characteristic of the Ce^{3+} ion which is in agreement with the emission spectra for Ce^{3+} in Figure 4. The additional broad peak beside the Ce^{3+} emission that was centered at 416 nm is assigned to the Eu^{2+} ions in SrF_2 (clearly shown in the inset graph of Figure 7a). The Eu^{2+} emission slightly increased before it decreased with increasing Eu concentration. In Figure 7a the emission spectrum of the SrF$_2$:Eu without Ce excited at 295 nm is also shown. It clearly shows no Eu^{2+} emission has occurred. The presence of Eu^{2+} emission under 295 nm excitation, in the co-doped samples, is therefore evidence of an energy transfer process from Ce^{3+} to Eu^{2+}. This process can occur in such material since the emission of Ce^{3+} overlaps the excitation spectra of Eu^{2+} (Figure 7b; SrF$_2$:Ce^{3+} (0.7 mol%), Eu (0.6 mol%)). Such spectral overlap is a necessary condition for the occurrence of the energy transfer from Ce^{3+} to Eu^{2+}. An efficient energy transfer from Ce^{3+} to Eu^{2+} in a fluoride crystal was previously demonstrated even for a very low concentration [34]. More evidence of energy transfer between Ce^{3+} and Eu^{2+} is shown in Figure 7c where the room temperature luminescence excitation spectra of SrF$_2$:Ce^{3+} (0.7 mol%), Eu (0.6 mol%) nanophosphors are plotted. The excitation spectrum of Eu^{2+} (dotted line) not only consists of the Eu^{2+}:4f^7→4f^{6}5d excitation transition but also the Ce^{3+} excitation band (clearly seen in the inset of the Figure 7c). All these results confirm the existence of energy transfer from Ce^{3+} to the Eu^{2+} ion.

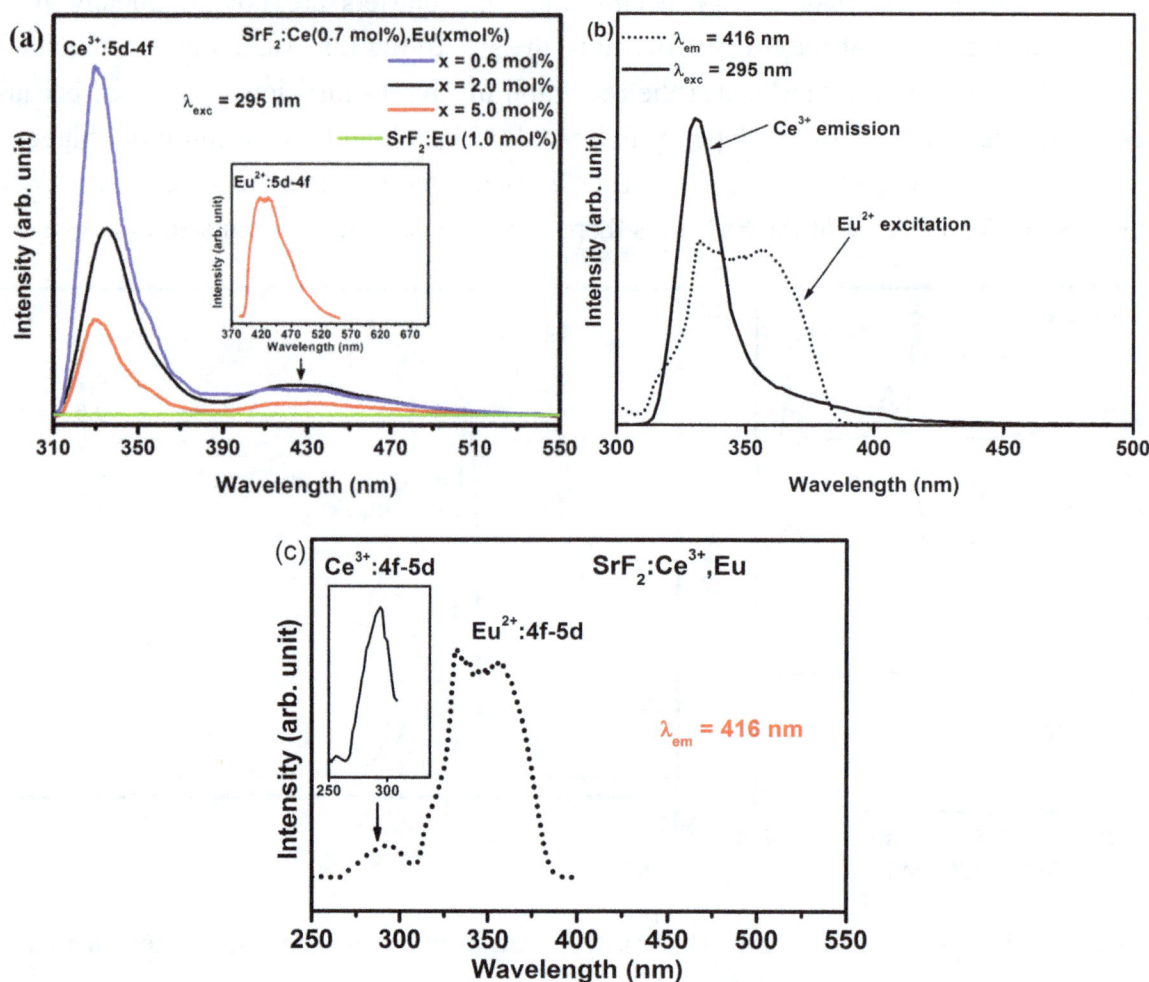

Figure 7. (a) PL emission spectra of Ce^{3+} and Eu^{2+} from SrF$_2$:Ce^{3+} (0.7 mol%) with different Eu doping concentration as well as from Eu^{2+} in SrF$_2$:Eu excited by an excitation wavelength of 295 nm; (b) Spectral overlap between Ce^{3+} emission and Eu^{2+} excitation and (c) excitation spectra of SrF$_2$:Ce^{3+} (0.7 mol%), Eu (0.6 mol%) nanophosphors measured at an emission wavelength of 416 nm. The inset in (a) is the enlarge spectrum of the Eu^{2+} emission ions and the inset in (c) is the enlarge Ce^{3+} excitation from SrF$_2$:Ce^{3+} (0.7 mol%), Eu (5.0 mol%).

Results obtained from the luminescence decay curves for Ce^{3+} emission also contributed further to the energy transfer process. The decay time of the donor ions does not change in the presence and absence of the acceptor ions if the radiative energy is dominant [35]. In the situation of non-radiative energy transfer the decay time of the donor ions gradually decreases with an increase in the acceptor concentration. The PL decay curves of Ce^{3+} with various Eu concentration are shown in Figure 8. The decay curve of the Ce^{3+} ions gradually decreased with an increase in the Eu concentration. The luminescence decay curve of Ce^{3+} singly doped SrF_2 nanoparticles can well be fitted into a single-exponential function, shown in the inset of Figure 8, whereas the decay curve of the entire co-doped concentrations were fitted with a bi-exponential decay model [35,36]:

$$I(t) = A_1 \exp(-t/\tau_1) + A_2 \exp(-t/\tau_2) \tag{1}$$

$I(t)$ is the luminescence intensity at time t; A_1 and A_2 are constants; and τ_1 and τ_2 are the short- and long-decay components, respectively. The average lifetime constant (τ^*) can be calculated from the following equation:

$$\tau^* = (A_1 \tau_1^2 + A_2 \tau_2^2)/(A_1 \tau_1 + A_2 \tau_2) \tag{2}$$

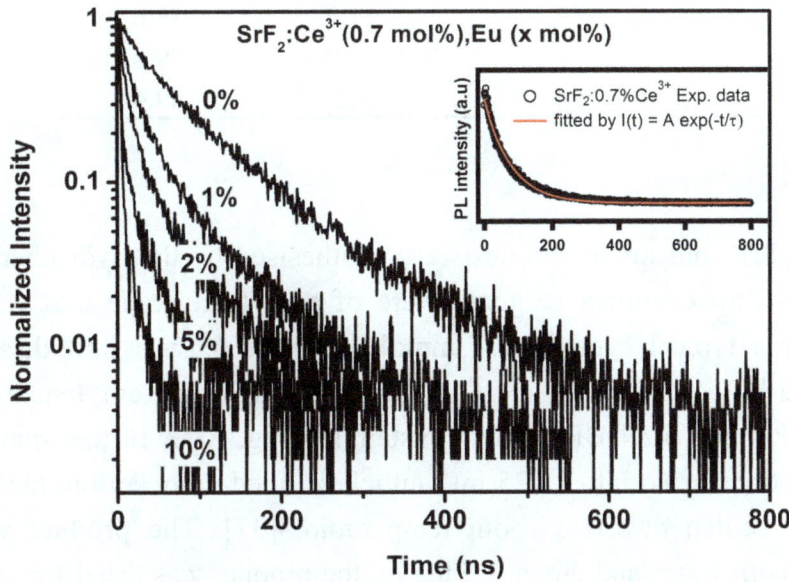

Figure 8. The decay lifetime of Ce^{3+} ions in the SrF_2 host with an increase in Eu concentration. The inset graph shows the decay curve of 0.7% Ce^{3+} in SrF_2 fitted to a single-exponential fitting function.

The lifetime of the Ce^{3+} doped SrF_2 is determined to be 77.15 ns. This value is in good agreement with the reported value of Ce^{3+} in SrF_2 [27]. In the Eu ions co-doped system, the average lifetime of the donor ion (Ce^{3+}) decreased up to 8.2 ns at 10 mol% Eu concentration. This results confirm that the excitation energy of Ce^{3+} ions was transferred to the Eu^{2+} ions. The lifetime results for the Ce^{3+} ions in the SrF_2 host strongly suggest that the energy transfer from Ce^{3+} to Eu^{2+} was non-radiative. The energy transfer efficiency from Ce^{3+} to Eu is defined by the following expression:

$$\eta_{ET} = 1 - \tau/\tau_0 \tag{3}$$

where τ and τ_0 are the average lifetime of Ce^{3+} in the presence and absence of the Eu ions, respectively. The corresponding lifetime and energy transfer efficiencies are tabulated in Table 2. From Table 2, the energy transfer of Ce^{3+} increased gradually with an increase in the Eu concentration. The maximum energy transfer efficiency is about 89.4% for the sample doped with 0.7 mol% Ce^{3+} and 10 mol% Eu. An efficient energy transfer has occurred from Ce^{3+} to Eu^{2+}. The emission of Eu^{2+} has slightly increased before it decreased with increasing Eu concentration due to the decrease of the Eu^{2+} ratio in the SrF_2 host. In our previous investigation of SrF_2:Eu the Eu^{2+} ion was, however, found to be unstable when irradiated by a YAG laser. The Eu^{2+} ion's PL emission intensity rapidly decreased with time and this result made the SrF_2:Eu nanophosphor an unsuitable candidate for several applications, such as white light-emitting diodes and wavelength conversion films for silicon photovoltaic cells [10].

Table 2. Lifetime of the 5d–4f transition of Ce^{3+} (330 nm) and the Ce^{3+}-Eu energy transfer efficiency (η_{ET}) in SrF_2 matrix.

Eu concentration (mol%)	τ (ns)	η_{ET} (%)
0	77.15	0
1	46.3	40
2	31.9	58.6
5	16.05	79.2
10	8.2	89.4

3. Experimental Section

Doped and un-doped SrF_2 phosphor samples were synthesised by the hydrothermal method. For the hydrothermal process, all chemical reagents were of analytical grade and were used without further purification. For a typical synthesis, 1 mmol of $Sr(NO_3)_2$ was first dissolved in 30 mL distilled water, followed by 5 mmol of $C_{10}H_{14}N_2O_8.2H_2O$ (Na$_2$EDTA, ethylenediamine tetraacetic acid disodium salt) and 2 mmol of $NaBF_4$ under constant stirring. After further magnetic stirring for 10 min the solution was transferred into a 125 mL autoclave lined with Teflon, heated at 160 °C for one hour and naturally cooled down to room temperature [37]. The product was collected by centrifugal and washed with water and ethanol. Finally, the product was dried for 10 h in an oven at 60 °C. Ce^{3+} and Eu co-doped SrF_2 samples were prepared by the same hydrothermal technique. $Eu(NO_3)_3(H_2O)_5$ and $Ce(NO_3)_3(H_2O)_6$ were used as sources for the Eu and Ce dopants, respectively.

The phosphors were characterized by X-ray diffraction (XRD) (Bruker AXS Gmbh, Karlsruhe, Germany) (Bruker Advance D8 diffractometer with Cu K_α radiation (λ = 0.154 nm)) to identify the crystalline structure of the powder. Auger spectra were collected with a PHI 700 Scanning Auger Nanoprobe (ULVAC-PHI Inc, Chanhassan, MN, USA) equipped with a scanning Auger microscope (SAM). The field emission electron gun used for the SAM analyses was set at: 2.34 A filament current; 4.35 kV extractor voltage and 381.4 μA extractor current. With these settings a 25 kV, 10 nA electron beam was obtained for the Auger analyses. The electron beam diameter was about 10 nm. An IonTof time of flight secondary ion mass spectrometer (TOF-SIMS) instrument (ION-TOF Gmbh, Muenster, Germany) equipped with a Bi primary ion source was used to characterize the nanophosphor materials for their chemical composition and dopants distribution. In spectroscopy mode, the system equipped with a DC current of 30 nA and a pulsed current of 1 pA at 30 kV with a heating current of 2.95 A and emission current of 0.8 µA was used. High resolution X-ray photoelectron spectroscopy (XPS) was obtained with a PHI 5000 Versaprobe system (ULVAC-PHI Inc, Chanhassan, MN, USA). A low energy Ar$^+$ ion gun and low energy neutralizer electron gun were used to minimize charging on the surface. A 100 µm diameter monochromatic Al Kα X-ray beam ($hv = 1486.6$ eV) generated by a 25 W, 15 kV electron beam was used to analyze the different binding energy peaks. The pass energy was set to 11 eV giving an analyzer resolution ≤0.5 eV. Multipack version 8.2 software (ULVAC-PHI Inc, Chanhassan, MN, USA) was utilized to analyze the spectra to identify the chemical compounds and their electronic states using Gaussian-Lorentz fits. Photoluminescence spectra (PL) were collected using a Cary Eclipse fluorescence spectrophotometer (Varian Ltd, Mulgrave Victoria, Australia) equipped with a xenon lamp and also with a He-Cd laser PL system with a 325 nm excitation wavelength. Luminescence decay curves were recorded by using a NanoLED with a 335 nm excitation wavelength and repetition rate of 1 MHz. All measurements were performed at room temperature.

4. Conclusions

As-prepared SrF2:Eu,Ce nanophosphors were successfully synthesised with the hydrothermal technique. The average crystallite size that was calculated by using Scherrer's equation was found to be 7.6 nm for the host sample. Dopant ions were intended to decrease the particle size of the host. The Auger spectra confirmed the presence of Sr, F, Eu and Ce elements in the host matrix. Photoluminescence properties of Ce^{3+} and Eu co-doped SrF2 nano-phosphor have been investigated. A possible efficient energy transfer from Ce^{3+} to Eu^{2+} ions was demonstrated. From the PL decay curves the energy transfer efficiency was calculated to be 89.4% for the SrF2: 0.7 mol% Ce^{3+}, 10 mol% Eu sample.

Acknowledgments

This work is based on the research supported by the South African Research Chairs Initiative of the Department of Science and Technology and National Research Foundation of South Africa. The financial assistance of the National Research Foundation (NRF) and the University of the Free State towards this research is hereby acknowledged.

Author Contributions

Hendrik C. Swart is the leader of the research group and supervisor of the PhD students and he helped with the data interpretation and writing of the paper. Elizabeth Coetsee is also one of the supervisors of the PhD students and she helped with the editing of the paper. Luyanda Noto helped with the TOF-SIMS analysis and discussion and Mubarak Y. A. Yagoub was mainly responsible for the planning, experimental part as well as writing the main part of the paper, the decay measurements were done with the help of Peber Bergman in his laboratory.

Conflicts of Interest

The authors declare no conflict of interest.

References

1. Ivanovskikh, K.V.; Pustovarov, V.A.; Krim, M.; Shulgin, B.V. Time-resolved vacuum ultraviolet spectroscopy of Er^{3+} ions in the SrF_2 crystal. *J. Appl. Spectrosc.* **2005**, *72*, 564–568.

2. Van der Ende, B.M.; Aarts, L.; Meijerink, A. Near-infrared quantum cutting for photovoltaics. *Adv. Mater.* **2009**, *21*, 3073–3077.

3. Ivanovskikh, K.; Pustovarov, V.; Smirnov, A.; Shulgin, B. Inter- and intraconfigurational luminescence of trivalent rare earth ions doped into strontium fluoride crystals under vacuum ultraviolet excitation. *Phys. Stat. Solidi C* **2007**, *4*, 889–892.

4. Kristianpoller, N.; Weiss, D.; Chen, R. Optical and dosimetric properties of variously doped SrF_2 crystals. *Radiat. Meas.* **2004**, *38*, 719–722.

5. Shendrik, R.; Radzhabov, E.A.; Nepomnyashchikh, A.I. Scintillation properties of pure and Ce^{3+}-doped SrF_2 crystals. *Radiat. Meas.* **2013**, *56*, 58–61.

6. Shendrik, R.; Radzhabov, E. Emission in CaF_2, BaF_2, SrF_2. *IEEE Trans. Nucl. Sci.* **2009**, *57*, 1295–1299.

7. Gao, D.; Zheng, H.; Zhang, X.; Fu, Z.; Zhang, Z.; Tian, Y.; Cui, M. Efficient fluorescence emission and photon conversion of $LaOF:Eu^{3+}$ nanocrystals. *Appl. Phys. Lett.* **2011**, *98*, 011907–011909.

8. Chung, P.; Chung, H.; Holloway, P.H. Phosphor coatings to enhance Si photovoltaic cell performance. *J. Vac. Sci. Technol. A* **2007**, *25*, 61–66.

9. Huang, X.Y.; Wang, J.X.; Yu, D.C.; Ye, S.; Zhang, Q.Y.; Sun, X.W. Spectral conversion for solar cell efficiency enhancement using $YVO_4:Bi^{3+},Ln^{3+}$ (Ln = Dy, Er, Ho, Eu, Sm, and Yb) phosphors. *J. Appl. Phys.* **2011**, *109*, 113526–113532.

10. Yagoub, M.Y.A.; Swart, H.C.; Noto, L.L.; O'Connell, J.H.; Lee, M.E.; Coetsee, E. The effects of Eu-concentrations on the luminescent properties of SrF2:Eu nanophosphor. *J. Lumin.* **2014**, *156*, 150–156.

11. Li, Y.C.; Chang, Y.H.; Lin, Y.F.; Chang, Y.S.; Lin, Y.J. Synthesis and luminescent properties of Ln^{3+} (Eu^{3+}, Sm^{3+}, Dy^{3+})-doped lanthanum aluminum germanate $LaAlGe_2O_7$ phosphors. *J. Alloys Compd.* **2007**, *439*, 367–375.

12. Jin, Y.; Qin, W.; Zhang, J. Preparation and optical properties of SrF_2: Eu^{3+} nanospheres. *J. Fluor. Chem.* **2008**, *129*, 515–518.

13. Huang, X.; Han, S.; Huang, W.; Liu, X. Enhancing solar cell efficiency: The search for luminescent materials as spectral converters. *Chem. Soc. Rev.* **2013**, *42*, 173–201.

14. Rakov, N.; Guimaraes, R.B.; Franceschini, D.F.; Maciel, G.S. Er:SrF_2 luminescent powders prepared by combustion synthesis. *Mater. Chem. Phys.* **2012**, *135*, 317–321.

15. Oprea, C.; Ciupina, V.; Prodan, G. Investigation of nanocrystals using TEM micrographs and electron diffraction technique. *Rom. J. Phys.* **2008**, *53*, 223–230.

16. Seed Ahmed, H.A.A.; Ntwaeaborwa, O.M.; Kroon, R.E. The energy transfer mechanism in Ce,Tb co-doped LaF_3 nanoparticles. *Curr. Appl. Phys.* **2013**, *13*, 1264–1268.

17. Wang, F.; Han, Y.; Lim, C.S.; Lu, Y.; Wang, J.; Xu, J.; Chen, H.; Zhang, C.; Hong, M.; Liu, X. Simultaneous phase and size control of upconversion nanocrystals through lanthanide doping. *Nature* **2010**, *463*, 1061–1065.

18. Childs, K.D.; Carlson, B.A.; LaVanier, L.A.; Moulder, J.F.; Paul, D.F.; Stickle, W.F.; Watson, D.G. *Handbook of Auger Electron Spectroscopy*, 3rd ed.; Physical Electronics: Eden Peairie, MN, USA, 1995.

19. Kroon, R.E.; Swart, H.C.; Ntwaeaborwa, O.M.; Seed Ahmed, H.A.A. Ce decay curves in Ce, Tb co-doped LaF3 and the energy transfer mechanism. *Phys. B* **2014**, *439*, 83–87.

20. Van Wijngaarden, J.T.; Scheidelaar, S.; Vlugt, T.J.H.; Reid, M.F.; Meijerink, A. Energy transfer mechanism for downconversion in the (Pr^{3+}, Yb^{3+}) couple. *Phys. Rev. B* **2010**, *81*, 155112–155117.

21. Vasquez, R.P. SrF_2 by XPS. *Surf. Sci. Spectr.* **1992**, *1*, 24–30.

22. Vercaemst, R.; Poelman, D.; van Meirhaeghe, R.L.; Fiermans, L.; Laflere, W.H.; Cardon, F. An XPS study of the dopants' valence states and the composition of $CaS_{1-x}Se_x$:Eu and $SrS_{1-x}Se_x$:Ce thin film electroluminescent devices. *J. Lumin.* **1995**, *63*, 19–30.

23. Lu, D.; Sugano, M.; Sun, X.Y.; Su, W. X-ray photoelectron spectroscopy study on $Ba_{1-x}Eu_xTiO_3$. *Appl. Surf. Sci.* **2005**, *242*, 318–325.

24. Zhang, J.; Yang, M.; Jin, H.; Wang, X.; Zhao, X.; Liu, X.; Peng, L. Self-assembly of $LaBO_3$:Eu twin microspheres synthesized by a facile hydrothermal process and their tunable luminescence properties. *Mater. Res. Bull.* **2012**, *47*, 247–252.

25. Lässer, R.; Fuggle, J.C.; Beyss, M.; Campagna, M. X-ray photoemission from Ce core levels of $CePd_3$, CeSe, $CeAl_2$ and $CeCu_2Si_2$. *Phys. B C* **1980**, *102*, 360–366.

26. Gamza, M.; Slebarski, A.; Rosner, H. Electronic structure of $Ce_5Rh_4Sn_{10}$ from XPS and band structure calculations. *Eur. Phys. J. B* **2008**, *63*, 1–9.

27. Zhang, C.; Hou, Z.; Chai, R.; Cheng, Z.; Xu, Z.; Li, C.; Huang, L.; Lin, J. Mesoporous SrF_2 and $SrF_2:Ln^{3+}$ (Ln = Ce, Tb, Yb, Er) Hierarchical microspheres: Hydrothermal synthesis, growing mechanism, and luminescent properties. *J. Phys. Chem. C* **2010**, *114*, 6928–6936.

28. Dorenbos, P. Energy of the first $4f^7 \rightarrow 4f^65d$ transition of Eu^{2+} in inorganic compounds. *J. Lumin.* **2003**, *104*, 239–260.

29. Dorenbos, P. Valence stability of lanthanide ions in inorganic compounds. *Chem. Mater.* **2005**, *17*, 6452–6456.

30. Baran, A.; Barzowska, J.; Grinberg, M.; Mahlik, S.; Szczodrwksi, K.; Zorenko, Y. Binding energy of Eu^{2+} and Eu^{3+} ions in β-Ca_2SiO_4 doped with europium. *Opt. Mater.* **2013**, *35*, 2017–2114.

31. Biswas, K.; Sontakke, A.D.; Sen, R.; Annapurna, K. Luminescence properties of dual valence Eu doped nano-crystalline BaF_2 embedded glass-ceramics and observation of $Eu^{2+} \rightarrow Eu^{3+}$ energy transfer. *J. Fluoresc.* **2012**, *22*, 745–752.

32. Pandey, C.; Dhopte, S.M.; Muthal, P.L.; Kondawar, V.K.; Moharil, S.V. $Eu^{3+} \leftrightarrow Eu^{2+}$ redox reactions in bulk and nano CaF_2:Eu. *Radiat. Eff. Defects Solids* **2007**, *162*, 651–658.

33. Wang, X.; Wu, N.; Shimizu, M.; Sakakura, M.; Shimotsuma, Y.; Miura, K.; Zhou, S.; Qiu, J.; Hirao, K. Space selective reduction of europium ions via SrF_2 crystals induced by high repetition rate femtosecond laser. *J. Ceram. Soc. Jpn.* **2011**, *119*, 939–941.

34. Caldino, U.G.; Gruz, C.D.; Monoz, G.H.; Rubio, J.O. $Ce^{3+} \rightarrow Eu^{2+}$ energy transfer in CaF_2. *Solid State Commun.* **1989**, *4*, 347–351.

35. Zhou, J.; Xia, Z.; You, H.; Shen, K.; Yang, M.; Liao, L. Synthesis and tunable luminescence properties of Eu^{2+} and Tb^{3+}-activated $Na_2Ca_4(PO_4)_3F$ phosphors based on energy. *J. Lumin.* **2013**, *135*, 20–25.

36. Katsumata, T.; Nabae, T.; Sasajima, K.; Komuro, S.; Morikawa, T. Effects of composition on the long phosphorescent $SrAl_2O_4 : Eu^{2+}$, Dy^{3+} Phosphor Crystals. *J. Electrochem. Soc.* **1997**, *144*, L243–L245.

37. Peng, J.; Hou, S.; Liu, X.; Feng, J.; Yu, X.; Xing, Y.; Su, Z. Hydrothermal synthesis and luminescence properties of hierarchical SrF_2 and $SrF_2:Ln^{3+}$ (Ln = Er, Nd, Yb, Eu, Tb) micro/nanocomposite architectures. *Mater. Res. Bull.* **2012**, *47*, 328–332.

3

Field Performance of Recycled Plastic Foundation for Pipeline

Seongkyum Kim and Kwanho Lee *

Department of Civil Engineering, Kongju National University, Cheonan 330-717, Korea;
E-Mail: tjdrua0614@kongju.ac.kr

* Author to whom correspondence should be addressed; E-Mail: kholee@kongju.ac.kr;

Academic Editor: Jung Ho Je

Abstract: The incidence of failure of embedded pipelines has increased in Korea due to the increasing applied load and the improper compaction of bedding and backfill materials. To overcome these problems, a prefabricated lightweight plastic foundation using recycled plastic was developed for sewer pipelines. A small scale laboratory chamber test and two field tests were conducted to verify its construction workability and performance. From the small scale laboratory chamber test, the applied loads at 2.5% and 5.0% of deformation were 3.45 kgf/cm^2 and 5.85 kgf/cm^2 for Case S1, and 4.42 kgf/cm^2 and 6.43 kgf/cm^2 for Case S2, respectively. From the first field test, the vertical deformation of the recycled plastic foundation (Case A2) was very small. According to the analysis based on the PE pipe deformation at the connection (CN) and at the center (CT), the pipe deformation at each part for Case A1 was larger than that for Case A2, which adopted the recycled lightweight plastic foundation. From the second field test, the measured maximum settlements of Case B1 and Case B2 were 1.05 cm and 0.54 cm, respectively. The use of a plastic foundation can reduce the settlement of an embedded pipeline and be an alternative construction method.

Keywords: field test; pipe deformation; recycled plastic foundation; sewer pipe

1. Introduction

Modern civilization has been developed in a city-oriented manner, and infrastructure has gained increasing importance to maintain the functions of cities. Sewage pipelines are crucial city infrastructure. Pipelines are analogous to blood vessels in our body; water pipes are like the main arteries, and sewage pipelines are like veins that act as the lifeline of the city. Once sewage pipelines are constructed, they are expected to remain in use for 20 years as a highly critical national infrastructure to carry wastewater and rainfall to sewage treatment plants or reservoirs [1].

The government of the Republic of Korea declared 2002, "the first year of sewer special maintenance", and has focused strong attention and investment on the sewage and infrastructure sectors. Because of this, the Ministry of the Environment has begun the Sewer BTL (Build-Transfer-Lease) Project and the Han River sewer improvement project, a sewer facility expansion project for upstream areas across the nation. They have also devoted more effort to ensure smooth progress of the sewer improvement project budget for the active sewer business.

In recent years, the use of flexible pipe including thermoplastic resin pipe and thermosetting plastic resin pipe has increased. The major advantages are easy use, excellent durability, flame-retardation and easy transportation. These sewer pipes are classified as flexible plastic pipe. The performance of flexible pipe highly depends on the interaction with the surrounding soils. Korea features several types of major damage to sewage pipelines: approximately 38% are joint defects, approximately 30% are extrusions of the connection part, approximately 30% are joint connection defects, 12% are due to the accumulation of impurities inside of the pipes, 10% are cracks and 10% are other types of damage. Joint defects, joint connection defects and other defects make up approximately 70% of the total damages and are related to quality management and the degree of the compaction of backfill materials that are executed around the pipes [2–4]. The age of the material, sink of pipe backfill materials and various other factors affect the damage to sewage pipelines. Furthermore, pipes buried underground are not visible, and risks are not visible. Therefore, some problems may result in serious damage, as depicted in Figure 1. More severe damage can occur in a short time frame due to heavy torrential rain caused by recent climate change, inadequate drainage for rainfall, the defects in connection joints and pipe damage [5,6]. RCA is a by-product of the construction and demolition activities of concrete structures. Concrete chunks are crushed into aggregates of variable sizes depending on the field of application. Various authors have reported on the geotechnical properties of RCA in geotechnical and pavement sub-base applications [7,8]. Several researchers have stated that appropriate design methods in pipe backfilling can be set in a way that can minimize or partly prevent contaminants [9–11]. Among them, Rahman et al. [8] investigated the suitability of recycled construction and demolition materials as alternative pipe backfilling materials for storm-water and sewer pipes. Three commonly found recycled construction and demolition waste materials—crushed brick, recycled concrete aggregate and reclaimed asphalt pavement—were investigated to assess their suitability as pipe backfilling material. The physical, geotechnical and chemical properties of these construction and demolition materials were compared to specifications from the local engineering and water authorities for typical quarried materials to assess their performance as a viable substitute for virgin quarried aggregates in pipe backfilling applications.

The objective of this study was to evaluate the performance of recycled plastic foundations for sewage systems. The use of plastic foundations could reduce damage to connections of sewage pipelines and improve their safety. In this research, two different types of tests (a small chamber test and a field test) were conducted. The purpose of the small chamber test is to verify the performance of PE pipe without and with plastic foundation. This result was used to setup the test combinations or the field test. Two types of field test were conducted. The first test checked the feasibility and constructability of plastic foundations. The second test simulated the field construction to check the deflection of PE pipe, which was installed on the manhole.

Figure 1. Failure of underground pipelines.

2. Results and Discussion

2.1. Testing Materials

2.1.1. PE Triple Wall Pipe

This study used a flexible PE triple wall pipe. The PE triple wall pipe is a flexible pipe molded by extrusion using high-density polyethylene (HDPE) [12]. This material is lightweight and easily-handled for field construction. In addition, the material is very resistant to acidic and alkaline substances. Care should be taken during construction for buoyancy and to compact the backfill materials around the pipeline [13]. Table 1 shows the standard specifications of PE pipe in Korea. In this research, the PE triple wall corrugated pipe used in this experiment featured a 250 mm inner diameter, 284 mm outer diameter, 20 mm thickness and 6000 mm length.

Table 1. Dimensions of polyethylene (PE) pipe in Korea.

Type	Inside (mm)	Outside (mm)	Thickness (mm)
D 150	150	180	15
D 200	200	232	16
D 250	250	284	17
D 300	300	340	20
D 350	350	398	24
D 400	400	460	30
D 450	450	510	30
D 500	500	570	35
D 600	600	694	47

2.1.2. Plastic Foundation

This study used a prefabricated plastic foundation with the injection molding of general plastic at high temperature and pressure, as shown in Figure 2. The injection-molded products are 1.5 m long and weighed 4 kg for ease of use. This process was simplified to meet the site conditions, and a cross section was fabricated to be applied in a straight section for testing purposes. The material used was HDPE. Injection molding products using 50% recycled plastic and 50% new plastic was applied with a 90 degree contact angle only.

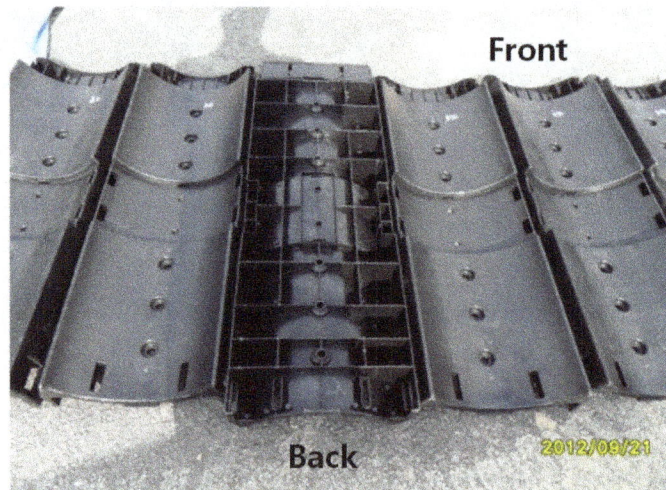

Figure 2. Plastic foundation by injection molding methods.

The properties of the composited plastic foundation were tested to fabricate the plastic foundations for sewage pipelines. The tested properties included density and the elastic modulus according to an impact resonance test and a uniaxial compressive test [1,14,15].

The material density was 0.93 t/m^3. The densities of the tested materials were low; specifically, this material was less dense than water. The elastic modulus was evaluated using an impact resonance testing machine. The measured elastic modulus value was 1764 N/mm^2. The uniaxial compressive test was conducted at 18 °C (normal temperature) and −15 °C (low temperature). Its yielded values of 21.22 N/mm^2 at room temperature and 21.22 N/mm^2 at low temperature, which were based on a uniaxial compressive test.

2.1.3. Backfill Materials

In general, natural sand is used a bedding and backfill material for sewer pipelines in Korea. The unit weight, internal friction angle and relative density were 15.46 kN/m^3, 43.5° and 80%, respectively. The natural sand is classified as SP by ASTM D 2487. In this study, recycled *in-situ* soil from a construction site near Cheonan City, Korea, was used. The characteristics of the *in-situ* soil were relatively uniform, and its water content was approximately 14%. This soil is classified as SC by ASTM D 2487.

2.2. Small Scale Lab Test

2.2.1. Properties of Sand Backfill

Two different testing conditions (Case S1 and Case S2) were used to verify the characteristics of deformation for buried PE pipes. The measured dry density and degree of compaction are shown in Table 2. The degrees of compaction at the top, middle and bottom were approximately 81.4%, 86.0% and 85.9%, respectively.

Table 2. Measured properties of sand backfill.

Items Case		Dry Density (g/cm³)		Degree of Compaction (%)	
		Average	Standard Deviation	Average	Standard Deviation
Case S1 (360° Sand Bedding)	Top	1.333	0.037	81.3	2.3
	Middle	1.394	0.019	85.0	1.1
	Bottom	1.408	0.031	85.8	1.9
Case S2 (50% Recycled Plastic Foundation	Top	1.336	0.013	81.5	0.8
	Middle	1.416	0.016	86.4	1.0
	Bottom	1.411	0.027	86.0	1.6

2.2.2. Vertical and Lateral Deformation of PE Pipe

Two different testing conditions (Case S1 and Case S2) were used to verify the characteristics of deformation due to the applied load that simulated traffic loading. The measured test results are shown in Table 3 and Figure 3. In Korea, the specification of allowable deformation for buried PE pipe is a 5% maximum deformation for the pipe diameter. For comparison, 2.5% and 5.0% of deformation were selected. The applied loads at 2.5% and 5.0% of deformation were 3.45 kgf/cm² and 5.85 kgf/cm² for Case S1, and 4.42 kgf/cm² and 6.43 kgf/cm² for Case S2, respectively. This means that the use of a 50% recycled plastic foundation (Case S2) can endure larger loading than that of sand bedding (Case S1).

In the case of lateral deformation, the measured deformations at 2.5% and 5% were 6.60mm and 12.51 mm for Case S1, and 5.90 mm and 11.30 mm for Case S2, respectively. Case S2 showed better resistance than Case S1.

Table 3. The measured vertical and lateral deformations of PE pipe.

Bedding Type		Case S1		Case S2	
Deformation Ratio According to Pipe Diameter		2.5%	5%	2.5%	5%
Vertical	Applied Stress (kgf/cm²)	3.45	5.85	4.42	6.43
Lateral	Deformation (mm)	6.60	12.51	5.90	11.30

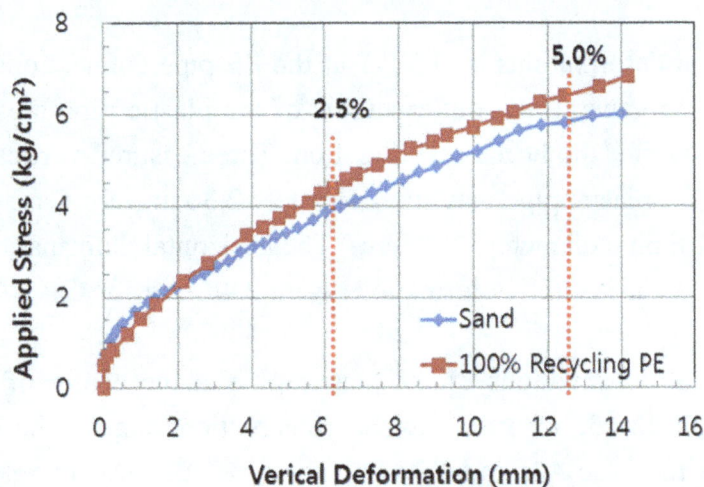

Figure 3. Vertical deformation of PE pipe.

2.2.3. Flatness of PE Pipe

The flatness is defined as the ratio of the vertical (compression) deformation to the lateral (tensile) deformation. A flatness 1.0 means that the pipe is a perfect circle. As the flatness number decreases below 1.0, the shape of the PE pipe is flattened. Table 4 and Figure 4 show the calculated flatness of the PE pipe for Case S1 and Case S2. In general, the calculated flatness of Case S2 is lower than that of Case S1, meaning that the PE pipe in Case S2 is more resistant than that of Case S1. The flatness of Case S1 and Case S2 at 6.0 kgf/cm^2 of applied stress was 0.896 and 0.921, respectively. The flatness number used to calculate the flowable flux.

Table 4. The measured flatness of PE pipe for each case at same applied stress.

Type	Case S1			Case S2		
Applied Stress (kgf/cm^2)	2.0	4.0	6.0	2.0	4.0	6.0
Flatness	0.988	0.949	0.896	0.988	0.960	0.921

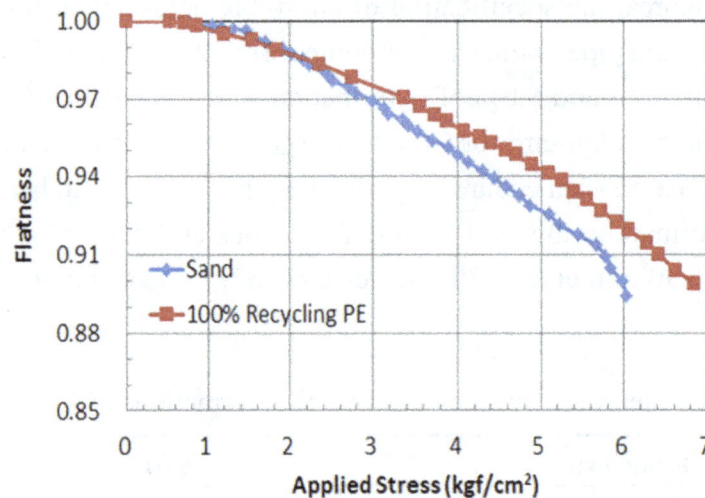

Figure 4. The flatness of PE pipe with applied stress.

2.3. First Field Test

The vertical and horizontal deformations by LVDT at the PE pipe connections (CN) of the plastic pipe are shown in Figure 5. In general, the deformations of the plastic pipe are compressive for the vertical direction and expansive for the horizontal direction. The measured vertical deformation at the PE pipe connection (CN) of the plastic pipe was approximately 2.36 mm for Case A1 and 2.35 mm for Case A2, or 8% of the total PE pipe diameter (300 mm). The horizontal deformations at the same point are related to tensile strain. The measured deformations were 2.49 mm for Case A1, and 1.82 mm for Case A2, respectively.

The vertical and horizontal deformations of the plastic pipe at the center of the pipe (CT) were measured and are shown in Table 5, according to the construction stages. The measured horizontal deformations were 0.39 mm for Case A1, and 1.34 mm for Case A2, respectively. In Case A1, there were some deformations of the sand bedding during the compaction stages of the backfill material, which induced the deformation of the PE plastic pipe for the vertical direction. This deformation is relatively large at the connection point and small at the center of the pipe. On the other hand, the vertical deformation of the recycled plastic foundation (Case A2) is very small. This means that the vertical pressure transferred to the horizontal direction. According to the analysis based on the PE pipe deformation at the connection (CN) and at the center (CT), the pipe deformations at each part for Case A1 were larger than those for Case A2, which adopted the recycled lightweight plastic foundation. These results mean that the use of recycled lightweight plastic foundation can reduce the differential deformation of PE pipe in the longitudinal direction.

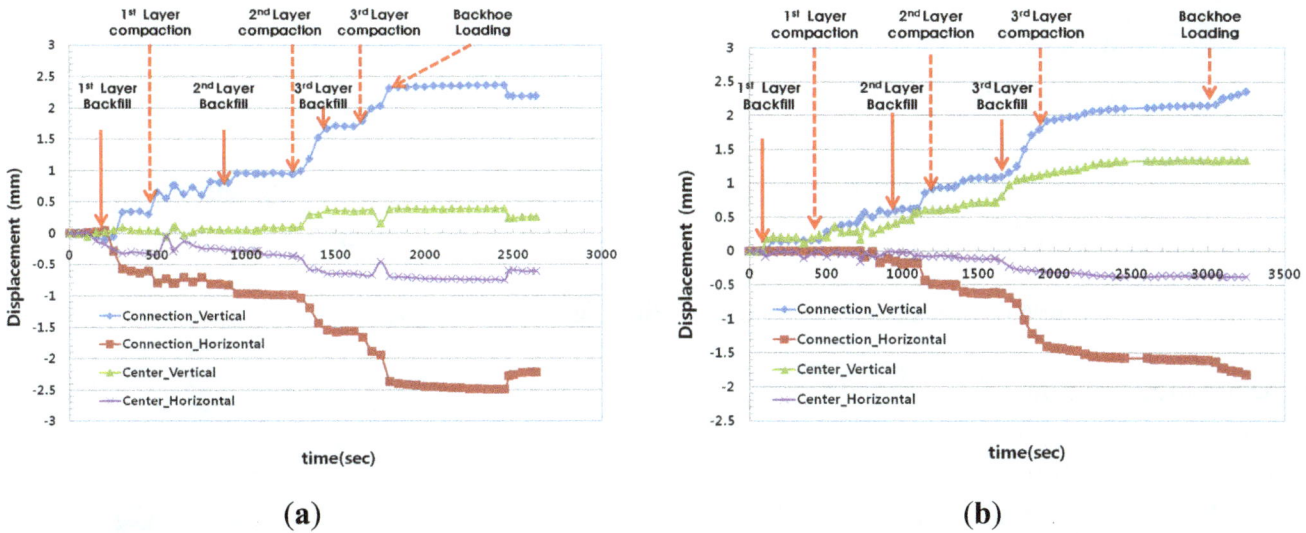

(a)   (b)

Figure 5. The displacement of each case with construction stages. (a) Case A1; (b) Case A2.

Table 5. Pipe Deformation at the Connection (CN) and the Center Position (CT).

| Case | At Connection (CN) | | | | At Center Position (CT) | | | |
| | Case A1 | | Case A2 | | Case A1 | | Case A2 | |
Construction Stage	Vertical	Horizontal	Vertical	Horizontal	Vertical	Horizontal	Vertical	Horizontal
After1st backfill	0.55	−0.73	0.21	−0.04	0.03	−0.05	0.29	−0.04
After 1st compaction	0.82	−0.82	0.24	−0.18	0.06	−0.24	0.41	−0.03
After 2nd backfill	0.94	−0.97	0.25	−0.02	0.04	−0.28	0.62	−0.08
After 2nd compaction	0.94	−0.99	0.39	−0.13	0.09	−0.37	0.72	−0.12
After 3rd backfill	1.18	−1.20	0.73	−0.45	0.29	−0.58	1.19	−0.32
After 3rd compaction	1.78	−1.67	1.46	−1.11	0.36	−0.67	1.30	−0.37
After backhoe loading	2.36	−2.49	2.34	−1.99	0.39	−0.75	1.34	−0.38

2.4. Second Field Test

The installed pipe length was 15 m from left to right. After installing the plastic foundation and the PE pipe, the backfill was dumped and compacted according to KS standard specifications. A dump truck with a full cargo was used as traffic loading right on the embedded pipeline. The settlements of the pipeline based on the construction level were measured using a level and are shown in Figure 6. The settlement of Case B2 with the recycled plastic foundation is smaller than that of Case B1 with the sand bedding. The measured maximum settlements of Case B1 and Case B2 are 1.05 cm and 0.54 cm, respectively. The use of a plastic foundation can reduce the settlement of the embedded pipeline.

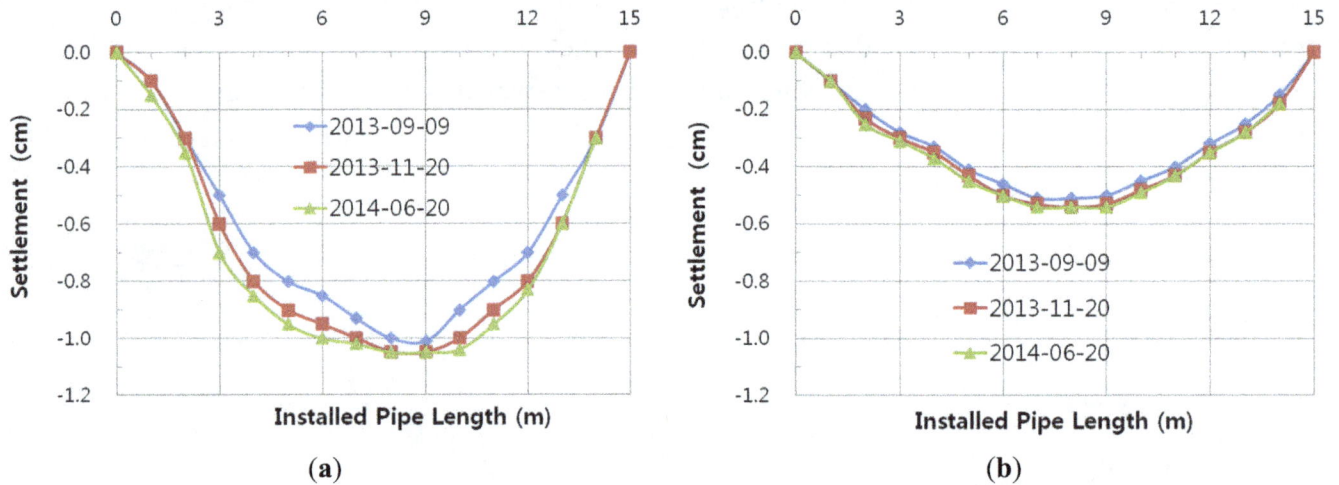

Figure 6. Measured settlement of Pipe. (**a**) Case B1; (**b**) Case B2.

3. Experimental Section

3.1. Small Scale Lab Test

The small-scale chamber test was performed to evaluate the performance of the plastic foundation for pipelines. The purpose of the small chamber test was to verify the performance of the PE pipe without and with a plastic foundation. This result was used to setup the test combinations or the field test. The dimensions of the small-scale chamber were 1.4 m × 0.6 m × 0.9 m, and it was reinforced with horizontal and vertical flat metal strips, as shown in Figure 7. As observed in the figures, a square metal plate with an elliptical hole was attached to the box with nuts and bolts. A rubber membrane sheet was placed between the chambers and the plate before tightening the plate, and a 30-cm circular hole was cut in the plate to insert the pipe. The membrane was used to ensure water tightness without affecting the behavior of the pipe.

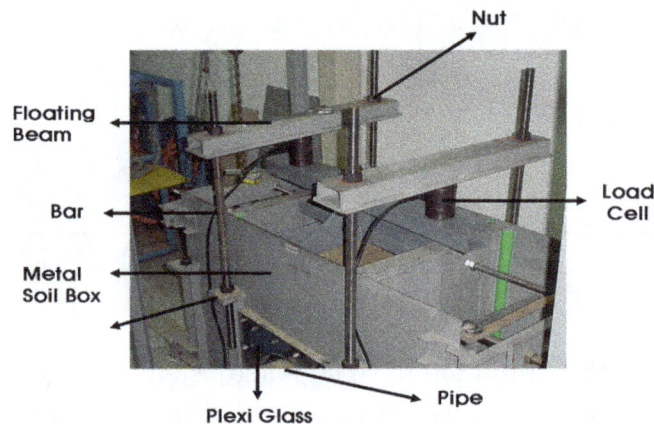

Figure 7. Small-scale chamber.

The loading system with a circular plate on top of the chamber and two metallic blocks over the plate is shown in Figure 8. A metal beam with a load cell was placed between the metal blocks and the beam. The maximum capacity of each load cell was two tons. The load cell was pre-calibrated using a 5-ton universal testing machine. A linear variable differential transformer (LVDT), shown in Figure 9, was installed to obtain the vertical and horizontal deflections of the pipe. An automated data acquisition system was used to collect the data.

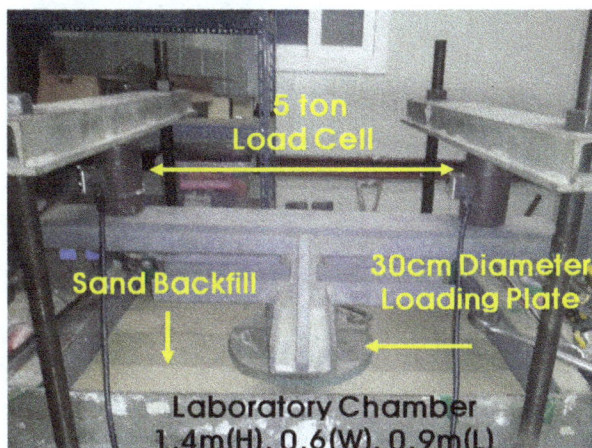

Figure 8. Loading system.

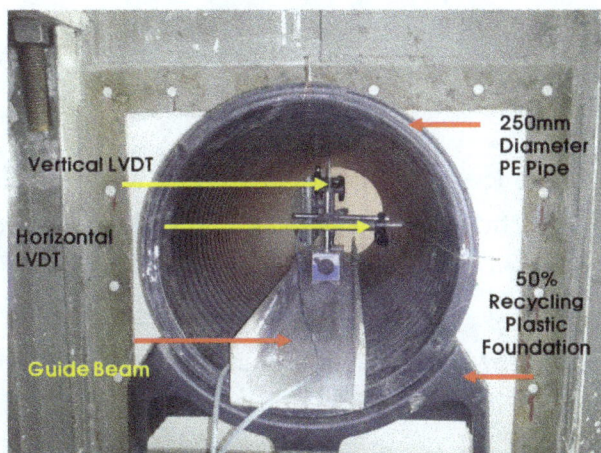

Figure 9. Setup of Linear Variable Differential Transducer (LVDT).

The two different conditions of the test are presented in Table 6 and Figure 10. Natural sand and 50% recycled plastic foundation were used as a bedding material. The natural sand was used as a backfill material. A PE pipe with a 25 cm diameter and 0.7 cm thickness was installed. In the laboratory chamber test, the chamber size was limited, which could affect the test results due to boundary conditions. To minimize the boundary effects, a vinyl sheet was attached to the chamber wall to reduce the friction effect. The length of the PE pipe was 1 m. In the field, the usual length of a PE pipe is over 6 m. The short length of the PE pipe induced a smaller deflection of the PE pipe at the center of the applied load. In this project, obtaining the real deformation or deflection of PE pipe in the field was not easy. However, we obtained the general trend of deformation and deflection of the PE pipe with and without the plastic foundation.

Table 6. Testing cases for small scale lab test.

Type	Bedding and Foundation	Backfill
Case S1	360° Sand Bedding	Natural Sand
Case S2	50% Recycled Plastic + 50% New Plastic Foundation	Natural Sand

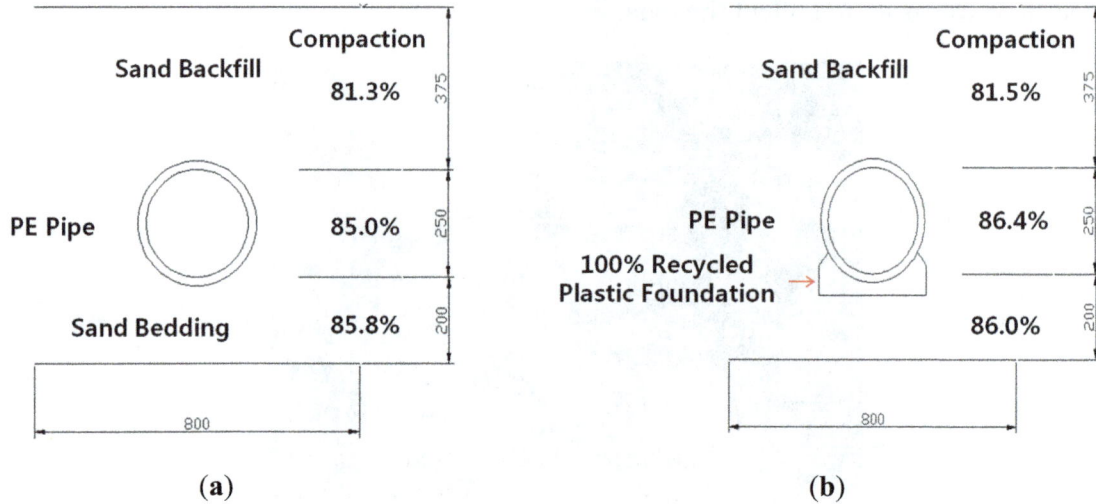

(a)   (b)

Figure 10. The two testing conditions, (**a**) Case S1; (**b**) Case S2.

3.2. First Field Test

3.2.1. Procedure of Field Construction

For the evaluation of field performance, two types of field tests with recycled plastic foundation for sewer pipelines were conducted. The first test checked the feasibility and constructability of plastic foundations. The second test simulated the field construction to check the deflection of PE pipe, which was installed on the manhole.

The process of field construction followed the Korean Standard Specification. As shown in Figure 11, the vertical trench was 1.5 m of deep and 1.0 m of wide.

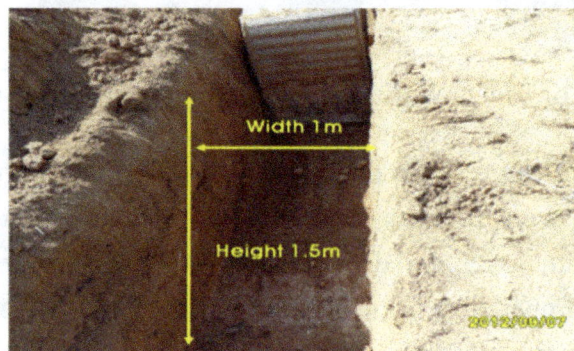

Figure 11. Trench excavation for the plastic foundation.

Three different cases (Table 7) with different bedding materials, foundations and backfill materials were used in this research. Figure 12 shows the field construction process, including the excavation of the trench, sand bedding, pipe installation, installation of measuring instruments, backfill and compaction. The density of each compaction layer was measured, and the average relative density was over 80%.

Table 7. Cases for the first field test.

Type	Bedding and Foundation	Backfill
Case A1	Sand	One Layer of Sand + Two Layers of Original Soil
Case A2	50% Recycled Plastic + 50% New Plastic Foundation	Three Layers of *In-Situ* Soil

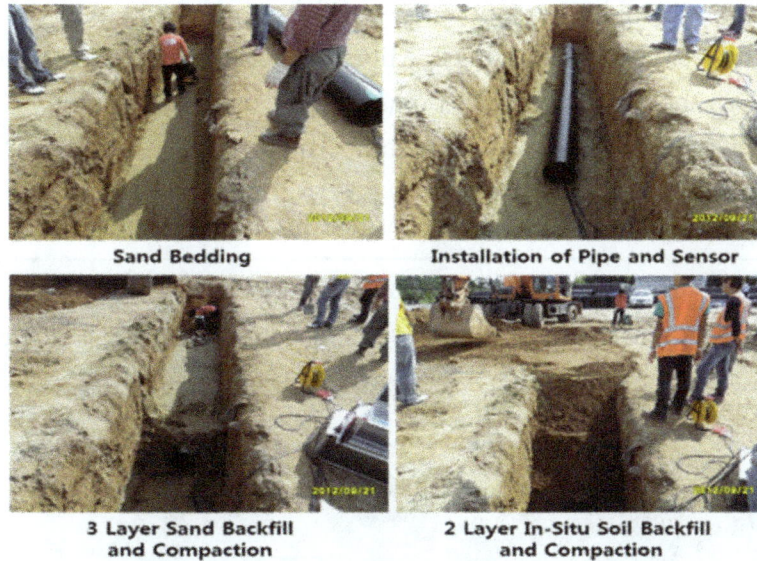

Sand Bedding   Installation of Pipe and Sensor

3 Layer Sand Backfill and Compaction   2 Layer In-Situ Soil Backfill and Compaction

Figure 12. Field construction process.

3.2.2. Measuring Instruments and Load

The measuring instruments are shown in Figure 13. To obtain the vertical and horizontal deformations of the PE pipe, two sets of LVDT for the vertical and lateral directions were installed at two different locations, such as at the PE pipe connection (CN) and the center of the PE pipe (CT). To measure the longitudinal strain of the PE pipe, three strain gauges were installed at the PE pipe connection (S1) to the center (S3) with 1 m of intervals. The backhoe was used to simulate real traffic loads in the roadway.

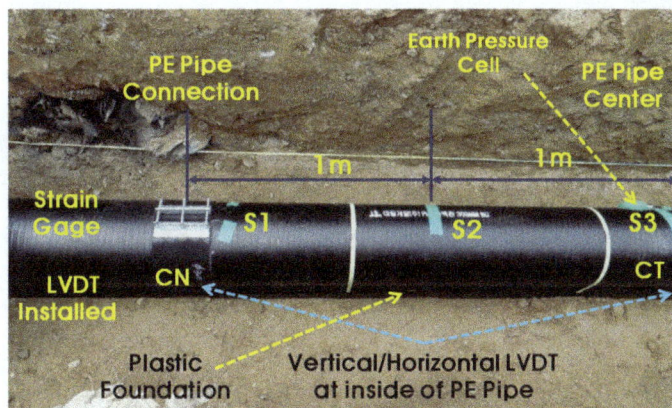

Figure 13. Measuring system.

3.3. Second Field Test

The second field test (Figures 14 and 15) was conducted to verify the performance of the recycled lightweight plastic foundation and to evaluate the deflection of the PE pipe. The stages of field construction were very similar to the first field test. The main different was the length of the PE pipeline, which varied from 6 to 15 m, with two connection parts. The PE pipe was installed using 15 m of manhole to manhole, which means that the pipe had fixed support at the beginning and ending positions. Table 8 shows the test cases of the second field test. Case B1 is a common process for installing PE pipelines with sand bedding in Korea. Case B2 is a brand new process using the recycled lightweight plastic foundation for sewer pipelines. After field construction, the truck load with fully occupied sand was located on the right top of the pipeline. Measurements were performed three times at two month intervals to check the variation of the installed pipeline elevation.

Table 8. Cases for the second field test.

Type	Bedding and Foundation	Backfill
Case B1	Sand Bedding	180° and one Layer Sand + two Layer Original Soil
Case B2	50% Recycled Plastic + 50% New Plastic Foundation	Three Layers of *In-Situ* Soil

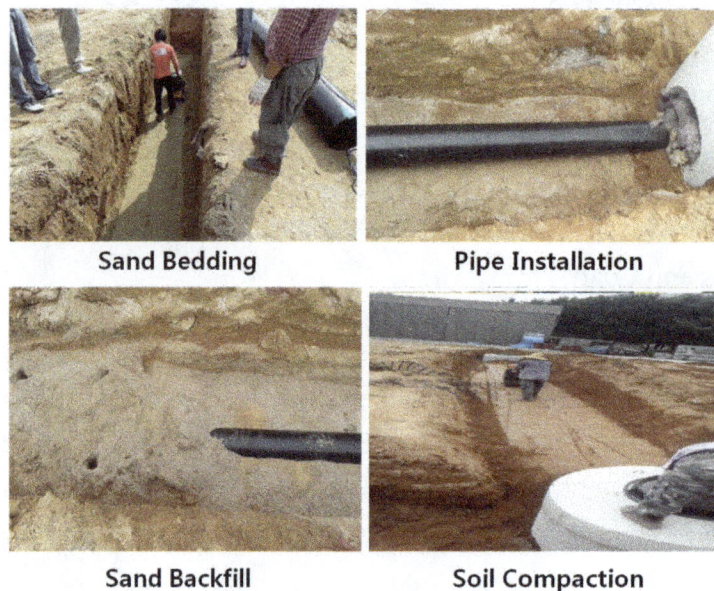

Sand Bedding   Pipe Installation

Sand Backfill   Soil Compaction

Figure 14. Second field construction for common process.

Installation of   Soil Compaction
Plastic Foundation

Figure 15. Second field construction for new process.

To measure the pipeline elevation after field construction, a level and ruler (Figure 16) were used. First, the ruler was set at the beginning position inside of the PE pipe. The elevation of the pipeline was measured at 1 m intervals of 15 m pipeline using a level.

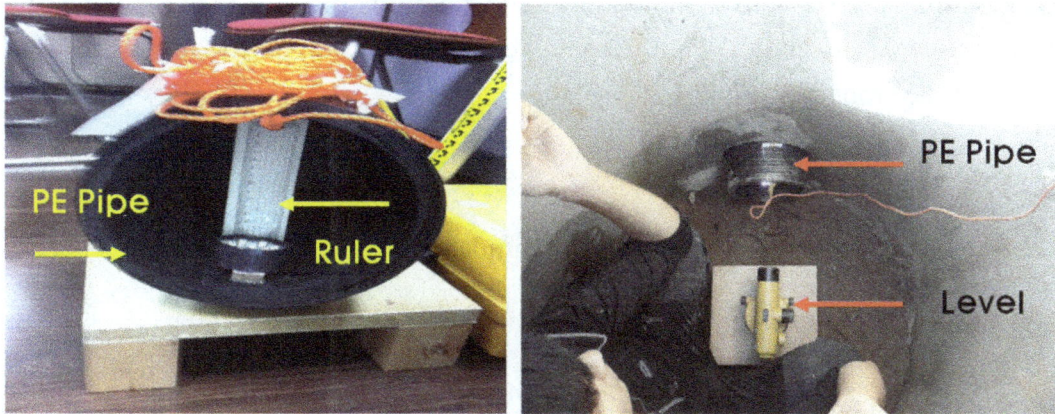

Figure 16. Measuring system of the second field construction.

4. Conclusions

The research presented in this paper aimed to evaluate the performance of recycled plastic foundations for embedded pipelines. A small scale laboratory chamber test and two different field tests were conducted. Despite the possible limitations of the number of tests, the following conclusions can be drawn:

1) From the small scale laboratory chamber test, the applied loads at 2.5% and 5.0% of deformation were 3.45 kgf/cm^2 and 5.85 kgf/cm^2 for Case S1, and 4.42 kgf/cm^2 and 6.43 kgf/cm^2 for Case S2, respectively. These results means that the use of 50% recycled plastic foundation (Case S2) can endure larger loading than that of sand bedding (Case S1). For the case of lateral deformation, the measured deformations at 2.5% and 5% were 6.60 mm and 12.51 mm for Case S1, and 5.90 mm and 11.30 mm for Case S2, respectively. Therefore, Case S2 showed better resistance than Case S1.

2) The calculated flatness of Case S2 is lower than that of Case S1, which means that the PE pipe of Case S2 is more resistant than that of Case S1. The flatness of Case S1 and Case S2 at 6.0 kgf/cm^2 of applied stress was 0.896 and 0.921, respectively.

3) From the first field test, the vertical deformation of the recycled plastic foundation (Case A2) was very small. According to the analysis based on the PE pipe deformation at the connection (CN) and at the center (CT), the pipe deformation at each part for Case A1 was larger than for Case A2, which adopted the recycled lightweight plastic foundation.

4) In general, as the load increased, the measured strain increased for all of the cases. In Case A1, the strain at S3 was smaller than at S1. In this case, the measured strains were not uniform at each measured location (S1, S2 and S3), which means that the applied overburden pressure, including the backfill and the backhoe load, was not uniform on the PE pipe due to the non-uniform compaction of sand bedding and backfill around the PE pipe. This non-uniform compaction induced the differential settlement. In Case A2, the measured strains at S1, S2 and S3 were relatively uniform. The use of a recycled lightweight plastic foundation uniformly supported the

installed PE pipe, meaning that the overburden pressure was relatively uniformly distributed on the PE pipe.

5) From the second field test, the settlement of Case B2 with the recycled plastic foundation was smaller than that of Case B1 with sand bedding. The measured maximum settlements of Case B1 and Case B2 were 1.05 cm and 0.54 cm, respectively. The use of a plastic foundation can reduce the settlement of the embedded pipeline.

Acknowledgments

This work was supported by Research Grant (2014) funded by Kongju National University.

Author Contributions

Seongkyum Kim conducted the laboratory test, and Kwanho Lee took charge of the whole project.

Conflicts of Interest

The authors declare no conflicts of interest.

References

1. Kim, S.K.; Lee, D.H.; Lee, K.H. Performance evaluation of plastic foundation for sewage pipeline. *ASTM J. Test. Eval.* **2014**, doi:10.1520/JTE20130127.

2. Park, J.S. Necessity of rehabilitation and current state of domestic sewage pipe network. *KSCE Mag.* **2007**, *55*, 135–143.

3. Korean Society of Water and Wastewater (KSWW). *A Study on Deformation Guideline of Sewage Pipeline*; KSWW: Seoul, Korea, 2011; p. 278.

4. Korea Ministry of Environment. *Sewer Facility Standard Construction Procedure*; Korea Ministry of Environment: Seoul, Korea, 2010; p. 590.

5. Kim, S.; Lee, K.; Jeon, S.; Hwang, C. Development of foundation for failure reduction of sewage pipeline. In Proceedings of 2011 Conference, Korean Society of Civil Engineers, Ilsan, Kyunggi Do, Korea, 2 November 2011.

6. Petersen, D.L.; Le, G.; Nelson, C.R.; McGrath, T.J. Analysis of live loads on culverts. In Proceeding of the 2008 Transportation Research Board Annual Meeting, Washington, DC, USA, 13–17 January 2008; pp. 3–5.

7. Arulrajah, A.; Piratheepan, J.; Disfani, M.M.; Bo, M.W. Geotechnical and geo-environmental properties of recycled construction and demolition materials in pavement sub-base applications. *J. Mater. Civil Eng.* **2013**, *25*, 1077–1088.

8. Rahman, M.A.; Imteaz, M.; Arulrajah, A.; Disfani, M.M. Suitability of recycled construction and demolition aggregates as alternative pipe backfilling materials. *J. Clean. Prod.* **2014**, *66*, 75–84.

9. Hellweg, U.; Fischer, U.; Hofstetter, T.B.; Hungerbühler, K. Site-dependent fate assessment in LCA: Transport of heavy metals in soil. *J. Clean. Prod.* **2005**, *13*, 341–361.

10. Susset, B.; Grathwohl, P. Leaching standards for mineral recycling materials—A harmonized regulatory concept for the upcoming German recycling decree. *J. Waste Manag.* **2011**, *31*, 201–214.

11. Disfani, M.M.; Arulrajah, A.; Bo, M.W.; Sivakugan, N. Environmental risks of using recycled crushed glass in road applications. *J. Waste Manag.* **2012**, *20*, 170–179.
12. Plastics Pipe Institute. New PE pipe material designation codes. In *Second Edition Handbook of PE Pipe*; Plastics Pipe Institute: Irving, TX, USA, 2007.
13. American Society for Testing Material (ASTM). *Innovations in Controlled Low-Strength Materials (Flowable Fill)*; ASTM STP 1459; American Society for Testing Material: West Conshohocken, PA, USA, 2004; p. 159.
14. Lee, D.H.; Lee, K.H. Laboratory loading test of light-weight prefabricated plastic foundation for sewage pipeline. *J. Korea Acad. Ind. Coop. Soc.* **2012**, *13*, 2757–2762.
15. Kang, S.Y.; Park, J.S.; Lee, K.H. *Study on Basement Modeling of Sewage Pipeline Based on Comparison of Finite Element Analysis Results with Experimental Data*; The Korea Academia-Industrial Cooperation Society: Cheonan, Korea, 2013; pp. 593–596.

4

Process Optimisation to Control the Physico-Chemical Characteristics of Biomimetic Nanoscale Hydroxyapatites Prepared Using Wet Chemical Precipitation

Piergiorgio Gentile *, Caroline J. Wilcock, Cheryl A. Miller, Robert Moorehead and Paul V. Hatton *

School of Clinical Dentistry, University of Sheffield, 19 Claremont Crescent, Sheffield S10 2TA, UK; E-Mails: mta07cw@sheffield.ac.uk (C.J.W.); c.a.miller@sheffield.ac.uk (C.A.M.); r.moorehead@sheffield.ac.uk (R.M.)

* Authors to whom correspondence should be addressed;
   E-Mails: p.gentile@sheffield.ac.uk (P.G.); paul.hatton@sheffield.ac.uk (P.V.H.);

Academic Editor: Andrew J. Ruys

Abstract: Hydroxyapatite nanoscale particles (nHA) were prepared by wet chemical precipitation using four different synthesis methods. Differences in physico-chemical properties including morphology, particle-size, and crystallinity were investigated following alteration of critical processing parameters. The nanoparticles were also studied using X-ray diffraction (XRD), Fourier Transform infrared spectroscopy in attenuated total reflectance mode (FTIR-ATR), and transmission electron microscopy (TEM) with energy dispersive X-ray (EDS) spectrometry. The results showed that the particles obtained were composed of nHA, with different morphologies and aspect ratios (1.5 to 4) and degrees of crystallinity (40% to 70% following calcination) depending on the different process parameters of the synthesis method used, such as temperature, ripening time and pH. This study demonstrated that relatively small adjustments to processing conditions of different wet chemical preparation methods significantly affect the morphological and chemical characteristics of nHA. For the predicable preparation of biomimetic nHA for specific applications, the selection of both production method and careful control of processing conditions are paramount.

Keywords: bioceramics; hydroxyapatite; nanoparticles; processing; wet-precipitation

1. Introduction

Hydroxyapatite (HA) is a synthetic biomaterial commonly used in bone tissue repair and augmentation on account of its recognised biocompatibility and surface active properties [1]. Its chemical composition ($Ca_{10}(PO_4)_6(OH)_2$) is comparable to the mineral component of natural bone, and associated bioactivity promotes bone tissue regeneration at the site of implantation [2]. Furthermore, newly deposited mineral in developing bone and tooth tissues is organised at the nanoscale [3], consisting of nanocrystalline HA that is measures less than 100 nm on all axes [4,5]. These observations suggest that the use of synthetic nanoscale HA (nHA) in medical devices and regenerative medicine represents a biomimetic strategy that has the potential to improve clinical performance.

It is generally held that synthetic nHA is defined as having a grain size less than 100 nm in at least one direction, and it may provide several advantages over micro- and macro-scale HA. Various studies report that ceramic biomaterials based on nano-sized HA exhibit enhanced resorbability [6,7] and greater bioactivity [8,9] than micron-sized HA (mHA). The release of calcium ions from nHA is also similar to that from biological apatite and significantly faster than that from coarser crystals. In addition, new models for nanoscale enamel and bone demineralization suggest that demineralization reactions may be inhibited when particle sizes fall into certain critical nanoscale levels [10]. Some studies have also reported that nHA possesses a significant capability of decreasing apoptotic cell death and hence improving cell proliferation and cellular activity related to bone growth [8,11]. The improved cell proliferation and differentiation may be due to the superior surface functional properties of nano-sized HA compared to its microphase counterpart; indeed nHA has higher surface area and surface roughness, resulting in better cell adhesion and cell–matrix interactions [12]. In fact, as reported by Huang *et al.* [13] in their study, the HA size influenced dramatically the osteogenic differentiation of rat bone marrow derived mesenchymal stem cells into osteoblasts. The authors observed that nHA, compared with the traditional mHA, induced: (1) a substantial increase in the transcriptional expression of the early osteoblast-related genes including core binding factor alpha 1, alkaline phosphatase and the alpha 1 chain of type I collagen, and (2) a higher osteoinductivity, implying that nHA changed the micro-environments of cell culture, adsorbed protein, formed a neo-matrix and significantly enhanced osteogenesis.

Several procedures for preparing nanoscale HA have been identified, involving hydrothermal, sol-gel, wet-chemical, and biomimetic deposition methods [14]. These methods are based on known chemical synthesis routes, in which accuracy is considered to be a key determinant of repeatability and outcome. Careful management of parameters such as pH, reaction time, temperature, and concentration of the reactants, together with the proper selection of the precursor materials, are therefore thought to be important during HA synthesis, as these parameters and materials may affect the composition and properties of the final product. However, the processing conditions that best reproduce the scale and morphology of natural nHA have not been reported. In this study, HA nanoparticles were therefore prepared by the wet chemical precipitation from aqueous solutions according to four protocols presented in literature, and compared to data available for nHA identified in mammalian bone. It is highly likely that biological response and therefore clinical performance is influenced by the chemistry and morphology of the HA nanoparticle, so there is now a greater need to better understand the nature of the nanoscale particles produced using the most promising methodologies. The aim of this systematic study was therefore to investigate the similarities and differences in physico-chemical properties of nHA powders prepared using different methods and conditions. This research will enable the scientific and medical device communities to make progress in the development of biomimetic nHA to improve clinical performance.

2. Results and Discussion

2.1. Results

2.1.1. The Influence of Processing Conditions on the Physico-chemical Characteristics of nHA before Calcination

XRD spectra of all the dried powders, precipitated by the four experimental methods (details given in Section 3.1), are shown in Figure 1. In all samples, only HA reflections were detected as confirmed with JCPDS file No. 09-432. Broad diffraction peaks were observed for the HA powders synthesised at low temperature (using BiancoB and Pang protocols). The increase of the synthesis temperature caused sharpened diffraction peaks, indicating increased crystallinity of nHA powders.

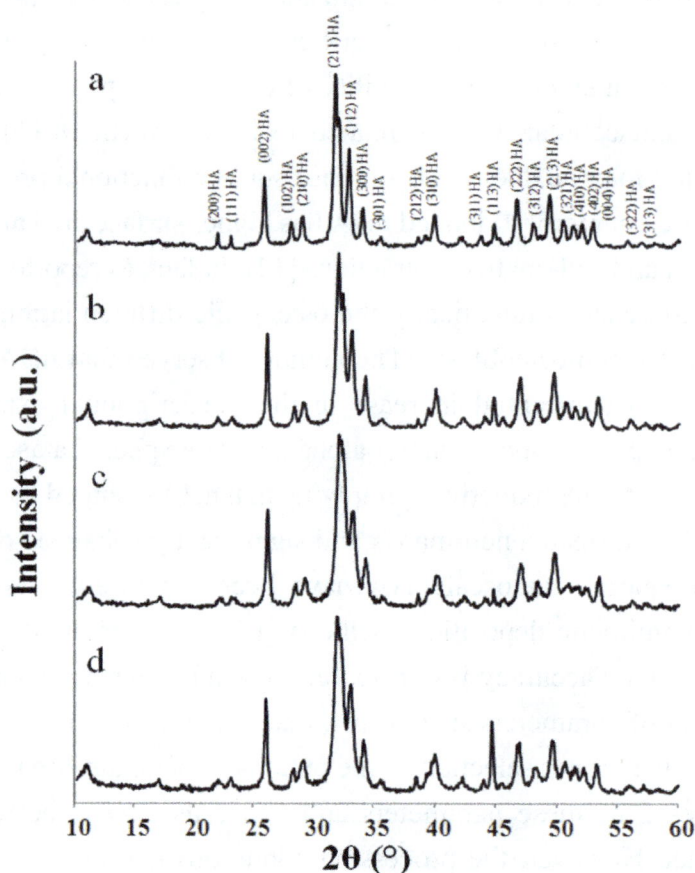

Figure 1. XRD patterns of as-precipitated hydroxyapatite (HA) powders synthesised following: (a) Prakash; (b) BiancoA; (c) BiancoB; and (d) Pang protocol.

The observed degrees of crystallinity and crystallite size for these samples were calculated using Rietveld refinement as described in Section 3.2 and summarised in Table 1. The crystallinity ranged between 21% ± 0.9% for Pang protocol synthesised at 40 °C up to about 68% ± 0.5% for Prakash protocol prepared at 95 °C. It was noticed that the increase of the crystallinity with the temperature was not linear.

There was no significant change in crystallinity when the synthetic temperature was lower than 70 °C. The crystallite sizes of the "as prepared" HA powders were less than 50 nm, and a slight increase in crystallite size of the HA powders with the increase of synthesis temperature was noted (Table 1).

Table 1. The χ_c and χ_s for the HA powders before and after calcination at 650 °C.

Protocols	Crystallinity χ_c (-)		Crystallite Size χ_s (nm)	
	Before calcination	After calcination	Before calcination	After calcination
Prakash	0.68 ± 0.05	0.70 ± 0.04	35.4 ± 2.2	25.1 ± 0.8
BiancoA	0.45 ± 0.08	0.61 ± 0.06	31.4 ± 1.1	28.5 ± 1.2
BiancoB	0.42 ± 0.09	0.47 ± 0.07	30.8 ± 1.5	29.1 ± 0.7
Pang	0.21 ± 0.09	0.39 ± 0.05	29.0 ± 1.9	27.0 ± 2.8

The FTIR-ATR analysis of all the HA powders before calcination is given in Figure 2.

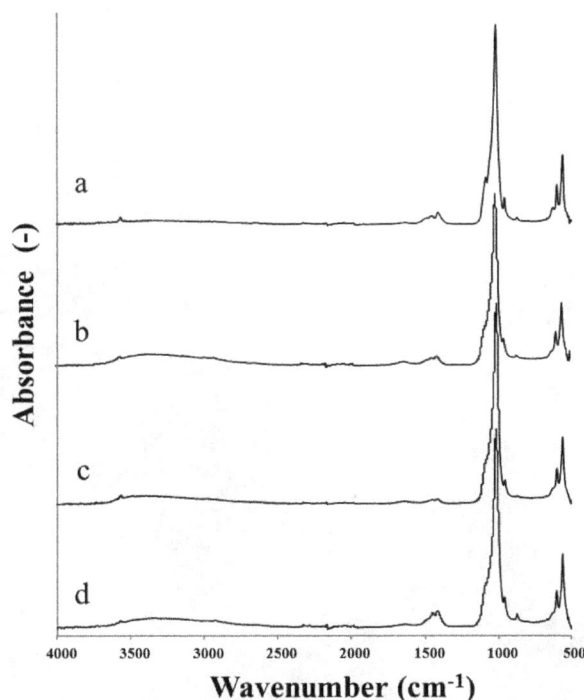

Figure 2. FTIR-ATR spectra of as-precipitated HA powders synthesised following: (a) Prakash; (b) BiancoA; (c) BiancoB; and (d) Pang protocol.

The results showed that all the HA powders exhibited the characteristic bands of hydrated partially carbonated hydroxyapatite [15]: ν_{OH} (3570 cm^{-1}) and δ_{OH} (633 cm^{-1}); $\nu_1(PO)4$ (962 cm^{-1}), $\nu_3(PO)4$ (broad band 1090–1040 cm^{-1}) and $\nu_4(PO)4$ (565 cm^{-1} and 603 cm^{-1}); $\nu_2(CO)3$ (875 cm^{-1}) and $\nu_3(CO)3$ (1420 cm^{-1} and 1457 cm^{-1}). The broad band at 3700–2500 and the sharp peak at 1637 cm^{-1} are most likely associated with the presence of either absorbed or combined water. The spectra contained bands attributed to carbonate substitution in HA, which probably resulted from the presence of $CaCO_3$ in the reactants and dissolved CO_2 from the atmosphere. The lack of the characteristic $\nu_3(CO)3$ peak at 1500 cm^{-1} suggested that only B-type carbonated hydroxyapatite was formed [16]. In the spectra obtained from the nHA produced using Prakash and BiancoB methods, the characteristic peaks (540–530 cm^{-1}, 855 cm^{-1}, 1130 cm^{-1} and 1210 cm^{-1}) of HPO_4 were not detected [15]. Moreover, the lack of bands in the range 750–700 indicates the absence of calcium carbonates, *i.e.*, calcite (712 cm^{-1}), aragonite (713 cm^{-1} and 700 cm^{-1}) and valerite (745 cm^{-1}). Finally, in BiancoA the absorption band at 1385 cm^{-1} ascribed to nitrate was also absent, due to the effective washing process.

Figure 3 shows the TEM micrographs of the HA powders. The appearance of these nanoparticles was quite different from each other, showing nano-sized needle-like with different aspect ratios. In particular, at low synthesis temperature the HA nanoparticles carried a needle-like morphology (30–50 nm in width and 80–120 nm in length with a shape factor ranging from 3 to 4), more irregular and with less clear contours, as observed for BiancoB and Pang methods (Figure 3c,d). On the other hand, increasing the reaction temperature (up to 70 °C for the BiancoA and 95 °C for the Prakash protocol) changed the crystal from a needle-like shape to more regular form with a lower aspect ratio (20–40 nm in width and 60–80 nm in length with a shape factor about 1.5–2), with clearer contours (Figure 3a,b).

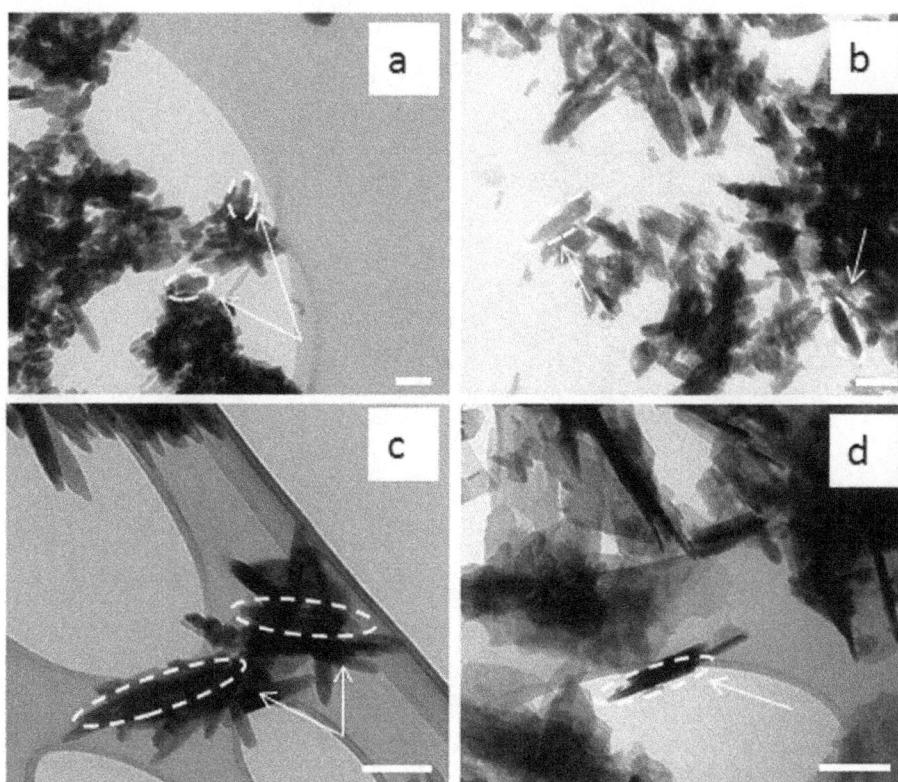

Figure 3. TEM micrographs of as-precipitated HA powders synthesised following: (**a**) Prakash; (**b**) BiancoA; (**c**) BiancoB; and (**d**) Pang protocol. Scale bars = 100 nm.

2.1.2. Effect of Calcination

The XRD diffraction patterns for the HA powders prepared following the different routes after calcination at 650 °C for 2 h are shown in Figure 4.

At 650 °C pure HA was present without any evidence of other phases of calcium phosphate or impurities. Moreover, all the calcinated HA powders showed much sharper diffraction peaks and higher crystallinity (X_c) compared with the corresponding "as prepared" powders, as shown in Table 1. The crystallinity ranged between 39% ± 0.5% for the Pang protocol synthesised at 40 °C up to about 70% ± 0.4% for the Prakash protocol. After calcination, the X_s values increase for the HA powders with lower crystallinity and smaller crystallite sizes but remain the same for those with high crystallinity and larger crystallite sizes.

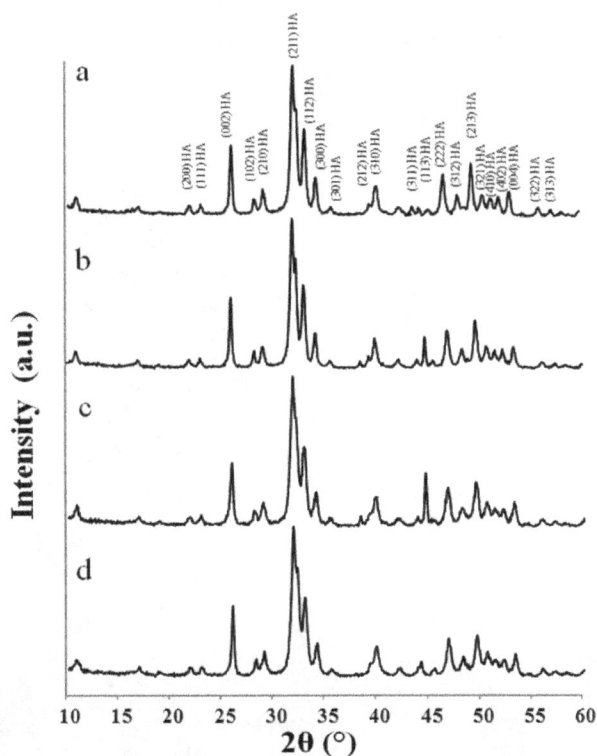

Figure 4. XRD patterns of calcinated HA powders synthesised following: (a) Prakash; (b) BiancoA; (c) BiancoB; and (d) Pang protocol.

FTIR-ATR spectra of calcinated samples (Figure 5) differed considerably from those of as-precipitated materials. The intensity of peaks associated to H_2O was noticeably decreased, indicating that the water loss had occurred during calcination.

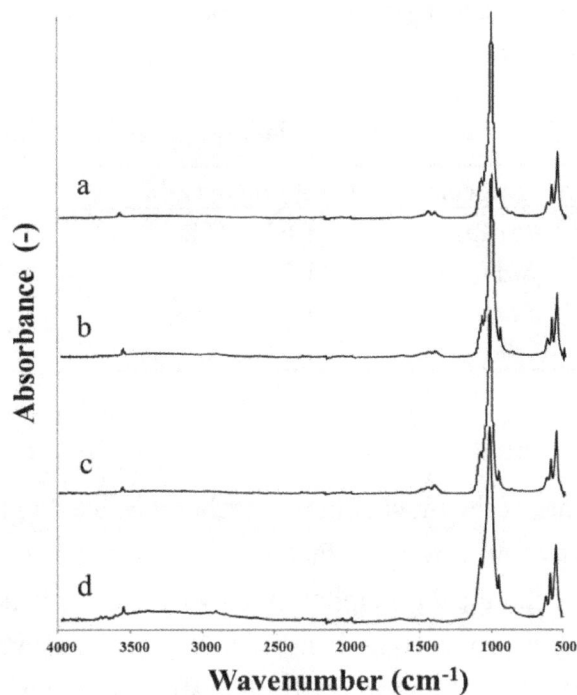

Figure 5. FTIR-ATR spectra of calcinated HA powders synthesised following: (a) Prakash; (b) BiancoA; (c) BiancoB; and (d) Pang protocol.

The morphology was again investigated using TEM (Figure 6). Following heat treatment, all particles showed a decrease in the size with a shape factor ranging from 1.25 to 2.5, in which a large amount of smaller new crystallites were formed during the calcination process.

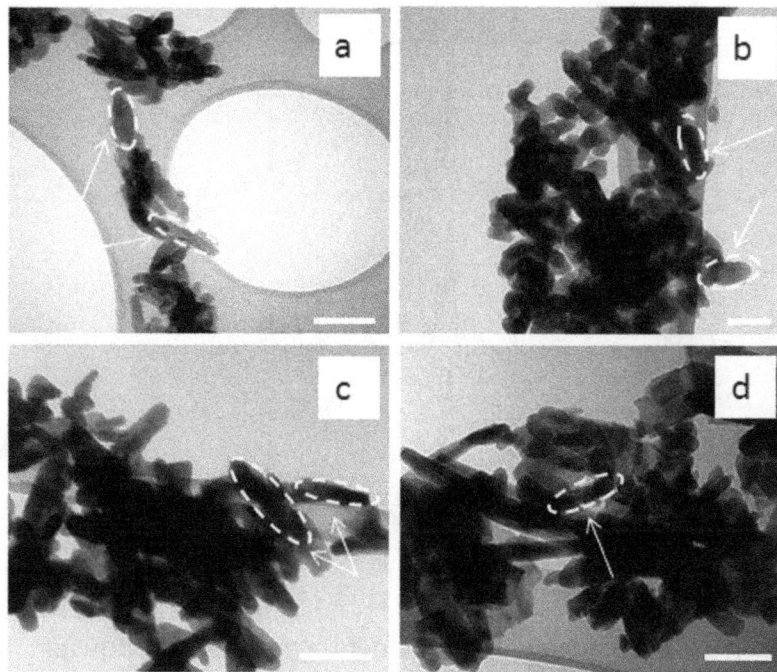

Figure 6. TEM micrographs of calcinated HA powders synthesised following: (**a**) Prakash; (**b**) BiancoA; (**c**) BiancoB; and (**d**) Pang protocol. Scale bars = 100 nm.

Furthermore, compositional analyses were performed on all powders to evaluate the Ca to P molar ratio. EDS analysis determined that synthesised hydroxyapatite particles had a stoichiometry close to the theoretical value for apatite, ranging from 1.69 to 1.77 (Table 2) [17].

Table 2. The Ca/P molar ratio for the HA powders after calcination.

Protocols	Ca/P molar ratio
Prakash	1.69 ± 0.02
BiancoA	1.72 ± 0.04
BiancoB	1.74 ± 0.04
Pang	1.77 ± 0.03

2.2. Discussion

Wet chemical precipitation is one of the most common techniques used in the synthesis of nanoscale hydroxyapatite due to its simplicity and low cost, factors that also make it suitable for adaptation to industrial production. In particular, chemical precipitation may be accomplished using various calcium- and phosphate- containing reagents as proposed in this study. A typical procedure involves the dropwise addition of one reagent to another under continuous and gentle stirring, while the molar ratio of elements (Ca/P) is kept at stoichiometry according to its ratio in HA (1.67). As the last step, the resultant suspension may be aged under atmospheric pressure or immediately washed, filtered, dried and crushed into a powder [1,9].

In this paper, nano-hydroxyapatite powders were synthesised through wet chemical precipitation following four protocols, differing through primary reagents composition, reactant addition rate, reaction temperature, reaction pH and post-synthesis aging time. The results of this study clearly demonstrated a strong effect of the process parameters on the physico-chemical characteristics of the HA products. In particular, the crystallinity was markedly affected by the calcination temperature, reaching a value of up to approximately 70% for the HA prepared following the Prakash protocol at 95 °C. However, it was noticed that the increase of the crystallinity with temperature was not linear. In fact, a substantially increased crystallinity was observed when the synthetic temperature was greater than 70 °C, as the crystalline activation energy of the HA was overcome at that temperature. Bouyer et al. reported a similar phenomenon [18], in which they found that 60 °C was a transition temperature. Below that temperature the HA crystals are monocrystalline, while above this temperature the HA crystals become multicrystalline.

Further evidence for the HA crystallinity may be acquired from the FTIR-ATR spectra by evaluating the intensities of the two hydroxyl absorption bands (the sharp peak at 3570 cm^{-1} and 1637 cm^{-1}) and the band at 940 cm^{-1} for phosphate group [19]. It was seen that the intensities of these three bands increased with the synthesis temperature. The result was consistent with the obtained XRD results. In addition, the broad band at 3700–2500 cm^{-1}, which is a reflection of the combined water in HA powders, also decreased with increasing synthesis temperature. The lower water content resulted from the higher crystallinity of HA powder, implying the crystalline HA powder had less affinity for water than its amorphous counterpart (as observed for the nHA prepared by the Prakash protocol).

The analysis of the HA morphology made an important contribution to this study, because it has been reported previously that small crystals (nano-size) were suitable due to potentially greater bioactivity when compared to coarse crystals (submicron-size) [20]. Moreover, the finer crystals also provided larger interfaces for osseointegration [21]. The morphological behaviour observed in all the precipitated powders, in which the appearance of these nanoparticles was quite different from each other and may be related with the synthesis temperature, which effects the nano-powders crystal shape. The change from needle-like (with irregular shape and contours) to a more regular shape (lower aspect ratio) is strictly related to the increase in the synthesis temperature, corresponding with an increase of the crystallinity of the HA nano-crystals. In effect, a more regular shape of the particles was observed when the powders had higher crystallinity, as shown in Figure 6a for Prakash protocol. [22]. Furthermore, it was noticed that the increase of the synthesis temperature increased slightly the HA nanoparticle size. The exact particle nucleation and growth mechanisms are not clear. Nucleation occurs probably by either hydrolytic reactions or a salting-out phenomenon. Growth could be via diffusive molecular deposition or an aggregation of primary particles due to increased precipitation temperatures [23].

Moreover, the influence of calcination on the physico-chemical characteristics of HA nanoparticles was also investigated. The calcination temperature of 650 °C was selected to avoid the formation of impurities that could be introduced at higher temperatures (i.e., tricalcium phosphate content above 900 °C) [24]. The calcinated HA powders showed much higher crystallinity and sharper diffraction peaks compared with the as-precipitated ones, as confirmed by the infrared spectra, that showed a decrease in the intensity of peaks associated to H_2O, indicating the water loss occurred during calcination. However, the small sharp peak at 3571 cm^{-1}, attributed to the hydroxyl group of HA, remained after calcination. These results suggested that no dehydration occurred within the HA molecules during calcination, and the water loss in the samples resulted from the loss of combined water in the HA powders [24].

Furthermore, it was interesting to notice that the change in crystallite sizes after calcination was dependent on the crystallinity and crystallite sizes of "as prepared" HA powders. After calcination, the X_s values increased for the HA powders with lower crystallinity and smaller crystallite sizes but remained the same for those with high crystallinity and larger crystallite sizes. This unusual change in crystallite size after calcination was also observed by Gibson et al. [25] for apatites after heating to temperatures between 650 °C and 750 °C for 2 h. Finally, after heat treatment at 650 °C, all the particles showed a decrease in the size with a shape factor ranging from 1.25 to 2.5. This was most likely due to a large amount of relatively smaller new crystallites forming during the calcination, as reported previously [22].

3. Experimental Section

3.1. Hydroxyapatite Nanophase Synthesis

HA powders were prepared by the direct precipitation from aqueous solutions according four different protocols. Many factors influence the HA properties since these different methods are characterised by different starting materials, pH (in a range from 6.5 to 12), temperature (ranging from 25 to 95 °C), and ripening time (from 24 to 96 h). In details:

Prakash protocol [26]. HA was precipitated using the reaction formula shown below:

$$10Ca(OH)_2 + 6H_3PO_4 \rightarrow Ca_{10}(PO_4)_6(OH)_2 + 18H_2O \tag{1}$$

$Ca(OH)_2$ and 85 wt% H_3PO_4 (>99% pure) was purchased from Sigma-Aldrich, St. Louis, MO, USA. The quantity of various reactants used are as follows: 1.85 g of $Ca(OH)_2$ powder in 250 mL of deionized (DI) water and 1.73 g of 85 wt% H_3PO_4 in 250 mL of DI water. The acid was added to the base at a rate approximately equal to 3.5 mL/min using a peristaltic pump, and prior to PO_4^{3-} ion addition to Ca^{2+} ion solution, the latter was heated to reaction temperature at 95 °C and stirred for 1 h at 400 rpm. The temperature was maintained within ±2 °C.

BiancoA and BiancoB protocols [22]. Nanoscale HA powders were prepared under magnetic stirring (around 400 rpm) following two different precipitation routes:

(i) from calcium nitrate (coded as BiancoA),
(ii) from calcium hydroxide (coded as BiancoB).

BiancoA Protocol. Stoichiometric volume of calcium nitrate tetrahydrate ($Ca(NO_3)_2 \cdot 4H_2O$, Sigma-Aldrich 99.2%) aqueous solution was added dropwise to diammonium hydrogen phosphate (($NH_4)_2HPO_4$, Sigma-Aldrich 99.2%) aqueous solution.

The following reaction occurred:

$$10Ca(NO_3)_2 + 6(NH_4)HPO_4 + 2H_2O \rightarrow Ca_{10}(PO_4)_6(OH)_2 + 12NH_4NO_3 + 8HNO_3 \tag{2}$$

During precipitation, the pH was continuously monitored and adjusted at 10 ± 0.1 by adding NH_4OH (Sigma-Aldrich). Precipitates were aged in situ for 24 h at 70 °C and then washed several times with NH_4OH aqueous solution (pH 10).

BiancoB Protocol. An aqueous suspension of calcium hydroxide ($Ca(OH)_2$, Sigma-Aldrich 99.5%) was titrated with phosphoric acid (H_3PO_4, Sigma-Aldrich 85%). The following reaction occurred as

Equation (1). The pH was finally adjusted at 9.4 by adding NH₄OH. As-precipitated powder was aged in situ for 24 h at room temperature and washed several times with NH₄OH aqueous solution (pH 9.4).

Pang protocol [24]. HA nanoparticles were prepared by chemical precipitation through aqueous solutions of the reactants. Calcium chloride and ammonium hydrogenphosphate (both supplied by Sigma-Aldrich, St. Louis, MO, USA) were first dissolved in deionised water to form 0.5 and 0.3 M aqueous solutions, respectively. Equal amounts of these two aqueous solutions were separately pre-heated to the synthesis temperature, and then mixed under vigorous stirring at 40 °C.

$$10CaCl_2 + 6(NH_4)HPO_4 + 8NH_4OH \rightarrow Ca_{10}(PO_4)_6(OH)_2 + 20NH_4Cl + 6H_2O \qquad (3)$$

Meanwhile, ammonium hydroxide (Sigma-Aldrich) was added immediately to adjust the reaction mixture to pH 10. The pH value was kept constant throughout the experiment. After ripening for a specified period of time (96 h), the precipitates were recovered by centrifugation and then washed with water. At least five cycles of washing and centrifuging were required to ensure complete removal of the by-product, ammonium chloride.

All the reactants used in this study were dissolved in deionized (DI) water produced using a Milli-Q unit (Millipore, Watford, UK) and the conductivity of the DI water was 0.05 µS/cm at 25 °C.

The calcination of HA powders synthesised under different temperatures and ripening times was carried out by first drying the samples at 70 °C for 24 h in oven and then calcination at 650 °C for 6 h at the ramp rate of 5.0 °C/min with a dwell time of 2 h in a tube furnace.

3.2. Physical, Chemical and Morphological Characterisation

X-ray powder diffraction (XRD) studies were carried out on a Philips RW1710 powder diffractometer, using Bragg-Brentano geometry with CuKα radiation and a solid-state Peltier cooled detector. All powder diffraction spectra were measured in continuous mode using the following conditions: 2θ angular range 10°–60°, tube power 45 kV and 40 mA, 2θ step size 0.02° and a scan rate of 1° 2θ/min. The diffraction spectra were compared with those in the Powder Diffraction Files (PDF) database of the Joint Committee on Powder Diffraction Standards (JCPDS). At least five spectra were recorded for each sample, and the results were averaged with standard deviation. The crystallinity degree (X_c), corresponding to the fraction of crystalline phase present in the examined volume, was evaluated as follows:

$$X_c = 1 - (V_{112/300} / I_{300}) \qquad (4)$$

where I_{300} is the intensity of (300) reflection of HA and $V_{112/300}$ is the intensity of the hollow between (112) and (300) reflections, which completely disappears in non-crystalline samples. In agreement with Landi et al. [27] a verification was done as follows:

$$B_{002} \sqrt[3]{X_c} = K \qquad (5)$$

where K is a constant found equal to 0.24 for a very large number of different HA powders, and B_{002} is the full width at half maximum (in degrees) of reflection (002).

Moreover, the peak broadening of XRD reflection can be used to estimate the crystallite size in a direction perpendicular to the crystallographic plane based on Scherrer's formula as follows [28]:

$$X_s = 0.9\lambda(FWHM \cdot \cos\vartheta) \qquad (6)$$

where X_s is the crystallite size (nm), λ the wavelength of monochromatic X-ray beam (nm) ($\lambda = 0.15406$ nm for CuKα radiation); FWHM the full width at half maximum for the diffraction peak under consideration (rad); and ϑ the diffraction angle (°). The diffraction peak at $2\theta = 26.04°$ was selected for calculation of the crystallite size since it is sharper and isolated from others. This peak assigns to (002) Miller's plane family and shows the crystal growth along the c-axis of the HA crystalline structure [29].

FTIR-ATR spectra were obtained using a Thermoscientific Nikolett spectrometer (Unicam Ltd., Cambridge, England). The spectra were recorded between 500 and 4000 cm^{-1} with a spectral resolution of 4 cm^{-1} averaging 64 scans. The HA powders used for FTIR-ATR measurement were dried in a vacuum oven at 70 °C for 1 week before testing.

The morphology and size distribution of the synthesized HA powders was characterised with a transmission electron microscope (TEM; JEM 100CX, Jeol, Tokyo, Japan), operating at 80 kV and a particle size analyzer (ImageJ, Maryland, Bethesda, MD, USA), respectively. An aspect ratio can be defined by the ratio length/width of the HA nanocrystals:

$$F_s = L / l \tag{7}$$

where F_s is the shape factor; L is particle length (nm) and l is particle width (nm). At least five images were recorded for each sample.

Finally compositional analysis by EDS (Philips EDAX 9100, EDAX Inc., Mahwah, NJ, USA) was performed on all powders to evaluate the Ca to P molar ratio. Three replicates were measured, and the results were averaged with standard deviation.

4. Conclusions

HA nanoparticles with a stoichiometry close to the theoretical value for apatite were successfully synthesised by wet-chemical precipitation using four different protocols. Products showed different physico-chemical properties depending upon the adjustment of process parameters as well as the choice of synthesis method. In particular, the synthesis temperature influenced nanoparticle shape, with lower temperatures producing a more needle-like morphology, characterised by a shape factor of 3-4 (BiancoB and Pang). Increasing the reaction temperature caused the crystal morphology to change from a needle-like shape to a more regular form with a lower aspect ratio (1.5 for Prakash and BiancoA). Furthermore, the synthesis temperature influenced crystallinity, ranging from between 40% for Pang up to about 70% for the nHA obtained following the Prakash protocol, and confirmed by XRD and FTIR-ATR studies.

Controlling the particle morphology of this biomaterial is of utmost importance because its shape and size are known to be useful to mimic nHA in mammalian bone and tooth. Though dimensions of biological bone apatite crystals reported in the literature vary due to different treatment methods and analytical techniques, it is generally at the nanoscale with values in the ranges of 30–50 nm (length), 15–30 nm (width) with an aspect ratio ranging from 1.5 to 2. The Prakash protocol may be considered to be the most suited to mimic bone apatite and to industrial scale-up of nHA, as it is based on a relatively simple reaction involving aqueous solutions of calcium hydroxide and phosphoric acid, with only water as a by-product. However, with regards to nHA found in dental tissues, typical apatite crystals in enamel are rod-like in shape with widths of 25–100 nm and undetermined lengths of greater than 100 nm along the c-axis. Therefore, BiancoB protocol is considered to be more appropriate to produce nHA that mimics that found in tooth tissue. Moreover, nHA with different aspect ratio may be used as inorganic filler for the preparation of scaffolds for bone tissue engineering in order to enhance the mechanical properties and to improve the biological activity. Specifically nHA with low aspect ratio can be deposited onto the walls of porous freeze-dried polymeric scaffolds, while nanoparticles with high aspect ratio can be dispersed longitudinally across the polymeric fibers in electrospun membranes. In conclusion, the aqueous wet-chemical precipitation methods investigated here provide a versatile platform for preparation of different nHA biomaterials for use in devices and regenerative medicine products, but great care should be exercised when optimising process parameters to generate nHA with desired physico-chemical properties.

Acknowledgments

The authors would like to thank the FP7-PEOPLE-2011-IEF-302315-NBC-ReGen4 Grant for providing financial support to this study and Clara Mattu of Politecnico di Torino, Italy, for the EDS analysis. This work is within the remit of membership of the UK EPSRC Centre for Innovative Manufacturing of Medical Devices-MeDe Innovation (EPSRC grant EP/K029592/1).

Author Contributions

Piergiorgio Gentile, Cheryl A. Miller and Paul V. Hatton conceived the study. Piergiorgio Gentile and Caroline J. Wilcock designed and performed the experiments; Piergiorgio Gentile performed the FTIR-ATR and analysed the EDS data; Caroline J. Wilcock performed and analysed the TEM data; Robert Moorehead analysed the XRD data; Piergiorgio Gentile, Cheryl A. Miller and Paul V. Hatton provided additional intellectual insight and prepared the manuscript.

Conflicts of Interest

The authors declare no conflict of interest.

References

1. Sadat-Shojai, M.; Khorasani, M.T.; Dinpanah-Khoshdargi, E.; Jamshidi, A. Synthesis methods for nanosized hydroxyapatite with diverse structures. *Acta Biomater.* **2013**, *9*, 7591–7621.
2. Salgado, A.J.; Coutinho, O.P.; Reis, R.L. Bone tissue engineering: State of the art and future trends. *Macromol. Biosci.* **2004**, *4*, 743–765.
3. Carter, D.H.; Hatton, P.V.; Aaron, J.E. The ultrastructure of slam-frozen bone mineral. *Histochem. J.* **1997**, *29*, 783–793.
4. Zhang, L.J.; Webster, T.J. Nanotechnology and nanomaterials: Promises for improved tissue regeneration. *Nano Today* **2009**, *4*, 66–80.
5. Dorozhkin, S.V. Nanodimensional and nanocrystalline apatites and other calcium orthophosphates in biomedical engineering, biology and medicine. *Materials* **2009**, *2*, 1975–2045.
6. Dong, Z.H.; Li, Y.B.; Zou, Q. Degradation and biocompatibility of porous nano-hydroxyapatite/polyurethane composite scaffold for bone tissue engineering. *Appl. Surf. Sci.* **2009**, *255*, 6087–6091.

7. Wang, Y.Y.; Liu, L.; Guo, S.R. Characterization of biodegradable and cytocompatible nano-hydroxyapatite/polycaprolactone porous scaffolds in degradation *in vitro*. *Polym. Degrad. Stab.* **2010**, *95*, 207–213.

8. Cai, Y.R.; Liu, Y.K.; Yan, W.Q.; Hu, Q.H.; Tao, J.H.; Zhang, M.; Shi, Z.L.; Tang, R.K. Role of hydroxyapatite nanoparticle size in bone cell proliferation. *J. Mater. Chem.* **2007**, *17*, 3780–3787.

9. Dorozhkin, S.V. Nanosized and nanocrystalline calcium orthophosphates. *Acta Biomater.* **2010**, *6*, 715–734.

10. Wang, L.J.; Nancollas, G.H. Pathways to biomineralization and biodemineralization of calcium phosphates: The thermodynamic and kinetic controls. *Dalton Trans.* **2009**, *15*, 2665–2672.

11. Li, B.; Guo, B.; Fan, H.S.; Zhang, X.D. Preparation of nano-hydroxyapatite particles with different morphology and their response to highly malignant melanoma cells *in vitro*. *Appl. Surf. Sci.* **2008**, *255*, 357–360.

12. Webster, T.J.; Ergun, C.; Doremus, R.H.; Siegel, R.W.; Bizios, R. Enhanced osteoclast-like cell functions on nanophase ceramics. *Biomaterials* **2001**, *22*, 1327–1333.

13. Huang, Y.; Zhou, G.; Zheng, L.S.; Liu, H.F.; Niu, X.F.; Fan, Y.B. Micro-/Nano-sized hydroxyapatite directs differentiation of rat bone marrow derived mesenchymal stem cells towards an osteoblast lineage. *Nanoscale* **2012**, *4*, 2484–2490.

14. Zakaria, S.M.; Sharif, S.H.; Othman, M.R.; Yang, F.; Jansen, J.A. Nanophase hydroxyapatite as a biomaterial in advanced hard tissue engineering: A review. *Tissue Eng. Part B Rev.* **2013**, *19*, 431–441.

15. Krajewski, A.; Mazzocchi, M.; Buldini, P.L.; Ravaglioli, A.; Tinti, A.; Taddei, P.; Fagnano, C. Synthesis of carbonated hydroxyapatites: Efficiency of the substitution and critical evaluation of analytical methods. *J. Mol. Struct.* **2005**, *744*, 221–228.

16. Apfelbaum, F.; Diab, H.; Mayer, I.; Featherstone, J.D.B. An FTIR study of carbonate in synthetic apatites. *J. Inorg. Biochem.* **1992**, *45*, 277–282.

17. Jean, A.; Kerebel, B.; Kerebel, L.M.; Legeros, R.Z.; Hamel, H. Effects of various calcium-phosphate biomaterials on reparative dentin bridge formation. *J. Endod.* **1988**, *14*, 83–87.

18. Bouyer, E.; Gitzhofer, F.; Boulos, M.I. Morphological study of hydroxyapatite nanocrystal suspension. *J. Mater. Sci. Mater. Med.* **2000**, *11*, 523–531.

19. Lim, G.K.; Wang, J.; Ng, S.C.; Gan, L.M. Formation of nanocrystalline hydroxyapatite in nonionic surfactant emulsions. *Langmuir* **1999**, *15*, 7472–7477.

20. Kokubo, T.; Kim, H.M.; Kawashita, M. Novel bioactive materials with different mechanical properties. *Biomaterials* **2003**, *24*, 2161–2175.

21. Kailasanathan, C.; Selvakumar, N.; Jeyasubramanian, K. Effect of calcination in synthesis of nano hydroxyapatite for bone grafting. *Mater. Manuf. Process.* **2011**, doi:10.1080/10426914.2011.577874.

22. Bianco, A.; Cacciotti, I.; Lombardi, M.; Montanaro, L.; Gusmano, G. Thermal stability and sintering behaviour of hydroxyapatite nanopowders. *J. Therm. Anal. Calorim.* **2007**, *88*, 237–243.

23. de Bruyn, J.R.; Goiko, M.; Mozaffari, M.; Bator, D.; Dauphinee, R.L.; Liao, Y.Y.; Flemming, R.L.; Bramble, M.S.; Hunter, G.K.; Goldberg, H.A. Dynamic light scattering study of inhibition of nucleation and growth of hydroxyapatite crystals by osteopontin. *PLoS One* **2013**, *8*, e56764.

24. Pang, Y.X.; Bao, X. Influence of temperature, ripening time and calcination on the morphology and crystallinity of hydroxyapatite nanoparticles. *J. Eur. Ceram. Soc.* **2003**, *23*, 1697–1704.

25. Gibson, I.R.; Rehman, I.; Best, S.M.; Bonfield, W. Characterization of the transformation from calcium-deficient apatite to beta-tricalcium phosphate. *J. Mater. Sci. Mater. Med.* **2000**, *11*, 799–804.

26. Prakash, K.H.; Kumar, R.; Ooi, C.P.; Cheang, P.; Khor, K.A. Apparent solubility of hydroxyapatite in aqueous medium and its influence on the morphology of nanocrystallites with precipitation temperature. *Langmuir* **2006**, *22*, 11002–11008.

27. Landi, E.; Tampieri, A.; Celotti, G.; Sprio, S. Densification behaviour and mechanisms of synthetic hydroxyapatites. *J. Eur. Ceram. Soc.* **2000**, *20*, 2377–2387.

28. Cernik, R.J. *X-ray Powder Diffractometry. An Introduction*; Series: Chemical Analysis, Vol. 138; von Jenkins, R., Snyder, R.L., Eds.; John Wiley & Sons: New York, NY, USA, 1997.

29. Li, Y.B.; Klein, C.P.A.T.; Zhang, X.D.; Degroot, K. Formation of a bone apatite-like layer on the surface of porous hydroxyapatite ceramics. *Biomaterials* **1994**, *15*, 835–841.

5

Investigation of the High Mobility IGZO Thin Films by Using Co-Sputtering Method

Chao-Ming Hsu [1], Wen-Cheng Tzou [2], Cheng-Fu Yang [3,*] and Yu-Jhen Liou [3]

[1] Department of Mechanical Engineering, National Kaohsiung University of Applied Science, No. 415 Chien Kung Road, Kaohsiung 807, Taiwan; E-Mail: jammy@kuas.edu.tw

[2] Department of Electro-Optical Engineering, Southern Taiwan University, No. 1, Nan-Tai Street, Yungkang Dist., Tainan City 710, Taiwan; E-Mail: wjtzou@mail.stust.edu.tw

[3] Department of Chemical and Materials Engineering, National University of Kaohsiung, No. 700 Kaohsiung University Road, Nan-Tzu District, Kaohsiung 811, Taiwan; E-Mail: sonic7838@hotmail.com

* Author to whom correspondence should be addressed; E-Mail: cfyang@nuk.edu.tw;

Academic Editor: Teen-Hang Meen

Abstract: High transmittance ratio in visible range, low resistivity, and high mobility of IGZO thin films were prepared at room temperature for 30 min by co-sputtering of $Zn_2Ga_2O_5$ (Ga_2O_3 + 2 ZnO, GZO) ceramic and In_2O_3 ceramic at the same time. The deposition power of pure In_2O_3 ceramic target was fixed at 100 W and the deposition power of GZO ceramic target was changed from 80 W to 140 W. We chose to investigate the deposition power of GZO ceramic target on the properties of IGZO thin films. From the SEM observations, all of the deposited IGZO thin films showed a very smooth and featureless surface. From the measurements of XRD patterns, only the amorphous structure was observed. We aimed to show that the deposition power of GZO ceramic target had large effect on the E_g values, Hall mobility, carrier concentration, and resistivity of IGZO thin films. Secondary ion mass spectrometry (SIMS) analysis in the thicknesses' profile of IGZO thin films found that In and Ga elements were uniform distribution and Zn element were non-uniform distribution. The SIMS analysis results also showed the concentrations of Ga and Zn elements increased and the concentrations of In element was almost unchanged with increasing deposition power.

Keywords: IGZO; co-sputtering method; deposition power; SIMS

1. Introduction

The typical plasma enhanced chemical vapor deposition (PECVD) hydrogenated amorphous silicon (α-Si:H) thin-film transistors (TFTs) are mainly applied for flat panel displays (FPDs), such as electronic papers (e-papers), organic light-emitting-diode displays (OLEDs), and liquid crystal displays (LCDs). Even α-Si:H TFTs have sub-threshold swing of 0.3~0.4 V/decade, off-state drain current (I_{Doff}) below 10^{-13} A, and on-to-off ratio about 10^7, they have the shortcoming of low field-effect mobility (μ_{eff}) of about 0.6~0.8 cm^2/V-s and poor transparency [1]. To address this issue, several n-type transparent amorphous oxide semiconductors (TAOSs), which exhibit high mobility, excellent uniformity, good transparency and applicability for the low-temperature process (for polymer or plastic substrate), and have potential to serve as active layer in TFTs [2,3]. Recently, conventional amorphous or polycrystalline transparent conduction oxide semiconductors (TCOs) have been proposed as alternative channel materials, because they exhibit excellent optical transparency and good TFTs performance in ambient conditions. However, the grain boundaries of TCOs could affect device properties, such as uniformity and stability, over large areas. For that, over the last several years, there has been great interest in TFTs made of TCOs-based TAOSs thin films. Recently, several n-type TAOSs thin films, such as ZnO [4], Al-Sn-Zn-O (ASZO) [5], and In-Ga-Zn-O (IGZO) [6], have received a considerable attention in the large-area FPD industry since they may overcome the difficulties encountered in the amorphous α-Si:H and polycrystalline silicon TFTs technologies [7].

This is mainly due to TAOSs' thin-film transistors having unique advantages, such as transparency in visible light region, large-area uniform deposition at low temperature, and high carrier mobility. IGZO thin films are a semiconducting material and they can be used as the TFTs' backplane of FPDs because IGZO-TFTs' mobility is higher than that of amorphous silicon. The α-IGZO thin films could be deposited on polyethylene terephthalate at room temperature and exhibited high Hall effect mobility [8]. IGZO-TFTs were first developed by Professor H. Hosono's group at Tokyo Institute of Technology and Japan Science and Technology Agency (JST) in 2003 for crystalline IGZO-TFTs [7] and in 2004 for amorphous IGZO (α-IGZO)-TFTs [8]. They found that an n-type amorphous In-Ga-Zn-O with a molar ratio 1:1:1 is preferred for fabricating electronic devices because it has a reasonably large Hall mobility (>15 cm^2/V-s). Zan *et al.* reported that they utilized self-organized polystyrene spheres with a diameter of 200 nm to fabricate a porous gate structure and Ar plasma treatment through the porous gate performed dot-like doping on α-IGZO channel region. They fabricated a top-gate self-aligneda-IGZO TFT with an effective field-effect mobility as 79 cm^2/V-s, they also reported that an intrinsic IGZO thin film had electron mobility as 39.6 cm^2/V-s [9]. Bak et al used IGZO to fabricate the top-gate structured TFTs and the mobility of IGZO thin films was in the range of 11.8 cm^2/V-s~14.8 cm^2/V-s under different bias voltage [10]. Therefore, IGZO-TFTs can improve operation speed, resolution, and size of FDPs, and they are also considered as one of the most promising TFTs to drive OLED displays. As we know, various techniques were investigated for growth of IGZO thin films, such as electron beam evaporation, ion beam assisted deposition, and ion implantation. In particular, Jeong *et al.* obtained IGZO thin films by co-depositing the Ga:In$_2$O$_3$ and Zn:In$_2$O$_3$ targets to deposit the Ga and Zn co-doped In$_2$O$_3$ electrode at room temperature [11].

In the past, K. Nomura *et al.* presented that IGZO thin films are composed of alternating stacks of InO$_2$$^-$ and GaO(ZnO)$^+$ layers [12]. They found that the In$_2$O$_3$ concentration in IGZO thin films has a large effect on the properties of IGZO thin films, especially in the electrical properties. In this study, the Zn$_2$Ga$_2$O$_5$ and In$_2$O$_3$ ceramic targets were prepared separately and the two ceramic targets were used to deposit IGZO thin films by co-sputtering method at room temperature on glass substrates. We believed that as the deposition power of In$_2$O$_3$ was fixed, the concentration of the In$_2$O$_3$ in IGZO thin films could be controlled by changing the deposition power of GZO ceramic target. We systematically examined the crystallization, optical, and electrical properties and surface and cross-section morphologies of IGZO thin films as a function of deposition power of GZO ceramic target. Importantly, we showed that as the deposition power of GZO ceramic target was changed, the co-deposited IGZO thin films had the high mobility of 11.0 cm^2/V-s~163.4 cm^2/V-s.

2. Experimental Section

Ga$_2$O$_3$ powder (99.99%) was mixed with ZnO powder (99.99%) to form the Ga$_2$O$_3$-2ZnO composition (abbreviated as GZO). After being dried and ground, the GZO powder was calcined at 800 °C for 1 h, and ground again. GZO powder and In$_2$O$_3$ powder were mixed with polyvinylalcohol (PVA) as binder, and then the mixed powders were uniaxially pressed into pellets of 5 mm thickness and 54 mm diameter using a steel die. After being debindered, the GZO pellet and In$_2$O$_3$ pellet were sintered at 1200 °C and 1250 °C, respectively, for 2 h. Glass substrates (Corning 1737) with an area of 2 × 2 cm^2 were cleaned ultrasonically with isopropyl alcohol (IPA) and deionized (DI) water, and dried under a blown nitrogen gas. Then GZO and pure In$_2$O$_3$ ceramic targets were used to co-deposit the IGZO thin films. Deposition power of In$_2$O$_3$ ceramic target was 100 W and deposition power of GZO ceramic target was changed from 80 W to 140 W, respectively, room temperature (RT) and 30 min were used as deposition temperature and deposition time. The base pressure of sputtering chamber was below 5 × 10^{-6} Torr and the working pressure was maintained at 3 × 10^{-3} Torr in pure Ar (99.99%) ambient. Thickness and surface morphology of IGZO thin films were measured using a field emission scanning electron microscopy (FESEM), and their crystalline structures were measured using X-ray diffraction (XRD) patterns with Cu Kα radiation (λ = 1.5418 Å). Energy dispersive spectrometer (EDS) and secondary ion mass spectrometry (SIMS) analyses were used to find the concentration variations of In, Ga, and Zn elements in the depth profile of IGZO thin films. Atomic Force Microscopy (AFM Analysis, Bruker, Germany) was used to measure surface topography and surface roughness of IGZO thin films. The optical transmission spectrum was recorded using a Hitachi U-3300 UV-V is spectrophotometer in the 250–1000 nm wavelength range. In the past, determination of the optical band gap (E_g) values was often necessary to develop the electronic band structure of a thin-film material. A Tauc plot is one method of determining the E_g values in semiconductors. However, as the Tauc plot is used, the E_g values of thin films are extracted from the data of absorption coefficient as a function of photon energy (*hv*). As the Tauc plot is used, the E_g values of IGZO thin films can be determined using the relation in Equation (1):

$$(\alpha hv)^2 = c(hv - E_g) \tag{1}$$

where α is the optical absorption coefficient, c is the constant for direct transition, h is Planck's constant, and v is the frequency of the incident photon [13]. While the Hall-effect coefficient of IGZO thin films was measured using a Bio-Rad Hall set-up.

3. Results and Discussion

As we know, FESEM could be used to observe the surfaces' crystallization of IGZO thin films. Figure 1 indicates that as deposition power was changed, the surface morphologies of IGZO thin films showed different results. As the deposition power of GZO ceramic target was 80 W or 100 W, the nano-crystalline grains were really observed on the surfaces of IGZO thin films. As the deposition power of GZO ceramic target was increased from 120 W to 140W, surface morphologies of IGZO thin films exhibited a very smooth surface regardless of deposition power of GZO ceramic target. However, most IGZO thin films showed stable and flat amorphous surface features. In order to achieve high performance TCOs-based TFTs or memory devices, the preparation of source and drain electrodes with a smooth surface morphology is very important because surface roughness of IGZO thin films will influence the leakage current between the semiconducting IGZO active layer and source/drain electrodes. The surface observation results suggest that the co-sputtering method is an acceptable method to deposit IGZO thin films, because all IGZO thin films have low roughness surfaces and can be used to fabricate the TCOs-based TFTs or memory devices with high performance.

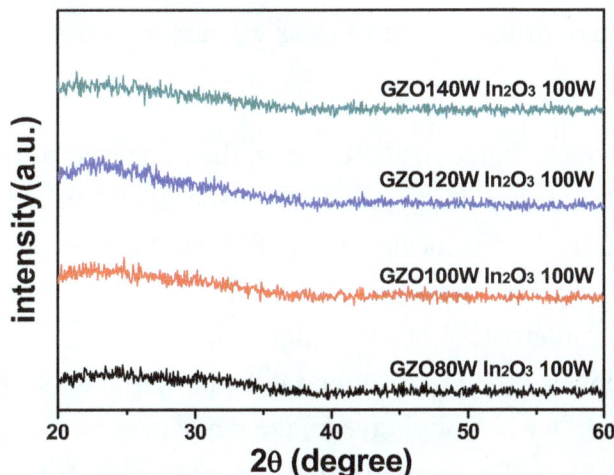

Figure 1. X-ray diffraction (XRD) patterns of In-Ga-Zn-O (IGZO) thin films as a function of deposition power of GZO ceramic target.

Figure 2 shows the cross-section observations of IGZO thin films as a function of deposition power of GZO ceramic target. As the results in Figure 2 show, thickness of IGZO thin films was around 103 nm, 138 nm, 149 nm, and 170 nm, as the deposition powers of GZO ceramic target was 80 W, 100 W, 120 W, and 140 W, respectively. Thickness of IGZO thin films increasing with deposition power of GZO ceramic target can be fairly expected, because more GZO particles will deposit onto glass substrates to form IGZO thin films. As the cross-session micrographs shown in Figure 2 were compared, there were different results as the deposition power of GZO ceramic target was changed. When the deposition power was 80 W, IGZO thin films grew irregularly. When 100 W and 120 W were used as the deposition powers, IGZO thin films grew like a densified aggregations of nano-laminations and nano-wires with random directions. AWhens the deposition power was 140 W, the aggregations of nano-laminations and nano-wires was changed to nano-wire-aggregated growths, and the nano-wires were highly oriented parallel to the substrate normal. In addition, there is no evidence of the segregation of GZO and In_2O_3 due to the uniform co-sputtering of GZO and In_2O_3 targets using tilted cathode guns. K. Nomura *et al.* reported that IGZO crystal is composed of alternating stacks of InO_2^- and $GaO(ZnO)^+$ layers and the concentration of In_2O_3 has a large effect on the crystallization of IGZO thin films [12].

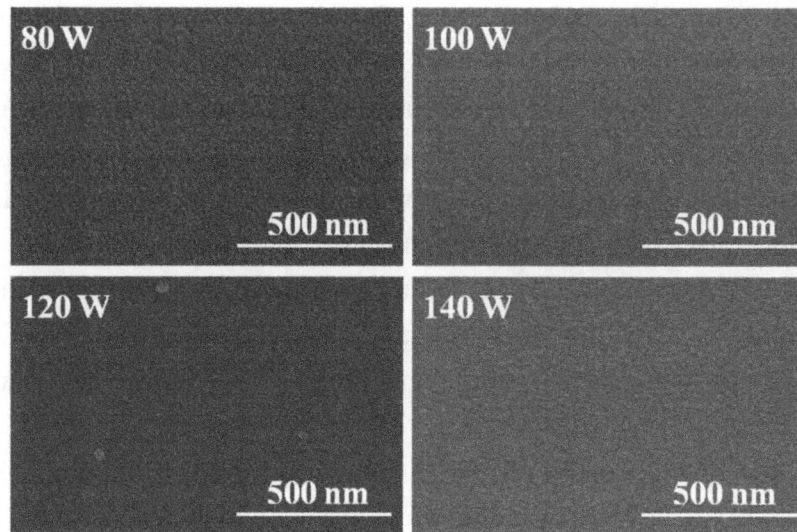

Figure 2. Surface morphology of IGZO thin films as a function of deposition power of GZO ceramic target.

From the standard XRD patterns revealed in JCPD cards, the main crystallization peak of In_2O_3 thin films is in the (222) plane [14] and the main crystallization peak of GZO thin films is in the (002) plane [15]. For the InO_2^- layer an In^{3+} ion is located at an octahedral site coordinated by six oxygen and for the $GaO(ZnO)^+$ layer Ga^{3+} and Zn^{2+} ions are located at triangle-bipiramidal sites and are each coordinated by five oxygen and alternately stacked along the (0001) direction. Those descriptions suggest that as the deposition power of GZO ceramic target in the co-sputtering method is changed, IGZO thin films have different surface morphology. As the deposition power of GZO target increases, the concentration of In_2O_3 decreases, the crystallization direction of GZO will dominate the growth direction of IGZO thin films, then the IGZO thin films have high c-axis orientation and are highly oriented parallel to the substrate normal. As Figure 2 shows, as the deposition power of GZO ceramic target is equal and higher than 120 W, the nano-wires parallel to the substrate normal suggest the stacked along the (002) direction. The structure of nano-laminations and nano-wires is changed to nano-wires parallel to the substrate normal, which proves that GZO thin films will dominate growth results of IGZO thin films. The results observed from the cross-session images of IGZO thin films shown in Figure 2 agree with the results of K. Nomura *et al.* [12] and Wang *et al.* [15].

AFM images of the two surfaces are presented in Figure 3 and the corresponding roughness values are measured using the described software. It can be noticed from Figure 3 that surface of IGZO thin films for GZO target's deposition power of 80 W (Figure 3a) is clearly much rougher than that of IGZO thin films for GZO target's deposition power of 140 W (Figure 3b) as indicated by the Root Mean Square (RMS) roughness values. The RMS roughness values of IGZO thin films' surface were obtained at five different locations and the average RMS roughness values were determined from the five data. The measured RMS roughness values for Figure 3a were in the range of 2.9 nm~4.5 nm and the average RMS value was 3.8 nm. The measured RMS roughness values for Figure 3b were in the range of 2.2 nm~3.2 nm and the average RMS value was 2.6 nm.

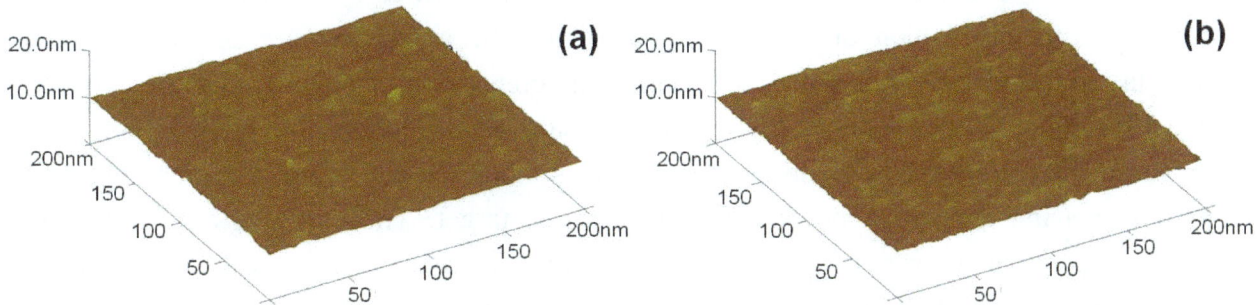

Figure 3. Surface Atomic Force Microscopy (AFM) morphology of IGZO thin films as a function of deposition power of GZO ceramic target. (**a**): 80 W (**b**): 140 W.

As the different sintering temperatures are used, differently crystalline phases will be formed in IGZO ceramic targets, and the multi-crystal phases are only observed in IGZO ceramic targets. Lo *et al.*, found cubic Ga_2ZnO_4 spinel and rhombohedral $InGaZnO_4$ phases are identified in the 1100 °C -sintered sample in addition to the as-prepared oxide powder phases of In_2O_3, Ga_2O_3, and ZnO [16]. However, most IGZO thin films will reveal the amorphous phase rather than the poly-crystal phases. For example, Jeong *et al.* co-deposited the $Ga:In_2O_3$ and $Zn:In_2O_3$ targets, and they obtained α-IGZO thin films rather than poly-crystal IGZO [11]. Jung *et al.* deposited IGZO thin films by using the facing targets sputtering (FTS) method at room temperature, also the deposited IGZO thin films revealed the amorphous phase [3]. As Figure 4 shows, only one weak and broad peak was assigned to the glass substrate, which proves that all deposited IGZO thin films exhibited the amorphous phase. However, the cubic Ga_2ZnO_4 and spinel rhombohedral (poly-crystal) $InGaZnO_4$ phases and the phases of precursor In_2O_3, Ga_2O_3, and ZnO were not observed in the Figure 4.

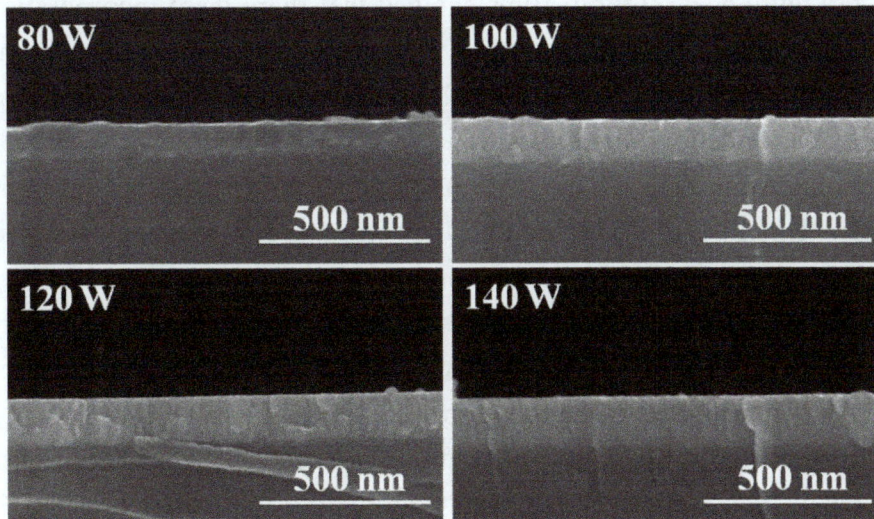

Figure 4. Cross-section observations of IGZO thin films as a function of deposition power of GZO ceramic target.

As we know, energy dispersive spectrometer (EDS) and secondary ion mass spectrometry (SIMS) are generally considered to be the qualitative techniques to find the large variation in ionization probabilities among different materials. For that, we used the two methods to analyze the IGZO thin films and to find the variations of atom ratios at the surface (EDS) and across the depth profile (SIMS) of IGZO thin films. Atomic ratio microanalysis in the FESEM is performed by measuring the energy or wavelength and intensity distribution of X-ray signal generated by a focused electron beam on the specimen. With the attachment of EDS, the precise elemental composition of materials can be obtained with high spatial resolution. Table 1 shows EDS analysis results as a function of deposition power of GZO ceramic target. The atom ratios of Zn and Ga elements increased and atom ratio of In element decreased with increasing deposition power of GZO ceramic target. Those results are expectable because as the deposition power of GZO ceramic target increases, more Ga_2O_3 and ZnO (or $Ga_2O_3 + 2\,ZnO$) molecules will be moved out from the surface of GZO ceramic target, then atom ratios of Zn and Ga increase and atom ratio of In increases. Table 1 shows important results: even the deposition power of the GZO target is higher than that of In_2O_3 target, and the atom ratio of the In element is higher than those of the Ga and Zn elements.

Table 1. Atom ratios of Zn, Ga as a function of deposition power of GZO target.

GZO Power	Zn	Ga	In
80 W	3.0	3.6	93.4
100 W	7.0	7.5	85.5
120 W	12.5	12.8	74.7
140 W	19.6	18.3	62.1

Because SIMS is a high sensitivity surface analysis technique for the determination of surface composition and contaminant analysis and for depth profile in the uppermost surface layers of a sample, it can detect very low concentrations of dopants and impurities. For that, the SIMS analysis was used to find the atomics' concentrations of the constituent elements (In, Ga, and Zn) as a function of the sample's depth to determine the elemental composition of the surface to a depth of about 120 nm, and the results are shown in Figure 5. IGZO thin films showed that there were incorporations of Zn, Ga, and In atoms in IGZO thin films, even the deposition process was proceeded at room temperature. The concentrations of In and Ga elements in the depth profile was almost unchanged and showed an uniformity distribution, independent of the deposition power of GZO target. However, the results in Figure 5 show that the concentration of Zn element in the depth profile was not uniform distribution. The concentration of Zn element first decreased and then increased as the analyzed depth increased, independent of the deposition power of GZO target.

As the results in Figure 5 are compared, the concentrations of Ga and Zn elements increased and the concentration of In element was almost unchanged as the deposition power of GZO ceramic target increased. The relative In concentration in the depth profile of IGZO thin films shown in Figure 5 is higher than that of the predicted values obtained from the used targets. Those results are very important because so far no SIMS analysis has been used to find the distribution of Zn, Ga, and In elements in the depth profiles of the deposited IGZO thin films. Figure 5 also shows that as the deposition power of GZO ceramic was changed from 80 W to 120 W, the relative concentration of In element was higher than those of Ga and Zn elements, those results are matched the analyzed results show in Table 1.

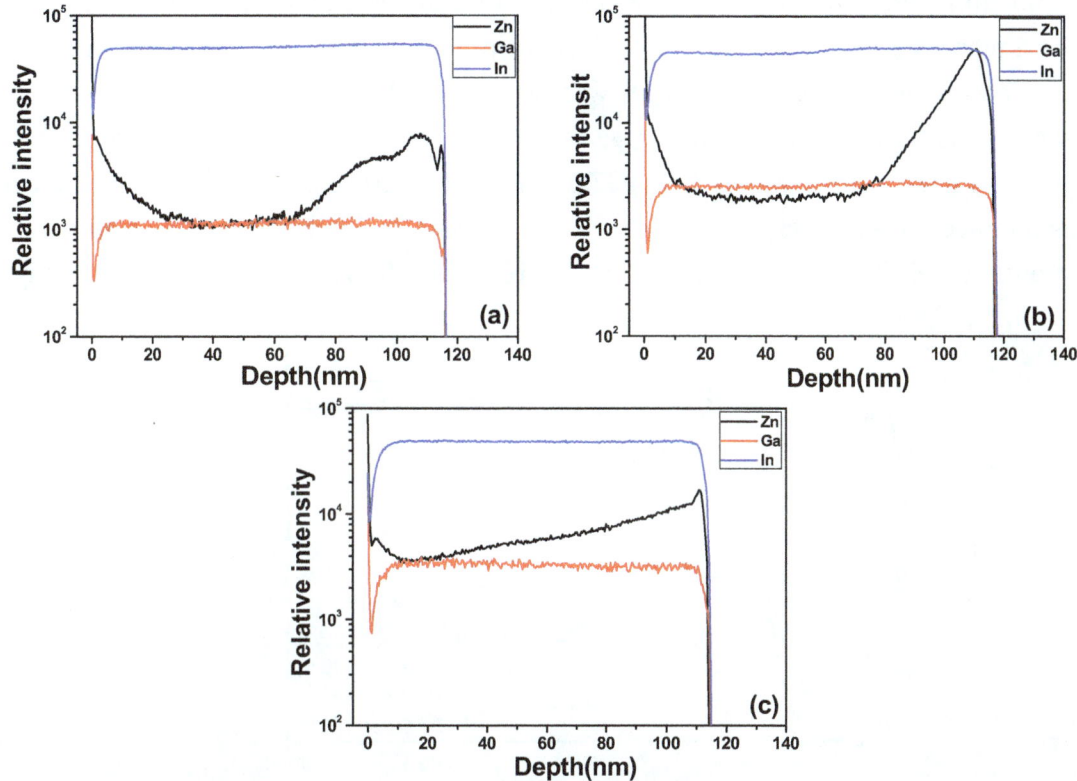

Figure 5. Second ion mass spectrometry analysis of IGZO thin film, the deposition power of GZO ceramic target was (**a**) 80 W (**b**) 100 W and (**c**) 120 W.

The reasons to cause the non-uniform distribution of Zn element in the depth profile are not really know. However, during the deposition process, the temperature significantly contributes to precursors decomposition and the growth mechanism and of thin films, and the growth mechanism is strongly depends on the reactor design and process parameters. Saha *et al.*, observed a significant decrease in growth rate and deteriorated structural of the ALD-ZnO films at 250 °C [17]. The higher growth rate might be because of precursor condensation due to their insufficient reactivity to the surface functional groups. Heo *et al.* used Zn as target to grow c-axis oriented ZnO thin films on c-plane Al_2O_3 via molecular beam epitaxy (MBE) using dilute ozone (O_3) as an oxygen source [18]. They found that for growth temperature higher than 350 °C; the rate dramatically decreased and for growth temperatures above 450 °C; continuous films were not realized. They also found that an increase in growth temperature causes a decrease of the sticking coefficient of Zn on the Al_2O_3 substrate which, subsequently, causes a decrease in the growth rate, even though the reactivity between Zn and the oxygen source is expected to increase with growth temperature. Those results suggest that using $Zn_2Ga_2O_5$ ceramic and In_2O_3 ceramic to co-deposit IGZO thin films at room temperature, the temperature on the glass substrates is higher, maybe higher than 300 °C; For that, the non-uniform distribution of Zn element in depth profile will be observed.

Figure 6 shows the transmittance ratios of IGZO thin films plotted against wavelengths in the region of 250–1000 nm, with deposition power of GZO ceramic target as the parameter. The results in

Figure 6 show that the transmittance ratios in the visible light region are apparently changed as the deposition power of GZO ceramic target is changed from 80 W to 140 W. The average transmittance ratio of IGZO thin films in the range of 400 nm~700 nm first increases with deposition power of GZO ceramic target and reaches a maximum value as the deposition power of GZO ceramic target is 120 W. As the deposition power of GZO ceramic target was 80 W, 100 W, 120 W, and 140 W, the average transmittance ratio of IGZO thin films in the range of 400 nm~700 nm was 77.3%, 77.5%, 91.4%, and 86.6%, respectively. Figure 6 also shows that IGZO thin films deposited on glass substrates had the maximum transmittance ratio of over 86.0%, 86.1%, 98.3%, and 96.3% in the range of 400~700 nm as the deposition power of GZO ceramic target was 80 W, 100 W, 120 W, and 140 W, respectively.

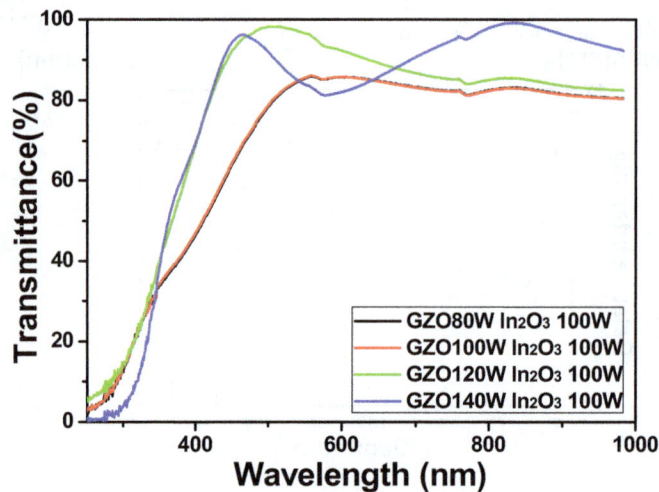

Figure 6. Transmittance spectrum of IGZO thin films as a function of deposition power of GZO ceramic target.

Those results suggest that as the co-sputtering method is used, we can deposit IGZO thin films with high transmittance ratio. From the results shown in Figure 1, the surfaces of all deposited IGZO thin films reveal a smooth structure and no agglomerated particles are observed, which are the reasons to cause IGZO thin films having high average transmittance ratio. For the transmission spectra shown in Figure 6, as the different deposition power of GZO ceramic target was used, the shift of the optical band edge was really observable and a greater sharpness was noticeable in the curves of the absorption edge. Those results suggest that the optical band gap (E_g) values will change as the co-sputtering method is used to prepare IGZO thin films.

The linear dependence of $(\alpha h v)^2$ on hv indicates that IGZO thin films are direct transition type semiconductors. In accordance with Equation (1), as Figure 7 shows, the calculated E_g values of IGZO thin films were 3.87 eV, 3.84 eV, 3.79 eV, and 3.71 eV as the deposition power of GZO target were 80 W, 100 W, 120 W, and 140 W, respectively. Because ZnO, Ga_2O_3, and In_2O_3 thin films have different E_g values, the variation in E_g values is believed to cause by the variation in the composition of IGZO thin films. The E_g values of ZnO [19], In_2O_3 [20], and intrinsic β-Ga_2O_3 thin films [21] are about 3.40 eV, 3.71 eV, and 4.90 eV, respectively. In general, the measured E_g values of IGZO thin films are consistent with and should be larger than that of ZnO and In_2O_3 thin films. However, the E_g values of IGZO thin films do not increase with the increase of deposition power of GZO target, even the atom concentration of Ga element (Ga_2O_3) increase, as Figure 7 shows.

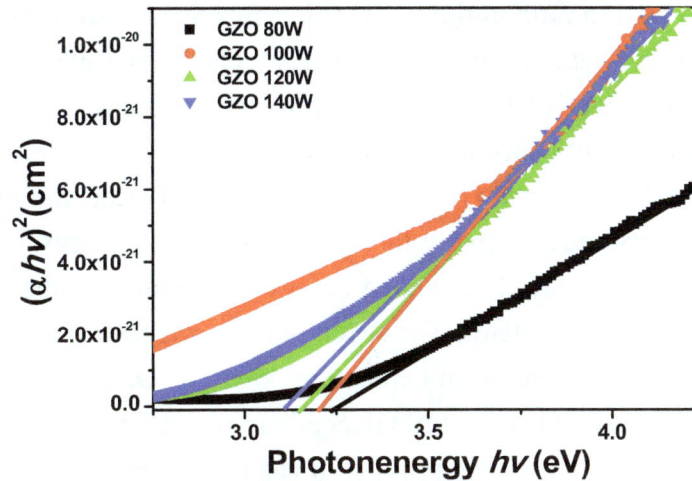

Figure 7. A^2 *vs.* hv-E_g Tauc plots of IGZO thin films as a function of deposition power of GZO ceramic target.

In the past, Nomura *et al.*, reported that the In^{3+} can provide extra carriers, when the TFTs devices is fabricated using the In-rich thin films the carrier concentration will be increased. Then, the devices will have larger drain current (I_{DS}) and better carrier mobility, and the needed off current will be increased [12]. Kim *et al.*, deposited α-IGZO thin films using the sol-gel method and the In:Ga:Zn mole ratio was controlled as 1:1:2, 3:1:2, and 5:1:2, respectively. They found that the transfer curves are shifted from the positive to the negative direction, *i.e.*, V_{th} of IGZO TFTs decreases from 15.84 V, 4.98 V, to −5.09 V as the In:Ga:Zn mole ratio increases from 1:1:2, 3:1:2, to 5:1:2 [22]. We believe the real In:Ga:Zn mole ratios are not 1:1:2, 3:1:2, and 5:1:2, respectively, but In concentration in IGZO thin films will affect their properties is un-doubtable. Kim *et al.* proved that the electronic concentration and mobility increased and resistivity decreased with increasing In/Ga ratio in IGZO thin films [22]. When IGZO thin films are deposited using the co-sputtering method, three reasons are believed to influence the carrier mobility of IGZO thin films. First, depositing at room temperature cannot provide enough energy to enhance the motion of plasma molecules. Then, the crystallization and grain size growth of IGZO thin films cannot be improved, the defects in IGZO thin films will generate during the deposition process. Second, if the agglomerated particles in IGZO thin films increase, that will cause the increase in the inhibiting of the barriers electron transportation and the mobility will decrease. Third, H. Hosono showed that the electron mobility and concentration evaluated from the Hall effects for α-IGZO thin films with different compositions, the mobility is primary determined by the fraction of In_2O_3 concentration and the highest value of ~40 cm^2 (V·s)$^{-1}$ is obtained around the samples containing the maximum In_2O_3 fraction [23]. From those reasons, the carrier mobility, carrier concentration, and resistivity of IGZO thin films are believed to be dependent on deposition power (or concentration) of GZO target.

In this study, at least five Hall-effect coefficients of IGZO thin films were measured for each deposition parameter, and the average values with the deviation ranges were shown in Figure 8. However, we obtained the different results as compared with those of Kim *et al.*, [22] and H. Hosono [23]. As Figure 8 indicates that as the deposition power of GZO ceramic target was 80 W, 100 W, 120 W, and 140 W, the carrier concentration was 6.45×10^{19} cm^{-3}, 2.34×10^{20} cm^{-3}, 7.30×10^{19} cm^{-3}, and 7.57×10^{18} cm^{-3}, and the carrier mobility was 163.4 cm^2/V-s, 11.0 cm^2/V-s, 17.6 cm^2/V-s, and 44.4 cm^2/V-s, respectively. There are two reasons are believed to cause IGZO thin films having a high mobility of

163.4 cm^2/V-s. The first is the high In ratio in the IGZO thin films formed for GZO ceramic target with a deposition power of 140 W. From the EDS and SIMS analyses results in Table 1, the In ratio decreased with the increase of deposition power of GZO target. Those results suggest that the concentration of In$_2$O$_3$ is the most important factor to influence the mobility of IGZO thin films and the results agree with the important results investigated by H. Hosono [23]. Generally, the field-effect mobility of semiconductor thin films of TFT devices is determined by many factors, including the energy band properties of the active layers and the interface states [24]. The related energy band states of the active layers involve deep states, band-tail states, and extended states. The second reason suggests that Hall mobility of IGZO thin films scarcely increases with the increase in deposition power of GZO ceramic target because the deep states and tail-like states in α-IGZO show little dependence on RF power. The resistivity of TCO thin films is proportional to the reciprocal of the product of carrier concentration N and mobility μ:

$$\rho = 1/Ne\mu \qquad (2)$$

Both the carrier concentration and the carrier mobility contribute to the conductivity. As the deposition power of GZO ceramic target was changed from 80 W to 140 W, the resistivity of IGZO thin films was linearly increased from 5.91×10^{-4} Ω-cm to 1.86×10^{-2} Ω-cm. The minimum resistivity of IGZO thin films at a deposition power of GZO ceramic target of 80 W is mainly caused by the carrier mobility at its maximum.

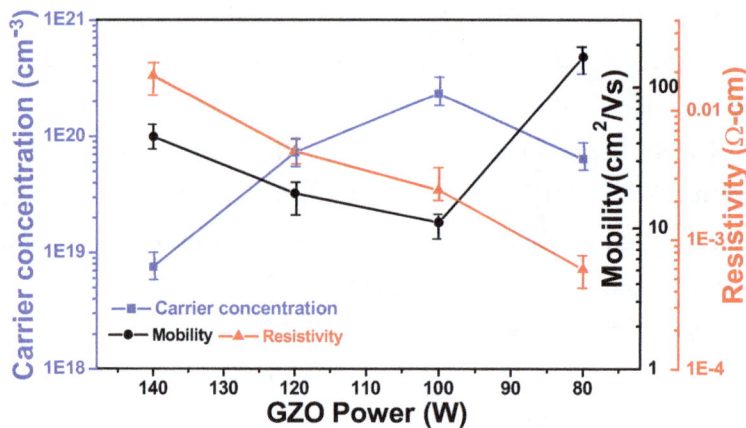

Figure 8. Hall mobility, carrier concentration, and resistivity of IGZO thin films as a function of deposition power of GZO ceramic target.

4. Conclusions

The characteristics of IGZO thin films prepared by Ga$_2$O$_3$-2 ZnO (GZO) and In$_2$O$_3$ co-sputtering method were well investigated in this study. As the deposition powers of GZO ceramic target was 80 W, 100 W, 120 W, and 140 W, the thickness of IGZO thin films was around 103 nm, 138 nm, 149 nm, and 170 nm; the average transmittance ratio of IGZO thin films in the range of 400 nm~700 nm was 77.3%, 77.5%, 91.4%, and 86.6%; and the calculated E_g values of IGZO thin films were 3.87 eV, 3.84 eV, 3.79 eV, and 3.71 eV, respectively. From the SIMS analysis results of IGZO thin films, the concentrations of In and Ga elements in the depth profile showed an uniformity distribution and the concentration of Zn element in the depth profile was not uniform distribution. As the deposition power of GZO thin films increased, the concentrations of Ga and Zn elements increased in the depth profile and the concentration of In element in the depth profile was almost unchanged. As the deposition power of GZO ceramic target increased from 80 W to 140 W, the carrier mobility of IGZO thin films was in the range of 11.0 cm^2/V-s~ 163.4 cm^2/V-s and the resistivity of IGZO thin films was linearly increased from 5.91 × 10^{-4} Ω-cm to 1.86 × 10^{-2} Ω-cm. The mobility of 163.4 cm^2/V-s is higher than those of most reported IGZO thin films.

Acknowledgments

The authors acknowledge financial supports of NSC 102-2221-E-218-036-, NSC 102-2622-E-390-002-CC3, and NSC 102-2221-E-390-027.

Author Contributions

Prof. Hsu and Prof. Tzou helped proceeding the experimental processes, measurements, and data analysis; Prof. Yang organized the paper and encouraged in paper writing; Mr. Liou helped proceeding the experimental processes and measurements.

Conflicts of Interest

The authors declare no conflict of interest.

References

1. Kanicki, J.; Libsch, F.R.; Griffith, J.; Polastre, R. Performance of thin hydro-genated amorphous silicon thin-film transistors. *J. Appl. Phys.* **1991**, *69*, 2339–2345.
2. Sato, A.; Shimada, M.; Abe, K.; Hayashi, R.; Kumomi, H.; Nomura, K.; Kamiya, T.; Hirano, M.; Hosono, H. Amorphous In-Ga-Zn-O thin-film transistor with coplanar homojunction structure. *Thin Solid Films* **2009**, *518*, 1309–1313.
3. Jung, Y.S.; Lee, K.H.; Kim, W.J.; Lee, W.J.; Choi, H.W.; Kim, K.H. Properties of In-Ga-Zn-O thin films for thin film transistor channel layer prepared by facing targets sputtering method. *Ceram. Int.* **2012**, *38S*, S601–S604.
4. Park, B.; Cho, K.; Kim, S.; Kim, S. Nano-floating gate memory devices composed of ZnO thin-film transistors on flexible plastics. *Nanoscale Res. Lett.* **2010**, *6*, doi:10.1007/s11671-010-9789-5.
5. Cong, Y.; Han, D.; Wu, J.; Zhao, N.; Chen, Z.; Zhao, F.; Dong, J.; Zhang, S.; Zhang, X.; Wang, Y. Studies on fully transparent Al-Sn-Zn-O thin-film transistors fabricated on glass at low temperature. *Jpn. J. Appl. Phys.* **2015**, *54*, doi:10.7567/JJAP.54.04DF01.
6. Jeong, S.K.; Kim, M.H.; Lee, S.Y.; Seo, H.; Choi, D.K. Dual active layer α-IGZO TFT via homogeneous conductive layer formation by photochemical H-doping. *Nanoscale Res. Lett.* **2014**, *9*, doi:10.1186/1556-276X-9-619.
7. Nomura, K.; Ohta, H.; Ueda, K.; Kamiya, T.; Hirano, M.; Hosono, H. Thin-film transistor fabricated in single-crystalline transparent oxide semiconductor. *Science* **2003**, *300*, 1269–1272.
8. Nomura, K.; Ohta, H.; Takagi, A.; Kamiya, T.; Hirano, M.; Hosono, H. Room-temperature fabrication of transparent flexible thin-film transistors using amorphous oxide semiconductors. *Nature* **2004**, *432*, 488–492.
9. Zan, H.W.; Tsai, W.W.; Chen, C.H.; Tsai, C.C. Effective mobility enhancement by using nanometer dot doping in amorphous IGZO thin-film transistors. *Adv. Mater.* **2011**, *23*, 4237–4242.

10. Bak, J.Y.; Kang, Y.; Yang, S.; Ryu, H.J.; Hwang, C.S.; Han, S.; Yoon, S.M. Origin of degradation phenomenon under drain bias stress for oxide thin film transistors using IGZO and IGO channel layers. *Sci. Rep.* **2015**, *5*, doi:10.1038/srep07884.

11. Jeong, J.A.; Kim, H.K. Transparent Ga and Zn co-doped In_2O_3 electrode prepared by co-sputtering of $Ga:In_2O_3$ and $Zn:In_2O_3$ targets at room temperature. *Thin Solid Films* **2011**, *519*, 3276–3282.

12. Nomura, K.; Kamiya, T.; Ohta, H.; Uruga, T.; Hirano, M.; Hosono, H. Local coordination structure and electronic structure of the large electron mobility amorphous oxide semiconductor In-Ga-Zn-O: Experiment and ab initio calculations. *Phys. Rev. B* **2007**, *75*, doi:10.1103/PhysRevB.75.035212.

13. Wang, F.H.; Kuo, H.H.; Yang, C.F.; Liu, M.C. Role of SiN_x barrier layer on the performances of polyimide Ga_2O_3-doped ZnO p-i-n hydrogenated amorphous silicon thin film solar cells. *Materials* **2014**, *7*, 948–962.

14. Sato, Y.; Otake, F.; Hatori, H. A dependence of crystallinity of In_2O_3 thin films by a two-step heat treatment of indium films on the heating atmosphere. *J. Mod. Phys.* **2010**, *1*, 360–363.

15. Wang, F.H.; Fu, M.Y.; Su, C.C.; Yang, C.F.; Tzeng, H.T.; Liu, H.W.; Kung, C.Y. Improve the properties of p-i-n α-Si:H thin-film solar cells using the diluted hydrochloric acid-etched GZO thin films. *J. Nanomater.* **2013**, *2013*, doi:10.1155/2013/495752.

16. Lo, C.C.; Hsieh, T.E. Preparation of IGZO sputtering target and its applications to thin-film transistor devices. *Ceram. Int.* **2012**, *38*, 3977–3983.

17. Saha, D.; Das, A.K.; Ajimsha, R.S.; Misra, P.; Kukreja, L.M. Disorder-driven carrier transport in atomic layer deposited ZnO thin films. *Mater. Sci.* **2013**, arXiv:1301.1172.

18. Heo, Y.W.; Ip, K.; Pearton, S.J.; Norton, D.P.; Budai, J.D. Growth of ZnO thin films on c-plane Al_2O_3 by molecular beam epitaxy using ozone as an oxygen source. *Appl. Surf. Sci.* **2006**, *252*, 7442–7448.

19. Wang, F.H.; Yang, C.F.; Liou, J.C.; Chen, I.C. Effects of hydrogen on the optical and electrical characteristics of the sputter-deposited Al_2O_3-doped ZnO thin films. *J. Nanomater.* **2014**, *2014*, doi:10.1155/2014/857614.

20. Beena, D.; Lethy, K.J.; Vinodkumar, R.; Detty, A.P.; Mahadevanpillai, V.P.; Ganesan, V. Photoluminescence in laser ablated nanostructured indium oxide thin films. *Optoelectron. Adv. Mater. Rapid Commun.* **2011**, *5*, 1–11.

21. Li, C.; Yan, J.L.; Zhang, L.Y.; Zhao, G. Electronic structures and optical properties of Zn-doped β-Ga_2O_3 with different doping sites. *Chin. Phys. B* **2012**, *21*, doi:10.1088/1674-1056/21/12/127104.

22. Kim, G.H.; Ahn, B.D.; Shin, H.S.; Jeong, W.H.; Kim, H.J.; Kim, H.J. Effect of indium composition ratio on solution-processed nanocrystalline InGaZnO thin film transistors. *Appl. Phys. Lett.* **2009**, *94*, doi:10.1063/1.3151827.

23. Hosono, H. Ionic amorphous oxide semiconductors: Material design, carrier transport, and device application. *J. Non-Cryst. Solids* **2006**, *352*, 851–858.

24. Shi, J.F.; Dong, C.Y.; Dai, W.J.; Wu, J.; Chen, Y.T.; Zhan, R.Z. The influence of RF power on the electrical properties of sputtered amorphous In-Ga-Zn-O thin films and devices. *J. Semicond.* **2013**, *34*, doi:10.1088/1674-4926.

Effect of Refractive Index of Substrate on Fabrication and Optical Properties of Hybrid Au-Ag Triangular Nanoparticle Arrays

Jing Liu [1,2], Yushan Chen [1], Haoyuan Cai [1], Xiaoyi Chen [3], Changwei Li [2] and Cheng-Fu Yang [4,*]

[1] School of Information Engineering, Jimei University, Xiamen 361021, China;
E-Mails: jingliu@jmu.edu.cn (J.L.); chenys@jmu.edu.cn (Y.C.); haoyuancai1@gmail.com (H.C.)
[2] China–Australia Joint Laboratory for Functional Nanomaterials, Xiamen University, Xiamen 361005, China; E-Mail: lichangweichangwei@gmail.com
[3] Department of Physics, National University of Singapore, Singapore 117551, Singapore;
E-Mail: a0123706@u.nus.edu
[4] Department of Chemical and Materials Engineering, National University of Kaohsiung, No. 700, Kaohsiung University Rd., Nan-Tzu District, Kaohsiung 811, Taiwan

* Author to whom correspondence should be addressed; E-Mail: cfyang@nuk.edu.tw;

Academic Editor: Teen-Hang Meen

Abstract: In this study, the nanosphere lithography (NSL) method was used to fabricate hybrid Au-Ag triangular periodic nanoparticle arrays. The Au-Ag triangular periodic arrays were grown on different substrates, and the effect of the refractive index of substrates on fabrication and optical properties was systematically investigated. At first, the optical spectrum was simulated by the discrete dipole approximation (DDA) numerical method as a function of refractive indexes of substrates and mediums. Simulation results showed that as the substrates had the refractive indexes of 1.43 (quartz) and 1.68 (SF5 glass), the nanoparticle arrays would have better refractive index sensitivity (RIS) and figure of merit (FOM). Simulation results also showed that the peak wavelength of the extinction spectra had a red shift when the medium's refractive index n increased. The experimental results also demonstrated that when refractive indexes of substrates were 1.43 and 1.68, the nanoparticle arrays and substrate had better adhesive ability. Meanwhile, we found the nanoparticles formed a large-scale monolayer array with the hexagonally close-packed structure. Finally, the hybrid Au-Ag triangular nanoparticle arrays were fabricated on quartz and SF5 glass substrates and their experiment extinction spectra were compared with the simulated results.

Keywords: discrete dipole approximation (DDA); nanosphere lithography (NSL); hybrid Au-Ag nanoparticles; substrate refractive index

---

1. Introduction

Surface plasmon resonances (SPRs) are surface electromagnetic waves that propagate in a direction parallel to the metal/dielectric (or metal/vacuum) interface. SPRs are used as the basis of many standard tools for measuring adsorption of material onto planar metal (typically gold and silver) surfaces or onto the surface of metal nanoparticles. However, in their simplest form, SPRs' reflectivity measurements can be used to detect molecular adsorption, such as polymers, DNA, proteins, *etc.* [1]. In a simple situation, such as that of nearly monodisperse spherical gold nanoparticle arrays in solution, the extinction spectrum exhibits a single peak known as the localized SPRs (LSPRs) [2]. LSPRs are collective electron charge oscillations in metallic plane or nanoparticle arrays that are excited by light and LSPRs exhibit enhanced near-field amplitude at the resonance wavelength. For that, LSPRs' spectroscopy of metallic nanoparticle arrays is a powerful technique for chemical and biological applications and different lab-on-a-chip sensors [3,4].

The discrete dipole approximation (DDA) is a method being used to compute scattering of radiation for particles having arbitrary shapes and having periodic structures [5]. The basic idea of the DDA algorithm was introduced in 1964 by DeVoe [6] who applied it to study the optical properties of molecular aggregation. Given a target of arbitrary geometry, one can calculate its scattering and absorption properties with the DDA method. Exact solutions to Maxwell's equations of LSPRs are known only for special geometries such as spheres, spheroids, or cylinders, so approximate methods are generally required. However, the DDA method employs no physical approximations and can produce accurately enough results, which can give sufficient computer power to simulate the LSPRs optical properties of nanoparticle arrays.

In 1909, Lorentz [7] showed that the dielectric and refractive properties of a substance could be directly related to the polarizabilities of the individual atoms of which it was composed. With a particularly simple and exact relationship, the Lorentz–Lorenz equation, also known as the Clausius–Mossotti relation and Maxwell's formula, the refractive index of a substance is related to its polarizability. Typical metals that support surface plasmons in the nanoparticle arrays' structure are silver (Ag) [8], gold (Au) [9], and chromium (Cr) nanoparticle arrays [10] in single-layer structure and hybrid Ag-Au [5] and hybrid Ni-Au [11] nanoparticle arrays in bi-layer structure. In addition, Cr can be used as the interlayer [5,12] to improve the adhesive effect of nanoparticle arrays.

As we know, the refractive index sensitivity (RIS) and figure of merit (FOM value, defined as the ratio of RIS/FWHM (full width at half maximum)) of the LSPR sensors [13,14] are sensitive to the used substrates. In the past, the electron-beam lithography (EBL) [15] and photolithography [16] methods can be used to fabricate the nanoparticle arrays for LSPRs' structures. Nanosphere lithography (NSL) [17,18] is one of the most low-cost and high-efficiency methods for producing periodically and geometrically tunable nanostructure arrays. In this paper, at first, the DDA method was used to simulate and find the RIS and FOM values of the hybrid nanostructure arrays on glasses with different refractive indexes. After finding glass refractive indexes had the better RIS and FOM values, the periodically hybrid Au-Ag triangular nanoparticle arrays were systematically grown on the glasses. The NSL method was also used to grow the hybrid Au-Ag triangular nanoparticle arrays. One important reason for using hybrid Au-Ag nanoparticle is that the Au can avoid Ag being oxidized or sulfurized. Glasses with different refractive indexes were used as the substrates, which were used to investigate the effect of different refractive indexes of the substrates on the optical properties of the hybrid Au-Ag triangular nanoparticle arrays.

2. Model Construction and Simulation

An implementation of electrodynamic theory called the discrete dipole approximation (DDA) can be used to model the experimentally measured extinction spectra. The DDA provides a convenient method for describing light scattering from nanoparticles or nanoparticle arrays of arbitrary shapes. Using the DDA algorithm, we could design and calculate the extinction spectra, RIS value, and FOM value of the hybrid Au-Ag triangular nanostructure arrays on different substrates. In the past, we had found that that the RIS value of the hybrid Au-Ag triangular nanoparticle arrays increased with the thickness of Cr interlayer from 4 to 12 nm and then decreased with further increase of Cr interlayer thickness to 20 nm [5]. Even the FOM value generally decreased with the increasing thickness of Cr thickness, we chose a suitable Cr interlayer thickness of 8 nm to improve the adhesion between Ag film and the substrates [5]. The corresponding schematic illustration of the hybrid nanoprism is shown in Figure 1: the structure was equilateral triangle with a side length of 180 nm. The thicknesses of Cr, Ag, and Au thin films were fixed on $h_{Cr} = 8$ nm, $h_{Ag} = 35$ nm, and $h_{Au} = 5$ nm, where h_{Au}, h_{Ag}, and h_{Cr} were defined in Figure 1b. The used substrates were quartz (with refractive index of 1.43), BAK1 glass (1.57), SF5 glass (1.68), SF10 glass (1.74), and SF6 glass (1.80), respectively.

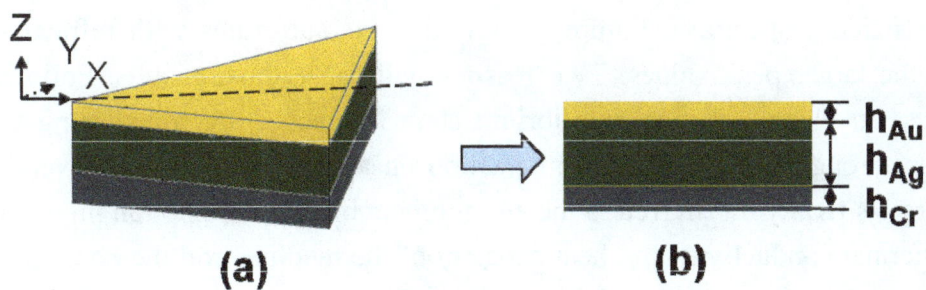

(a)   **(b)**

Figure 1. Schematic view of a single hybrid Au-Ag nanoparticle **(a)** In 3D view **(b)** Side cross-section view.

Extinction efficiency is a parameter used in physics to describe the absorption and scattering of electromagnetic radiation. Any changes in the parameters of metal nanoparticles or nanoparticle arrays could lead to the optical drift in extinction spectrum and thus influence the optical applications in practice. As Figure 2 shows, the peak value of the LSPR wavelengths had no apparent change but the peak intensity had large variation as the substrates' refractive index was changed. However, the extinction intensity did not gradually change with gradual change in refractive index of the substrate and the higher extinction intensities were revealed in using quartz and SF5 glass as the substrates. SPR is the resonant oscillation of conduction electrons at the interface between a negative and positive permittivity material stimulated by incident light. Surface plasmon polaritons are surface electromagnetic waves that propagate in a direction parallel to the metal/dielectric and/or metal/vacuum interfaces. The resonance condition is established when the frequency of incident photons matches the natural frequency of surface electrons oscillating against the restoring force of positive nuclei. Since the wave is on the boundary of the metal and the external medium (air in this paper), these oscillations are very sensitive to any change in this boundary, such as the adsorption of molecules to the metal surface. The peak value of LSPR wavelengths had no apparent change because the peak value is mainly caused by the Ag film, which will be proven later. When quartz and SF5 glass are used as substrates, their maximum extinction efficiencies are higher than the values using other glasses as substrates. We believe that as the refractive indexes of 1.43 (quartz) and 1.68 (SF5 glass) are used as the substrates, the frequency of incident photons will match the natural frequency of surface electrons on the two substrates. For that, when the quartz and SF5 glass are used as the substrates, the extinction intensities will be higher.

In addition, as the quartz and SF5 glass were used as the substrates, their FWHM value of the extinction efficiency spectra were smaller than the FWHM values of the extinction efficiency spectra using other glasses as substrates. In this study, the main extinction peak of the hybrid Au-Ag triangular nanoparticle arrays was observed at around 700 nm. In our previous work [19], we had found that as only silver film was deposited as the nanoparticle arrays, the resonance peak of the extinction spectrum was around 700 nm. If the gold film was deposited on the surface of silver film to form the hybrid Au-Ag films, as the thickness of gold film in the hybrid Au-Ag nanoparticle arrays increased, the corresponding resonance extinction peak of the extinction spectrum had a blue shift. However, the hybrid nanoparticle arrays are composed of gold and silver films and the thickness of gold film is only 5 nm, which is very thin. For that, the silver film is the mainly sensitive medium and this result suggests that the main extinction peak around 700 nm is mainly caused by Ag plasmon.

The extinction efficiency spectra in Figure 2 show that the substrates with refractive indexes of 1.43 and 1.68 had the larger peak values. Two reasons will cause those results. Following the pure dephasing processes, an electron lattice equilibrium state is reached by transferring the electronic energy to the lattice through electron-phonon interaction on a subpico-second to several pico-second time scale. The energy is finally transferred to the environment by phonon-phonon interaction, which is dependent on the thermal conductivity and heat capacity of the medium and the coupling between the nanoparticle arrays and the surrounding mediums [20]. At first, when the quartz (refractive index 1.43) and SF5 (1.68) are used as substrates, high density grain boundaries with dense, high-frequency molecular type vibrations are present, which are effective in removing the energy of the excited electrons in the nanoparticle arrays and in transforming the energy of incident light into the phonon and thermal energy. Then, the ionic plasmas can absorb more resonance energy. Second, when the quartz and SF5 are used as substrates, the coupling effect between substrates and metal nanoparticle arrays will be enhanced, and the enhancement between photo fields will also enhance the local filed.

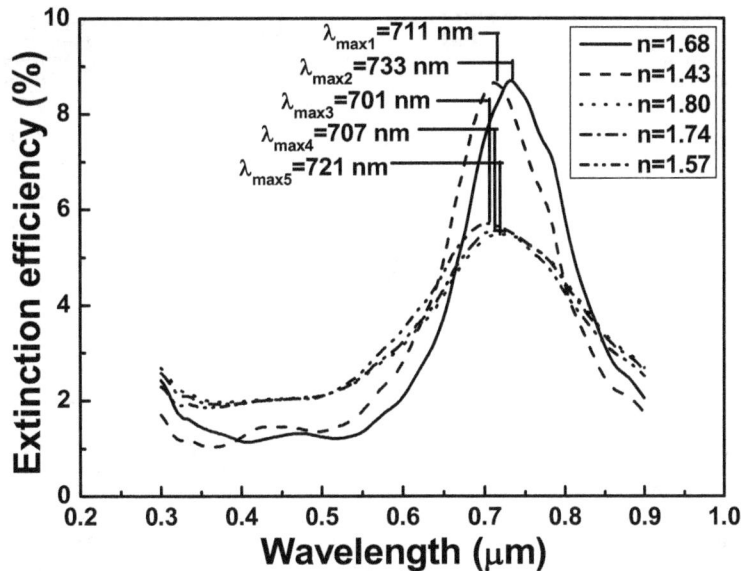

Figure 2. The extinction spectra results of compound nanostructure arrays as a function of substrate with different refractive indexes.

Because the substrates with refractive index of 1.43 (quartz) and 1.68 (SF5 glass) had the higher extinction efficiency, as Figure 2 shows, they were used to investigate the effect of the substrates' refractive indexes on the sensitivity of the hybrid Au-Ag nanostructure arrays. For that, the extinction efficiency spectra of the effective refractive index of the medium surrounding the nanostructure arrays were calculated. The refractive index sensitivity (RIS) is defined as $m = \Delta\lambda/\Delta n$ [5], where $\Delta\lambda$ denotes the change in peak value of the wavelength of the extinction efficiency spectra and Δn denotes the change in refractive index, respectively. As Figure 3a shows, as the substrate's refractive index was 1.43 and the medium's refractive index was increased from 1.0 to 1.15, the FWHM value and peak value of the extinction spectra had no apparent changes, the wavelength to reveal the peak value of the extinction spectra was shifted from 733 to 806 nm; however, as the substrate's refractive index was 1.68 and the medium refractive index was increased from 1.0 to 1.15, the FWHM value and peak value of the extinction spectra also had no apparent changes, the wavelength to reveal the peak value of the extinction spectra was shifted from 711 to 796 nm, respectively.

The results shown in Figure 3a–e suggest that the peak wavelength of the extinction spectra had a red shift when the medium's refractive index n increased. Figure 3f shows the relationships between the peak wavelengths of the extinction efficiency spectra and the medium's refractive indexes, which could be used to investigate the RIS values. As results in Figure 3a–e are compared, the extinction efficiencies of using quartz (1.43) and SF5 glass (1.68) as substrates are higher than those of using other glasses as substrates. In addition, the FWHM values of using quartz and SF5 glass as substrates are smaller than those using other glasses as substrates. As the substrates' refractive indexes are 1.43, 1.57, 1.68, 1.74 and 1.8, the RIS values of the hybrid Au-Ag triangular nanoparticle arrays are 560, 398, 486, 408, and 380 nm/RIU (refractive index unit), respectively. Because of having large extinction efficiencies and small FWHM values, only the FOM values of substrate's refractive indexes of 1.43 and 1.68 were calculated, respectively, and the calculated values were 3.06 and 2.63, which are compared in Table 1.

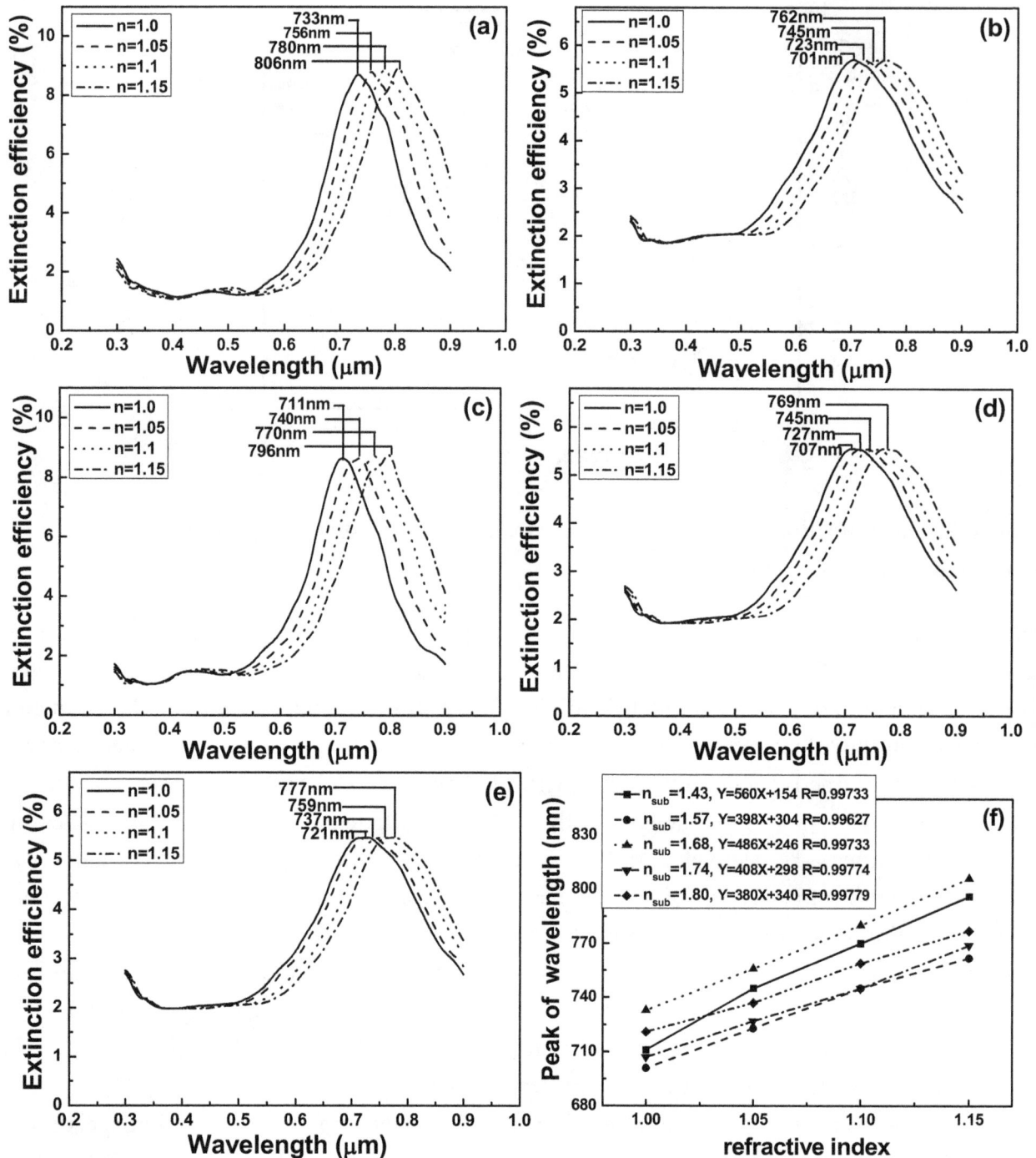

Figure 3. Effect of the substrates' refractive indexes on the sensitivity of the hybrid nanostructure arrays. Extinction spectra in different media for substrate with refractive index of (**a**) 1.43; (**b**) 1.57; (**c**) 1.68; (**d**) 1.74; and (**e**) 1.80; and (**f**) for refractive index sensitivity curves.

Table 1. Comparisons of refractive index sensitivity (RIS), width at half maximum (FWHM), and figure of merit (FOM) using two different substrates.

Feature/Characteristic	$n_{substrate} = 1.43$	$n_{substrate} = 1.68$
RIS (nm/RIU)	560	486
FWHM (nm)	183	185
FOM	3.06	2.63

3. Fabrication of Hybrid Au-Ag Triangular Nanoparticle Arrays

In this study, the quartz and glass with different refractive indexes were used as substrates to study the effect of refractive index on the change of the LSPRs' peak value and optical property of the hybrid Au-Ag nanoparticle arrays. Figure 4 illustrates the schematic view for the fabrication process of nanosphere lithography (NSL) method. In order to fabricate the hybrid nanoparticle arrays with the NSL method the polystyrene (PS) nanospheres were used as a deposition mask. The PS nanospheres with a mean diameter of 360 ± 10 nm and a concentration of 10 wt% in solution were purchased from Suzhou Nano-Micro Bio-Tech Co. Ltd. (Suzhou, China). The details for the preparation of the PS nanospheres were revealed in reference [5]. The depositions of Cr, Ag, and Au metals (all with 3N purity) were performed in a self-built thermal evaporator at a pressure of 5.0×10^{-4} Pa. The quartz and glass substrates were rotated at a speed of 16.5 rpm during the deposition process. The power for heating-up of the source materials was carefully increased in order to achieve homogeneous deposition. The deposition rate was about 4.0 nm/s for Cr thin film and the deposition rates were about 2.5 nm/s for both Au and Ag thin films, respectively. The thicknesses of the deposition thin films were monitored using a Dektak 3 Series surface profiler (Bruker, Billerica, MA, USA) to achieve an identical depth with a low reflectance. The deposition thicknesses of Cr, Ag, and Au thin films on PS nanospheres were about $h_{cr} = 8$ nm, $h_{Ag} = 35$ nm, and $h_{Au} = 5$ nm, respectively, by controlling the deposition time. After depositions of Cr, Ag, and Au thin films, the PS spheres were lifted off by immersing in absolute ethanol for about 5 s. The PS spheres were also removed by sonication (B3500S-MT, 140 W, 42 kHz, Branson, Danbury, CT, USA) in absolute ethanol to examine the adhesive ability of the hybrid Au-Ag nanoparticles on different substrates. The achieved PS mask and the structures of the achieved hybrid Au-Ag nanoparticle arrays on different substrates were characterized by scanning electron microscope (LEO-1530, Zeiss, Oberkochen, Germany). Ultraviolet visible (UV-Vis) spectra are obtained on a Cary 5000UV-Vis-NIR (175–3300 nm, Varian, New York, NY, USA) spectrophotometer.

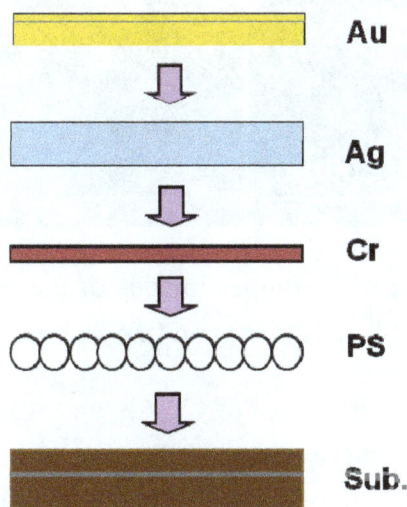

Figure 4. Schematic illustration for the fabrication process of the hybrid Au-Ag nanoparticle arrays.

4. Results and Discussion

Figure 5 shows the top morphologies of the hybrid Au-Ag triangular nanoparticle arrays with the different substrates. Top morphologies of the deposited hybrid Au-Ag nanoparticle arrays, as Figure 5a,c show, exhibited a hexagonally arranged disc structure with triangular structures at the six corners. The hybrid Au-Ag nanoparticle arrays were regular and well-defined independent in the substrates, and no tiny cracks were observed in the structure. Those results suggest that the cohesive force between the PS nanospheres and the quartz and SF5 glass substrates seems to be strong enough. However, the results shown in Figure 5 revealed different results as the substrates were different. As SF5 glass substrates were used and the results in Figure 5a,b were compared, the as-deposited hybrid Au-Ag periodically nanoparticle arrays had a sharp angle in the triangular structure. The annealed Au-Ag periodically nanoparticle arrays were changed to quasi-half-ball structure and were shrunk together, the Cr interlayer was also observed.

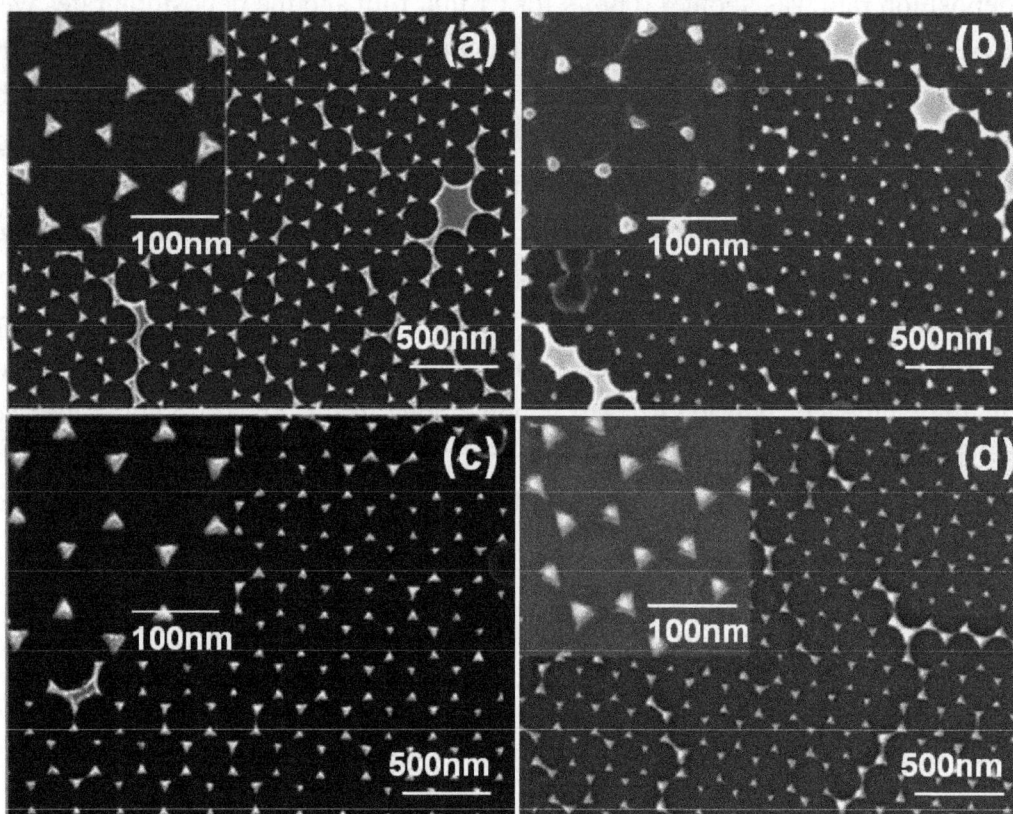

Figure 5. SEM (scanning electron microscope) images of the hybrid Au-Ag nanoparticle arrays deposited on two different substrates: (**a**) and (**b**): SF5, (**c**) and (**d**): quartz.

The reason is caused by the deposition processes of Cr, Ag, and Au thin films, which are carried out at a higher temperature. Because of this, more defects will exist in the Cr thin film and interaction or collision mean free path between Cr and hybrid Au-Ag increases, and more Ag, Au atoms will be diffused into the Cr thin films. When the deposition processes of Ag and Au thin films are stopped, then the substrates will persist a quasi-quench process. According to thermodynamic theory, this process will cause the instability in the atoms' stacking state. Therefore, from the dynamics theorem the Ag and Au have a lower separation speed, the quasi-quench process will keep the metal in the high-temperature state. For that, the Ag and Au can be uniformly deposited onto the Cr thin films. As the annealing duration persists for a period, the deposition multi-layer thin films will attain a condition of thermodynamic stability, which is changing from the non-equilibrium state to the equilibrium state. At this time, the separation condition will happen between Cr thin film and hybrid Au-Ag thin films. Then, the structure of Ag, Au thin films changes from triangle to quasi-half-ball because of the decrease of surface tension.

When the quartz was used substrates, the as-deposited hybrid Au-Ag periodically nanoparticle arrays had a sharp angle in the triangular structure, and the annealed nanoparticle arrays had no apparent change even the annealing process was used to treat on them, as Figure 5c,d were compared. As for the characteristics of the metal nanoparticles, the different morphologies after annealing can be attributed to that the Cr will diffuse into the SF5 glass substrate during the annealing process. As a result, the hybrid Au-Ag originally closed to the Cr will become adjacent to glass substrate. This phenomenon will lead to that the melting temperature of the Au-Ag nanoparticle arrays going lower than that of the Au-Ag-Cr nanoparticle arrays. However, the quartz substrate is a compact material, which can keep in stable even a higher temperature annealing process is treated, for that the Cr will not diffuse into the quartz substrate and the surfaces of the nanoparticle arrays are almost unchanged.

Figures 6 and 7 compare the extinction spectra of the experimental results for $n_{medium} = 1.0$ with those of the simulated results extrapolated from the data in Figure 3a,b. For the results of the DDA calculation, the wavelength with the maximum extinction efficiency was 711 nm for substrate with refractive index of 1.68 and was 733 nm for substrate with refractive index of 1.43. Thus, for the measured results, the wavelength with the maximum extinction efficiency was 675 nm for substrate with refractive index of 1.68 and was 688 nm for substrate with refractive index of 1.43, respectively. Figures 6 and 7 prove that the experimental results are generally in agreement with the calculated results. The major difference between with the calculation values and experiment values was that the wavelength with the maximum extinction efficiency had the blue shifts of 36 and 45 nm for using quartz and SF5 glass as substrates, respectively.

There are many reasons for the differences between the calculation values and experimental values, including the effects of different substrate [21] and the fabrication error caused by the uniformities of both size and shape of the hybrid Au-Ag nanoparticle arrays. In the DDA calculated model, the edge of the triangular nanoparticle arrays are straight while in real experimental results it will have a little warping and curving. Thus, effective medium theories can be proposed to explain the spectra of partially embedded objects [22], because the relative volumes of the two media have not been defined, except for the simple case of a half-embedded object. The exposed area and effective medium ideas are deficient because they fail to account for the change in LSPRs' wavelength that arises from the asymmetric environment and distorting nanoparticle arrays. However, the comparisons of the calculated and simulated results of the 2D hexagonally arranged hybrid Au-Ag triangular nanoparticle arrays show that the DDA calculated model is suitable method for our experimental fabrication.

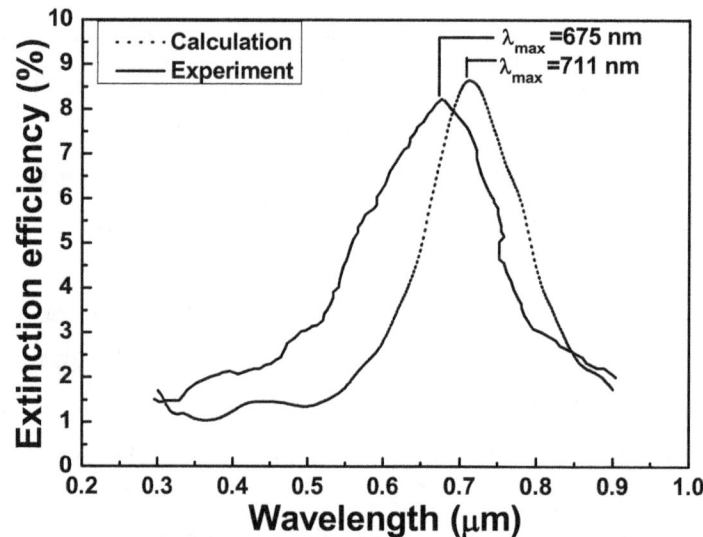

Figure 6. Extinction spectra of experiment and calculation with 1.43 refractive index substrate.

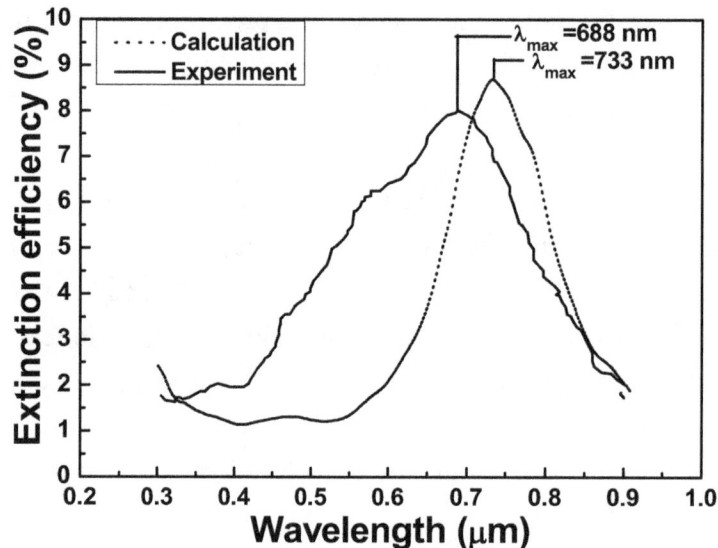

Figure 7. Extinction spectra of experiment and calculation with 1.68 refractive index substrate.

5. Conclusions

From the simulated results of the DDA numerical method, when the substrates' refractive indexes were 1.43 (quartz) and 1.68 (SF5 glass), as the medium's refractive index was increased from 1.0 to 1.15, the full width at half maximum of the LSPRs' spectra and the wavelength to reveal the maximum extinction efficiency of the LSPRs' spectra had no apparent changes. The wavelength to reveal the peak value of the LSPRs' spectra was shifted from 733 to 806 nm for the substrate's refractive index of 1.43 and was shifted from 711 to 796 nm for the substrate's refractive index of 1.68, respectively. As the substrates' refractive indexes were 1.68 and 1.43, the refractive index sensitivities of 486 and 560 nm/RIU were obtained in the hybrid Au-Ag triangular nanoparticle arrays. When the annealed process was used, the as-deposited hybrid Au-Ag periodically nanoparticle arrays on SF5 glass substrates were changed from a sharp angle in the triangular structure to quasi-half-ball structure and were shrunk together, and the Cr interlayer was also observed. The as-deposited hybrid Au-Ag periodically nanoparticle arrays on the quartz substrates had a sharp angle in the triangular structure and no apparent change even the annealing process was used to treat on them. For the results of the DDA calculation and substrates with refractive index of 1.68 and 1.43, the wavelengths to reveal the peak values of the LSPRs' spectra were 711 and 733 nm. For the measured results and substrates with refractive index of 1.68 and 1.43, the wavelengths to reveal the peak values of the LSPRs' spectra were 675 and 688 nm, respectively. However, the comparisons of the calculated and simulated results of the 2D hexagonally arranged hybrid Au-Ag triangular nanoparticle arrays show that the DDA calculated model is suitable to investigate the hybrid Au-Ag triangular nanoparticle arrays.

Acknowledgments

This work was supported by Fujian Provincial Natural Science Foundation under Grant No. 2015J01265, 2015J01266, 2015R0063, and 2015J0102, Science and Technology Plan Projects of Xiamen City under Grant No. 3502Z20143020. The plan of college youth outstanding research talents in Fujian Province (2015). Thanks are due to Xianfang Zhu for assistance with the experiments.

Author Contributions

Jing Liu organized the experimental procedure and results and paper writing. Yushan Chen, Haoyuan Cai, Xiaoyi Chen and Changwei Li helped the experimental procedure and analyzed the measured results. Cheng-Fu Yang analyzed the measured results helped in the paper writing.

Conflicts of Interest

The authors declare no conflict of interest.

References

1. Ha, M.H.; Endo, T.; Kerman, K.; Chikae, M.; Kim, D.K.; Yamamura, S.; Takamura, Y.; Tamiya, E. A localized surface plasmon resonance based immunosensor for the detection of casein in milk. *Sci. Technol. Adv. Mater.* **2007**, *8*, 331–338.
2. Malinsky, M.D.; Kelly, K.L.; Schatz, G.C.; Duyne, R.P.V. Nanosphere lithography: Effect of substrate on the localized surface plasmon resonance spectrum of silver nnanoparticles. *J. Phys. Chem. B* **2001**, *105*, 2343–2350.
3. Sekhon, J.S.; Malik, H.K.; Verma, S.S. DDA simulations of noble metal and alloy nanocubes for tunable optical properties in biological imaging and sensing. *RSC Adv.* **2013**, *3*, 15427–154343.
4. Yonzon, C.; Duyne, R.P.V. Localized and propagating surface plasmon resonance sensors: A study using carbohydrate binding protein. *Mater. Res. Soc. Symp. Proc.* **2005**, *876*, doi:10.1557/PROC-876-R7.3.
5. Liu, J.; Cai, H.; Kong, L.; Zhu, X. Effect of chromium interlayer thickness on optical properties of Au-Ag nanoparticle array. *J. Nanomater.* **2014**, *2014*, doi:10.1155/2014/650359.
6. DeVoe, H. Optical properties of molecular aggregates. I. Classical model of electronic absorption and refraction. *J. Chem. Phys.* **1964**, *41*, 393–400.
7. Lorentz, H.A. *Theory of Electrons*; Teubner: Leipzig, Germany, 1909.

8. Huang, S.; Yang, Q.; Zhang, C.; Kong, L.; Li, S.; Kang, J. Structural anomalies induced by the metal deposition methods in 2D silver nanoparticle arrays prepared by nanosphere lithography. *Thin Solid Films* **2013**, *536*, 136–141.

9. Jenkins, J.A.; Zhou, Y.; Thota, S.; Tian, X.; Zhao, X.; Zou, S.; Zhao, J. Blue-shifted narrow localized surface plasmon resonance from dipole coupling in gold nanoparticle random arrays. *J. Phys. Chem. C* **2014**, *118*, 26276–26283.

10. Shyu, J.H.; Lin, Y.C.; Lee, H.M.; Hsieh, C.T.; Huang, C.Y.; Wu, J.C. Tunable surface plasmon resonances based on chromium disk array containing liquid crystals. *Jpn. J. Appl. Phys.* **2011**, *50*, doi:10.1143/JJAP.50.09MG03.

11. Fu, Q.; Wong, K.M.; Zhou, Y.; Wu, M.; Lei, Y. Ni/Au hybrid nanoparticle arrays as a highly efficient, cost-effective and stable SERS substrate. *RSC Adv.* **2015**, *5*, 6172–6180.

12. Huang, S.; Kong, L.; Zhang, C.; Wu, Y.; Zhu, X. Effect of chromium interlayer deposition on periodic silver nanoparticle array structure fabricated by nanosphere lithography. *Phys. Lett. A* **2011**, *375*, 3012–3016.

13. Spackova, S.; Homola, J. Sensing properties of lattice resonances of 2D metal nanoparticle arrays: An analytical model. *Opt. Express* **2013**, *21*, 27490–27502.

14. Martinsson, E.; Shahjamali, M.M.; Enander, K.; Boey, F.; Xue, C.; Aili, D.; Liedberg, B. Local refractive index sensing based on edge gold-coated silver nanoprisms. *J. Phys. Chem. C* **2013**, *117*, 23148–23154.

15. Kim, S.; Marelli, B.; Brenckle, M.A.; Mitropoulos, A.N.; Gil, E.S.; Tsioris, K.; Tao, H.; Kaplan, D.L.; Omenetto, F.G. All-water-based electron-beam lithography using silk as a resist. *Nat. Nanotechnol.* **2014**, *9*, 306–310.

16. Barbillon, G. Plasmonic nanostructures prepared by soft UV nanoimprint lithography and their application in biological sensing. *Micromachines* **2012**, *3*, 21–27.

17. Haynes, C.L.; Haes, A.J.; Duyne, R.P.V. Nanosphere lithography: Synthesis and application of nanoparticles with inherently anisotropic structures and surface chemistry. *MRS Proc.* **2001**, *635*, doi:10.1557/PROC-635-C6.3.

18. Hong, Y.; Huh, Y.M.; Yoon, D.S.; Yang, J. Nanobiosensors based on localized surface plasmon resonance for biomarker detection. *J. Nanomater.* **2012**, *2012*, doi:10.1155/2012/759830.

19. Liu, J.; Cai, H.; Kong, L. Effect of gold layer thickness on optical properties and sensing performance of triangular silver nanostructure array. *Optoelectron. Adv. Mater. Rapid Commun.* **2014**, *8*, 1085–1090.

20. Huang, W.; Qian, W.; El-Sayed, M.A.; Ding, Y.; Wang, Z.L. Effect of the lattice crystallinity on the electron-phonon relaxation rates in gold nanoparticles. *J. Phys. Chem. C* **2007**, *111*, 26276–26283.

21. Chaumet, P.C.; Rahmani, A.; Bryant, G.W. Generalization of the coupled dipole method to periodic structures. *Phys. Rev. B* **2003**, *67*, doi:10.1103/PhysRevB.67.165404.

22. Kreibig, U.; Vollmer, M. *Optical Properties of Metal Clusters*; Springer-Verlag: Heidelberg, Germany, 1995.

7

Nanomaterials-Based Fluorimetric Methods for MicroRNAs Detection

Ming La [1], Lin Liu [2,*] and Bin-Bin Zhou [1,2,*]

[1] College of Chemistry and Chemical Engineering, Pingdingshan University, Pingdingshan 467000, Henan, China, E-Mail: xrc1202@gmail.com

[2] College of Chemistry and Chemical Engineering, Anyang Normal University, Anyang 455000, Henan, China

* Authors to whom correspondence should be addressed;
E-Mails: liulin@aynu.edu.cn (L.L.); gyp025809@gmail.com (B.-B.Z.);

Academic Editor: Jun-ichi Anzai

Abstract: MicroRNAs (miRNAs) are small endogenous non-coding RNAs of ~22 nucleotides that play important functions in the regulation of many biological processes, including cell proliferation, differentiation, and death. Since their expression has been in close association with the development of many diseases, recently, miRNAs have been regarded as clinically important biomarkers and drug discovery targets. However, because of the short length, high sequence similarity and low abundance of miRNAs *in vivo*, it is difficult to realize the sensitive and selective detection of miRNAs with conventional methods. In line with the rapid development of nanotechnology, nanomaterials have attracted great attention and have been intensively studied in biological analysis due to their unique chemical, physical and size properties. In particular, fluorimetric methodologies in combination with nanotechnology are especially rapid, sensitive and efficient. The aim of this review is to provide insight into nanomaterials-based fluorimetric methods for the detection of miRNAs, including metal nanomaterials, quantum dots (QDs), graphene oxide (GO) and silicon nanoparticles.

Keywords: microRNAs; biomarker; fluorescence; metal nanomaterials; quantum dots; graphene oxide; silicon nanoparticles

1. Introduction

MicroRNAs (miRNAs) are endogenous non-coding RNAs of ~22 nucleotides that regulate gene expression by translational repression or degradation of messenger RNA. The human genome may encode over 1000 miRNAs, which may target about 60% of mammalian genes and are abundant in many human cell types. Recently, miRNAs have been identified as diagnostic and prognostic biomarkers and predictors of drug response for many diseases, including a broad range of cancers, heart disease, and neurological diseases [1–3]. Since miRNAs are small molecular, high sequence homologous, and in low abundance in cancer cells (as low as a few molecules per cell), high sensitivity and specificity of a miRNAs detection system are very important for diagnosis and therapy of the cellular disease [4,5]. At present, Northern blotting technology, microarray and real-time PCR are being widely used for miRNAs analysis. However, these methods possess tedious sample preparation or require a thermal cycler. Recently, the simple and signal-amplified methods with the aid of nanomaterials have attracted great attention in bioassays. Nanomaterials contributed their high sensitivity, long life, and large surface area to the fabrication of different kinds of optical and electrochemical miRNAs biosensors [6]. Moreover, nanopore-based nucleotide analysis is an emerging technique that obviates the use of labeling, enzymatic reaction or amplification [7–10]. The studies in detecting cancer-derived miRNAs with nanopore implied that, in the next few years, the nanopore-based miRNAs technique may be validated for noninvasive and early diagnosis of diseases with the improvement of throughput. Progress in the fabrication of nanopore- and nanomaterials-based electronic miRNAs biosensors have been reviewed recently [6,11,12]. Fluorimetric methodologies in combination with nanotechnology are especially rapid, sensitive and efficient. Usually, nanomaterials can be used as the fluorophores or quenchers and as carriers for loading a large amount of probes to enhance the detection signal in the fluorescent sensing analysis. The recent advances in the development of nanomaterials-based fluorescent methods for miRNAs detection are summarized in this work, including metal nanomaterials, quantum dots (QDs), graphene oxide (GO), and silicon nanoparticles.

2. Metal Nanomaterials

2.1. Silver Nanoclusters

As promising alternatives to common fluorophores like organic dyes, few-atom metal nanoparticles (NPs) with strong and robust fluorescence emission have recently attracted considerable research interest. In particular, the creation of fluorescent silver nanoclusters (AgNCs) as new, bright, and photostable labels has received significant attention in the area of bioassays. When significantly small (less than 100 atoms) to exclude continuous density of states, discrete transitions between energy levels are possible for emission to occur. In order to achieve the creation of these small silver clusters and to avoid aggregation into larger nonemissive particles, a myriad of different scaffolds have been used. Typically, cytosine-rich DNA-templated fluorescent AgNCs (DNA/AgNCs) show great potential as fluorescent probes for biochemical applications due to their advantages of ultrafine size, low toxicity, good biocompatibility, outstanding photophysical properties, as well as facile integration with nucleic acid-based target-recognition abilities and signal amplification mechanisms [13–16]. Herein, we first summarized the development of AgNCs-based fluorescent probe design and its successful applications in detecting miRNAs.

Based on the fluorescence properties of DNA/AgNCs, Yang's group synthesized a DNA/AgNCs probe that can detect the presence of target miRNAs and investigated the effect of a range of diverse salts, organic solvents and buffer on the analytical performance [17–19]. The red fluorescence of the DNA/AgNC probe is diminished upon the presence of target miRNAs without pre- or post-modification, addition of extra enhancer molecules, and labeling. When target miRNAs are present, the emission of the DNA-nanosilver clusters (DNA/AgNCs) probe is significantly lower *versus* the case when no target miRNAs or other miRNAs are present. Additionally, to further advance the method toward multiplex miRNAs detection in solution, they also presented the design of three DNA/AgNCs probes showing green, red, and near-infrared (NIR) fluorescence [20]. Besides, Ye and co-workers reported the quantification of miRNAs expression levels in cell lysates by target assisted isothermal exponential amplification (TAIEA) coupled with fluorescent DNA-scaffolded AgNCs [16,21]. As shown in Figure 1, the unimolecular template involves five regions (AXAXB). Two repeat sequences of A are complementary to the target miRNAs. Also, two repeat sequences of X represent the "heart" of the template, upon their replication; the complementary strand includes the specific sequence for nicking by Nt·BstNBI. The sequence of B is complementary to the reporter oligonucleotide R acting as a scaffold for the synthesis of fluorescent AgNCs. After the amplification, the reporter oligonucleotide R was acting as a scaffold for the synthesis of fluorescent silver nanoclusters in the presence of Ag^+ through the reduction of $NaBH_4$.

Molecular beacon (MB) composing of a hairpin-like DNA stem-loop structure has been widely used for nucleic acid detection with excellent sensitivity and selectivity. Usually, the hairpin DNA probe was coupled with a fluorophore-quencher pair at its 5'- and 3'-termini. Upon hybridization with the target, the MB will undergo a spontaneous conformational reorganization, forcing the stem apart and causing the fluorophore and the quencher to separate from each other. As a result, the restoration of fluorescence occurs for detection assay. Recently, a few MB-based molecular detection platforms have been developed for miRNAs detection by using AgNCs as the fluorophores [22–24]. For example, Xia *et al.* found that the short single-stranded oligonucleotide probe with only six bases (5'-TCCCCC-3') can serve as the scaffold for the preparation of DNA/AgNCs. On the basis of this finding, a hairpin DNA probe with 5'-TCC/CCC-3' overhangs has been utilized for miRNAs assay. As shown in Figure 2A, the overhangs of 5'-TCC/CCC-3', which were bought into close proximity by intramolecular self-hybridization, served as the template for the synthesis of fluorescent AgNCs (a). Upon hybridization with target miRNAs (b), the hairpin-shaped oligodeoxynucleotide (ODN) was destroyed, resulting in the separation of the 5'-TCC/CCC-3' overhangs and the quenching of the AgNCs fluorescence. Moreover, Qiu *et al.* reported a DNA/AgNCs-based miRNAs detection platform based on hybridization chain reaction (HCR) using two complementary nucleic acid hairpin oligmers (MB1 and MB2) (Figure 2B) [24]. At first, MB1 and MB2 were in the closed form and spatially separated without the target miRNAs (let-7a). After the addition of $AgNO_3$ followed by the reduction with $NaBH_4$, highly fluorescent AgNCs were formed with the 6C loop in MB1 as the template. However, upon hybridization with let-7a, the hairpin structure of MB1 was opened, exposing a new single-stranded region that induced the opening of hairpin structured MB2, and the exposing of a new single strand of MB2 (identical in sequence to let-7a). This process led to the formation of a nicked double helix (polymer).

Thus, the 6C-loop of MB1 was completely hybridized with the 6G sticky end of MB2, inhibiting the formation of fluorescent AgNCs.

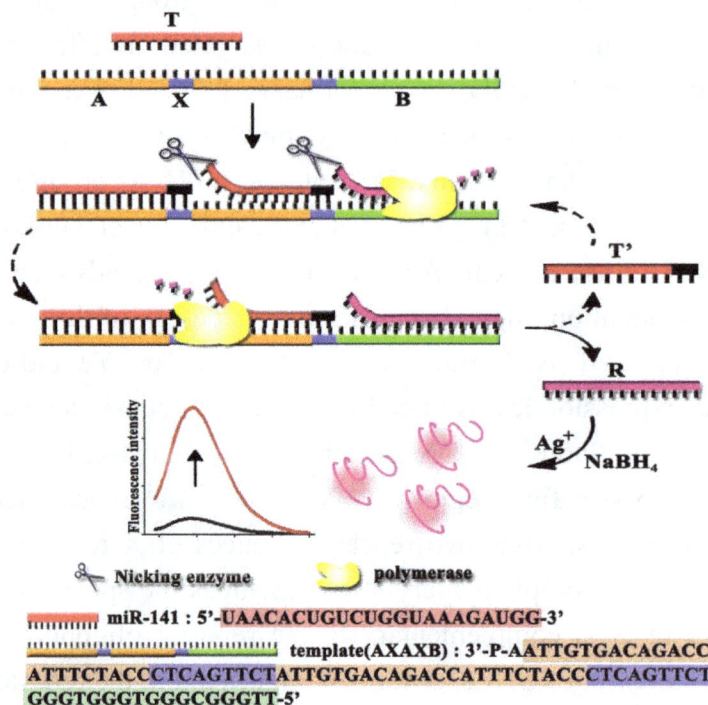

Figure 1. Detection of miRNA with attomolar sensitivity based on target-assisted isothermal exponential amplification (TAIEA) coupled with fluorescent DNA scaffolded AgNC probe. Reprinted with permission from [21]. Copyright 2012 American Chemical Society.

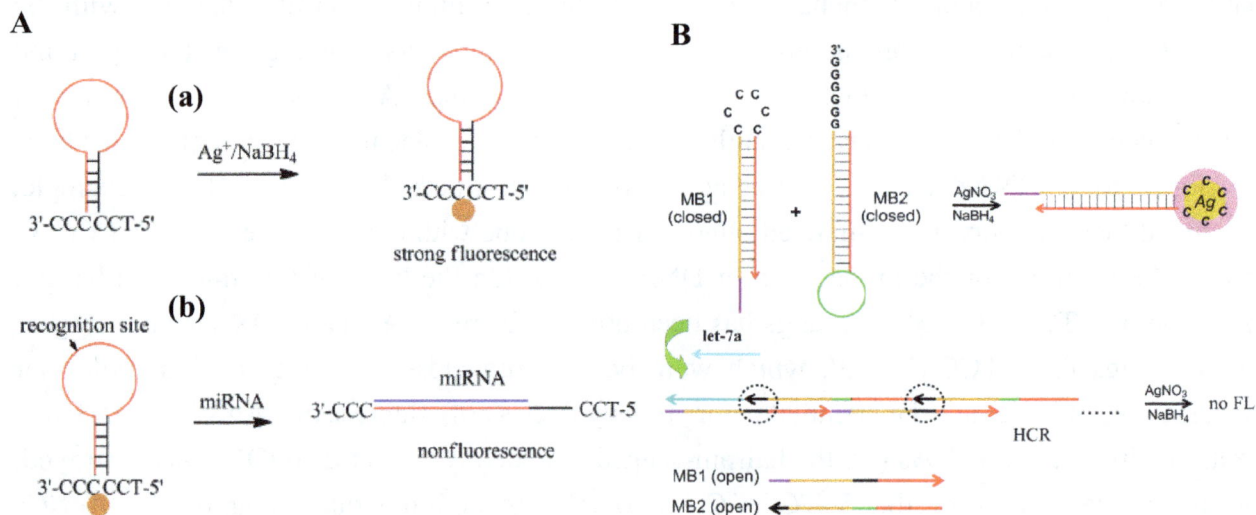

Figure 2. (A) Schematic representation of the creation of hairpin-shaped ODN probe/AgNCs for miRNA assay. (a) Template synthesis of AgNCs with strong fluorescence. (b) Hairpin-shaped ODN probe/AgNCs exhibited nonfluorescence or weak fluorescence after hybridization with target miRNA. Reprinted with permission from [22]. Copyright 2014 Elsevier. **(B)** Schematic illustration of HCR modulated fluorescent DNA-hosted Ag nanoclusters for the detection of the let-7a. Reprinted with permission from [24]. Copyright 2014 Elsevier.

2.2. Copper Nanoclusters

Compared with the other existing fluorescent metal nanoparticles, copper nanoclusters (CuNCs) or copper nanoparticles (CuNPs) are a type of newly emerged functional biochemical fluorescent probe [25–27]. In contrast to DNA/AgNCs, the newly emerging CuNCs, selectively formed on a DNA duplex, offer excellent potential for "on the spot" testing with a rapid and simple "mix-and-measure" format. For example, dsDNA-templated CuNCs (dsDNA-CuNCs) can be facilely prepared by reducing Cu^{2+} ions with ascorbic acid within fifteen minutes and the Cu^{2+} ions are soluble in many detection environments and thus have no precipitation phenomena like the Ag^+ ions. For these views, Wang et al. presented a label-free method for miRNAs detection using fluorescent dsDNA-CuNCs as signal indicators [28]. In this method, the miRNAs targets were transferred to the oligonucleotide reporters and acted as the scaffold for the synthesis of fluorescent CuNCs via an isothermal exponential amplification reaction, in which the unimolecular DNA designed for the miRNAs target is used as the amplification template and polymerases and nicking enzymes were used as mechanical activators [28]. Furthermore, Xu et al. reported the detection of miRNAs using concatemeric dsDNA-templated CuNPs (dsDNA-CuNPs) by introducing the rolling circle replication (RCR) technique into CuNPs synthesis [29]. In this strategy, the circular DNA template contained two functional regions (R and H) (Figure 3). A part of the recognition region R is complementary to the primer. The hybridization region H was designated to template the formation of CuNPs in its dsDNA form. In the presence of phi29 polymerase and dNTPs, the primer triggered the RCR process with a continual replication of the circular template. As a result, a short oligonucleotide primer was extended to a long concatemeric ssDNA with periodically repeated complementary parts of regions R and H (named as R' and H'). Through ensuing hybridization of the H' region with complementary DNA H, a long concatemeric dsDNA scaffold comprising two distinct alternating regions of R' ssDNA and H/H' dsDNA was obtained to synthesize concatemeric dsDNA-CuNPs after the addition of copper ions and ascorbic acid. In comparison with monomeric dsDNA-CuNPs, the sensitivity of concatemeric dsDNA-CuNPs was greatly improved with ~10,000-fold amplification.

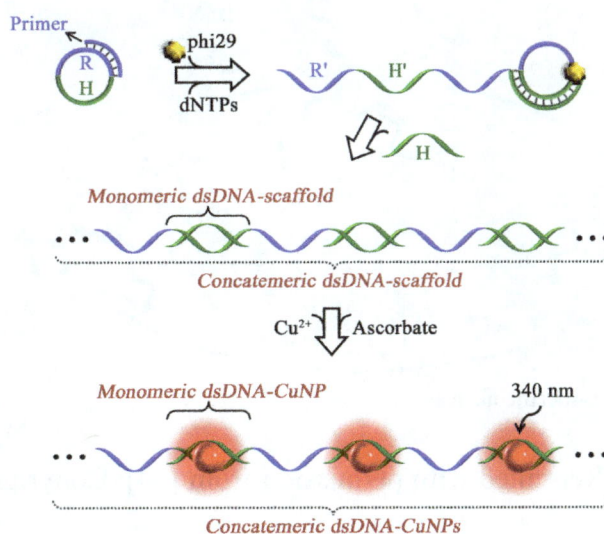

Figure 3. Schematic showing the principle of the RCR-mediated concatemeric dsDNA-CuNPs strategy. Reprinted with permission from [29]. Copyright 2014 American Chemical Society.

2.3. Gold Nanopartilces

The attractive characteristics of gold nanoparticles (AuNPs) such as high surface-to-volume ratio, high extinction coefficients, unique size-dependent optical properties and good conductance properties have proven to be of high utility in biomedical applications [12,30]. Recently, AuNPs-based fluorescent assays for miRNAs detection have been reported based on the ultraefficient fluorescence quenching efficiency of AuNPs [31,32]. For example, Tu *et al.* has demonstrated the detection of miRNAs by coupling AuNPs distance-dependent fluorescence quenching with a conformation-switched hairpin-structured probe that was labeled with thiol and a fluorophore [33]. The hairpin-structured probe DNA was anchored onto AuNPs surface via the formation of the Au–S bond. The stem-loop feature of the probe brings the fluorophore close to the AuNP surface to silence the fluorescence and no detectable fluorescence signal was observed. The miRNAs targets opened the loop of the probe by hybridization, leading to the fluorophore to be away from the AuNPs surface, the follow-up recovery of the fluorescence and the production of a readily detectable signal. The detection limit of this method was found to be 0.01 pM. Furthermore, Degliangeli *et al.* introduced a fluorescent method for the absolute quantification of miRNAs based on enzymatic processing of DNA-functionalized AuNPs [34]. Specifically, fluorescently labeled DNA probes were immobilized on the passivation layer of PEGylated AuNPs, inducing the efficient fluorescence quenching by the vicinity of the fluorophores to the AuNPs surface (Figure 4). In presence of target miRNAs, DNA-RNA heteroduplexes were formed, followed by hydrolyzation by the duplex-specific nuclease (DSN) enzyme. As a result, fluorophores were released in solution, leading to the appearance of a fluorescence signal. Note that the DSN is a highly stable, nonspecific endonuclease with can selectively cleave double stranded DNA and DNA in DNA-RNA heteroduplexes with little activity toward single-stranded nucleic acids and RNA-RNA duplex helixes [35]. Thus, miRNAs strands remained intact in this process and would be released back to the sample solution for recycling, making signal enhanced significantly. The detection limit was found to be 5~8 pM (or 0.2~0.3 fmol) after 5 h incubation.

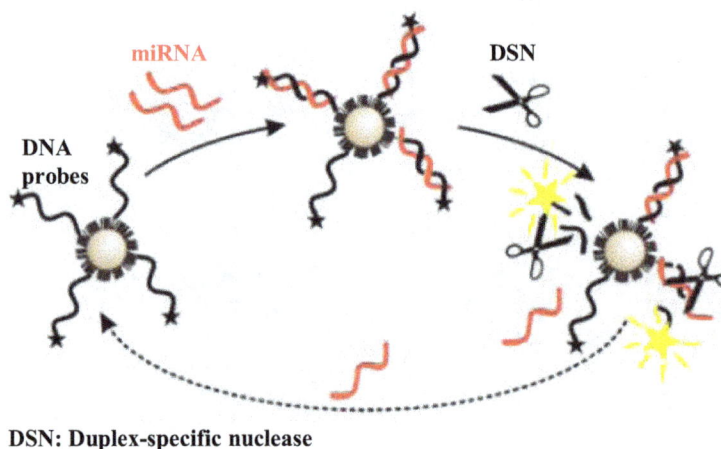

DSN: Duplex-specific nuclease

Figure 4. Assay strategy. Reprinted with permission from [34]. Copyright 2014 American Chemical Society.

Based on the fluorescence quenching efficiency of AuNPs, Baptista's group suggested that AuNPs functionalized with a fluorophore labeled hairpin-DNA (Au-nanobeacon) can be used to follow the synthesis and inhibition of RNA [36]. Furthermore, they proposed an innovative Au-nanobeacon-based theranostic approach for the detection and inhibition of sequence-specific miRNA *in vitro* [37]. The proposed method allows real-time detection of the beacon's signal while yielding a quantifiable fluorescence directly proportional to the level of gene silencing.

3. Quantum Dots

Quantum dots (QDs) are novel semiconductor nanocrystals with unique optical properties, including size-tunable photoluminescence spectra and relatively high quantum yield, and have been widely used as fluorescent markers in biological labeling, fluorescence imaging, and drug delivery [38–41]. In particular, QDs hold great promise as fluorescence resonance energy transfer (FRET) donors in the biosensing applications to homogeneously detect nucleic acids, proteins, and small molecules [42]. The use of QDs as the FRET donors offers several unique spectroscopic properties unmatched in any available organic fluorophores, including improved FRET efficiency as a result of coupling multiple acceptors around a single QD, tunable spectral overlap between the QD and the acceptor, minimization of direct acceptor excitation, and multiplex FRET configurations [43]. Recently, the fluorescent detection techniques in combination with QDs have been used to quantify miRNAs with high sensitivity [44,45]. Typically, Zeng *et al.* developed a QD-based miRNAs nanosensor for a point mutation assay using primer generation-mediated rolling circle amplification (PG-RCA) [46]. Zhang *et al.* reported a miRNAs detection method based on the two-stage exponential amplification reaction (EXPAR) and a single-QD-based nanosensor [44]. As shown in Figure 5, the two-stage EXPAR involves two templates and two-stage amplification reactions under isothermal conditions. The first-stage reaction (a, b) is an exponential amplification with the involvement of the X'-X' template, which can enable the amplification of miRNAs. The second-stage reaction (c) is a linear amplification with the involvement of the X'-Y' template, which can enable the conversion of miRNAs to the reporter oligonucleotide Y. It should be noted that different miRNAs can be converted to the same reporter oligonucleotides, which can be detected with the same set of capture and reporter probes without the need for resynthesis of the specific DNA probes for each new target miRNAs or reoptimization of the assay conditions. After amplification, the reporter oligonucleotide Y is sandwiched by a biotinylated capture probe and a Cy5-labeled reporter probe (d). This sandwich hybrid is then assembled on the surface of a 605QD to form the 605QD/reporter oligonucleotide Y/Cy5 complex through specific biotin-streptavidin binding (e). When this complex was excited by a 488 nm argon laser, the fluorescence signals of 605QD and Cy5 is observed simultaneously due to FRET from 605QD to Cy5 (f). Additionally, Su *et al.* reported a versatile "signal-off" strategy for miRNAs detection using CdTe/CdS core-shell QDs capped with 3-mercaptopropionic acid (MPA) (Figure 6) [45]. Surface-bound short-chain MPA molecules were first substituted by thiolated DNA via ligand exchange, leading to programmable DNA modification at the surface of QDs. The resulting DNA-conjugated QDs were used as fluorescence nanoprobes for miRNAs detection with strong photoluminescence and robust stability. Without the addition of a target miRNAs sequence, the organic quencher (BHQ$_2$)-labeled DNA had almost no influence on the DNA-QDs conjugate. With the addition of target miRNAs and BHQ2-labeled DNA sequences, sandwiched hybrids were formed. As a result, the fluorescence intensity of DNA-QDs decreased obviously with the addition of target miRNAs due to the energy transfer from QDs to BHQ2. Moreover, Jou *et al.* reported a two-step miRNAs detection platform with semiconductor CdSe/ZnS QDs modified by FRET quencher-functionalized DNA. Specifically, the DNA sequence on the QDs surface included the recognition sequences for target miRNAs and telomerase primer sequences for the second step of the analytical platform. In the presence of DSN, subjecting the DNA probe-modified QDs to target miRNAs induced the formation of miRNA/DNA duplex helixes. The follow-up DSN-mediated cleavage could lead to the regeneration of target miRNAs. The DSN-induced cleavage of the quencher units resulted in the activation of the fluorescence of the QDs, thus allowing for the fluorescent detection of miRNAs (the first step). The DNA residues associated with the QDs after cleavage of the DNA probe by DSN acted as primers for telomerase. The subsequent telomerase/dNTPs stimulated elongation of the primer units formed G-quadruplex telomer chains. Incorporation of hemin in the resulting G-quadruplex telomer chains yields horseradish peroxidase-mimicking DNAzyme units, catalyzing the generation of chemiluminescence in the presence of luminol/H_2O_2 (the second step) [47].

It has been suggested that each CdSe nanocrystal with an average diameter of 5.15 nm contains 2051 Cd^{2+} ions and every 1 nm increase in nanocrystal diameter represents 2200 more Cd^{2+} ions enclosed [48]. The large numbers of encapsulated ions can be released by a fast and gentle process, the cation exchange reaction. For this consideration, Li *et al.* reported the quantification of miRNAs in biological samples based on the signal amplification of CdSe QDs [49]. As shown in Figure 7, the capture probes are conjugated onto the magnetic beads and the detection probes are coupled with the nonfluorescent CdSe nanocrystals. The capture probe with a molecule beacon structure opened up and exposed the binding site for the detection strand upon hybridization of target miRNAs. Then, the nanocrystals were immobilized onto the magnetic beads. Afterwards Cd^{2+} ions were released to react with the Rhod-5N dye to produce intense fluorescent signals.

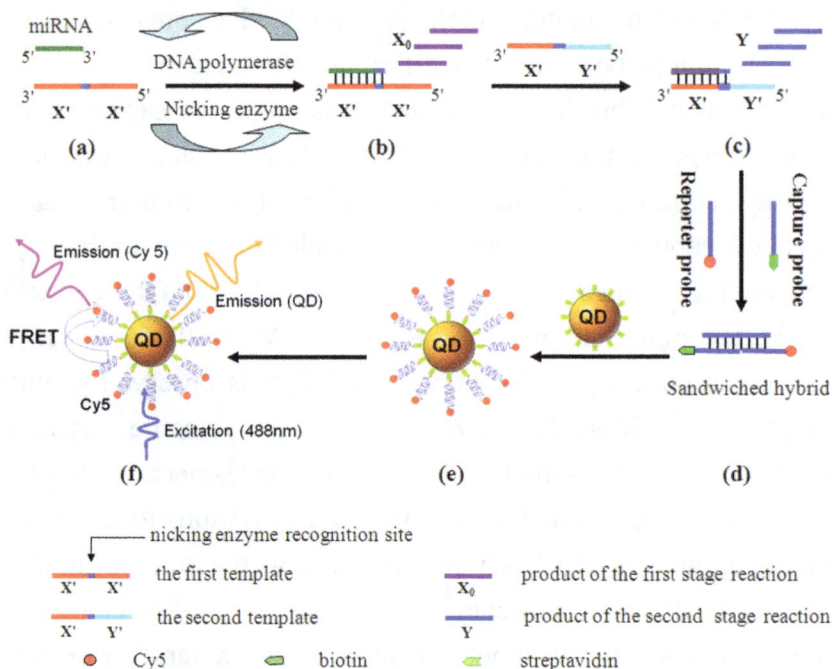

Figure 5. Scheme of the miRNA assay based on the two-stage EXPAR and single-QD-based nanosensor. Reprinted with permission from [44]. Copyright 2012 American Chemical Society.

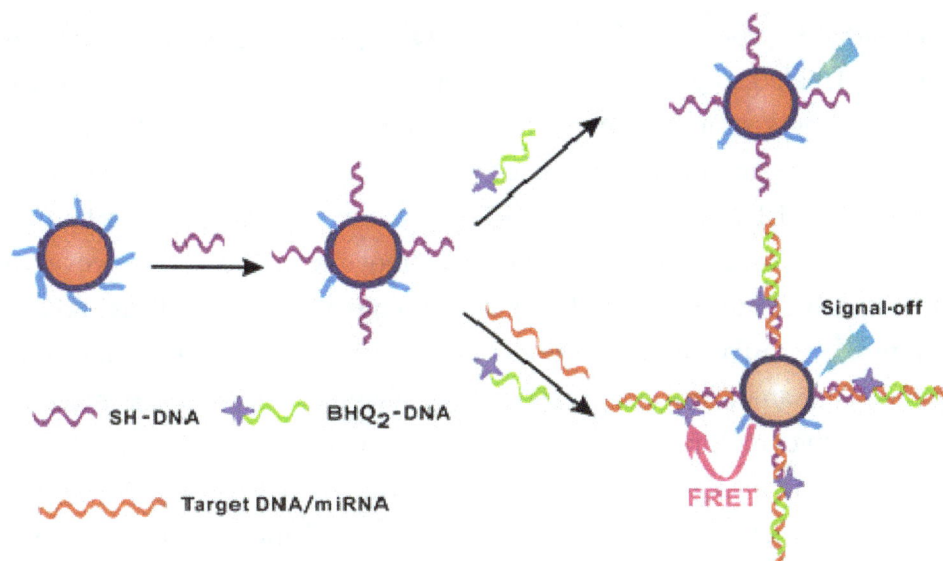

Figure 6. Schematic representation of the designed nanosensors for detection of DNA/miRNA based on the FRET system. Reprinted with permission from [45]. Copyright 2012 American Chemical Society.

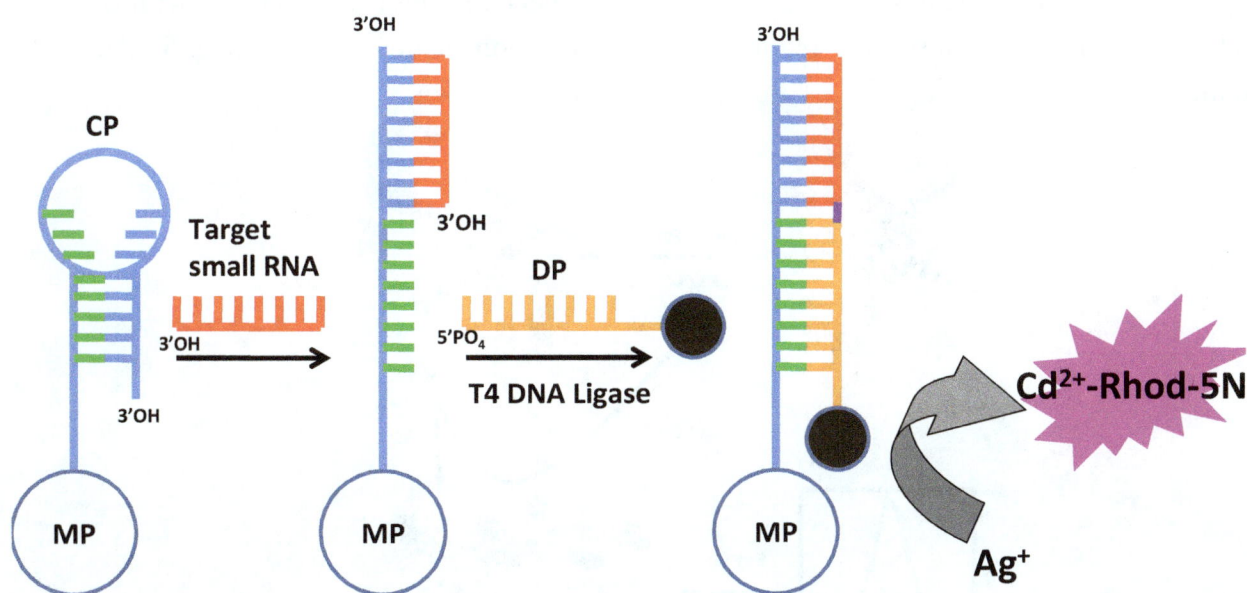

Figure 7. Schematic presentation of the small RNA detection assay using CXFluoAmp. MP: magnetic particles; CP: capture probe; DP: detection probe. Reprinted with permission from [49]. Copyright 2009 American Chemical Society.

4. Graphene Oxide

Graphene oxide (GO), a single-atom-thick, two-dimensional carbon nanomaterial, has become extremely popular in biological applications due to its unique characteristics, such as good water-solubility, flexible modification and super fluorescence quenching ability. In particular, it has been proven that GO can adsorb single-stranded nucleic acids via non-covalent π–π stacking interactions between the hexagonal cells of graphene and the ring structure of nucleobases, but has less affinity toward rigid double-stranded nucleic acids [50,51]. GO has been used as a platform for the detection of nucleic acids because of its remarkable properties [52,53] and proteins [54,55]. It has also been suggested that GO sheets could bind dye-labeled single-stranded DNA probes and efficiently quench the fluorescence of the labeled fluorescent-single-strand DNA. In the presence of target miRNAs, the labeled probes can be released from GO due to the formation of miRNA/DNA duplex helixes that disturb the interaction between the labeled DNA probes and GO. Then, the fluorescence can be recovered. This simple method could be used for assay of miRNAs at the nanomolar level [56,57]. However, the miRNAs content is at the attomolar to femtomolar level in many biological samples. Thus, great efforts have been made to develop more sensitive GO-based fluorescent sensors for the detection of low-abundance miRNAs with different signal amplification strategies [58]. Typically, Dong *et al.* presented a simple, highly sensitive, and selective multiple miRNAs detection method based on the GO fluorescence quenching and isothermal strand-displacement polymerase reaction (ISDPR) [58]. As shown in Figure 8, the high fluorescent quenching efficiency of GO by a FRET-based mechanism in combination with strong interaction between ssDNA and GO made the probes labeled with fluorescent dye exhibit minimal background fluorescence. Upon the recognition of specific target miRNAs, an ISDPR was triggered to produce numerous massive specific miRNA/DNA duplex helixes, and a strong emission was observed due to the weak interaction between the DNA-miRNA duplex helix and GO. The large planar surface of GO made it possible to simultaneously quench several DNA probes with different dyes and obtain a multiple biosensing platform for the detection of different target miRNAs in the same solution.

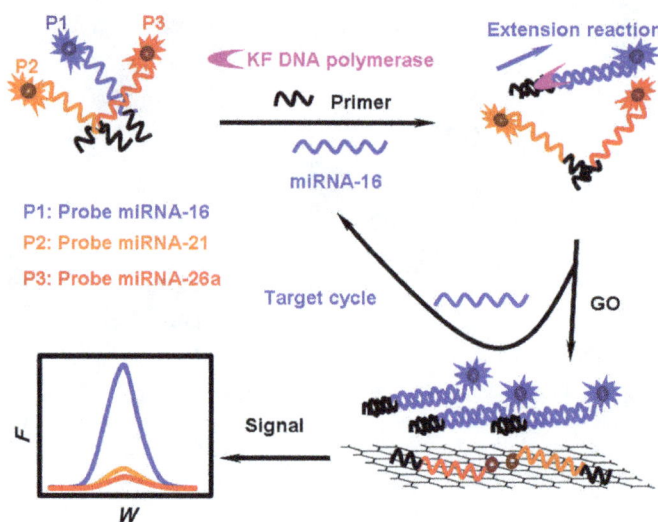

Figure 8. Illustration of the GO fluorescence quenching and ISDPR-based multiple miRNA analysis. Reprinted with permission from [58]. Copyright 2012 American Chemical Society.

Hybridization chain reaction (HCR) is enzyme-free amplification method that has shown great potential in nucleic acid detection. Yang *et al.* developed an enzyme-free signal amplified method for miRNAs detection using HCR coupled with a GO surface-anchored fluorescence signal readout pathway (Figure 9) [59]. In this method, miRNAs initiated HCR between two species of fluorescent hairpin probes in solution. When GO was added into the solutions after HCR, both of the excess hairpin probes and the HCR products were anchored onto the GO surface via the π–π stacking interaction. The fluorescence of the hairpin probes was completely quenched by GO, whereas the HCR products maintained strong fluorescence.

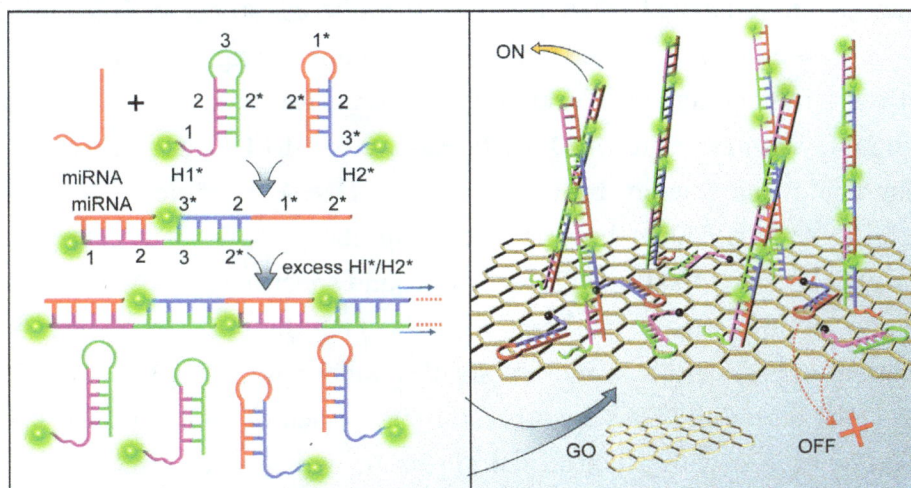

Figure 9. Schematic illustration of the proposed HCR/GO platform for miRNA detection. Reprinted with permission from [59]. Copyright 2012 American Chemical Society.

RsaI endonuclease is a site-specific endonuclease that recognizes the duplex symmetrical tetranucleotide sequence 5′-GTAC-3′ and catalyzing cleavage between T and A bases. Tu *et al.* reported the miRNAs assay by coupling the fluorescence quenching efficiency of GO with site-specific cleavage of RsaI endonuclease for improving selectivity (Figure 10) [60]. The designed single-stranded probe DNA carries both a binding region (44 bases) and a sensing region (22 bases). The binding region provides an anchoring function to facilitate the interaction between GO and the probe, inducing fluorescence quenching of the 5′-terminus-labeled fluorophore (6-carboxyfluorescein, FAM). The sensing region specifically recognizes the target miRNAs and hybridizes with it to form a duplex, which contains the specific sequence recognized by RsaI endonuclease. In the absence of a specific target, no fluorescence signal was detected. In the presence of target miRNAs, however, the formed duplex was subject to be released from the GO surface under the cleavage of RsaI endonuclease, resulting in the recovery of fluorescence of the fluorophore and producing a readily detectable signal. Moreover, Guo *et al.* reported a fluorescent sensing platform for miRNAs detection by combining the fluorescence quenching efficiency of GO and DSN-induced target recycling [61]. In the absence of target miRNAs, fluorophore-labeled DNA probes would be adsorbed by GO, leading to fluorescence quenching. In the presence of target miRNAs, the DSN cleaved the labeled DNA in the DNA-RNA hybrid duplex into small fragments and the miRNAs was released from the duplex for recycling, producing numerous small fluorophore-labeled DNA fragments that could not adsorb onto the GO surface.

It has been demonstrated that ssDNA and single-stranded RNA (ssRNA) probes were effectively protected from enzymatic digestion by nuclease after noncovalent adsorption onto GO surface due to a steric hindrance effect of GO that prevents nuclease from binding to the DNA and RNA [62]. Besides combining the extraordinary fluorescence quenching property of GO with the unique ssDNA/GO interaction to elaborately design miRNAs sensors in homogeneous solution, researchers have also been inspired to develop new types of GO-based miRNAs sensors. For example, DNase I is an endonuclease that nonspecifically cleaves DNA (single- and double-stranded DNA, chromatin and

DNA stranded in RNA/DNA complex) but not RNA and releases di-, tri- and oligonucleotide products; Cui *et al.* reported a signal amplification platform for multiplex analysis of miRNAs by combining DNase I and a single-labeled DNA fluorescent probe adsorbed on GO [63]. Specifically, single-stranded DNA was promptly adsorbed onto GO forming strong molecular interactions that prevented DNase I from approaching the constrained ssDNA. However, when hybridized with the target miRNAs, the double-stranded miRNA/DNA bound weakly to GO and released into solution, where the DNA probe could immediately be hydrolyzed by DNase I, while the miRNAs remained intact. The released miRNAs could then bind to another probe on GO to initiate a next round of cleavage. Moreover, Liu *et al.* developed a stable, sensitive, and specific miRNAs detection method on the basis of cooperative amplification combining with the GO fluorescence switch-based circular exponential amplification (CEA) and the multimolecules labeling of SYBR Green I (SG) [64]. As shown in Figure 11, the target miRNAs adsorbed on the surface of GO protected it from enzyme digest. If the miRNAs hybridized with a partial hairpin probe acting as a primer to initiate a strand displacement reaction to form a complete duplex, universal DNA fragments would be released under the action of nicking enzyme and used as triggers to initiate next reaction cycle, constituting a new circular exponential amplification. In the proposed strategy, a small amount of target miRNAs can be converted to a large number of stable DNA triggers, leading to a remarkable amplification for the target. Moreover, compared with labeling in a 1:1 stoichiometric ratio, multimolecules binding of intercalating dye SG to double-stranded DNA (dsDNA) can enhance the fluorescence signal significantly, improving the detection sensitivity.

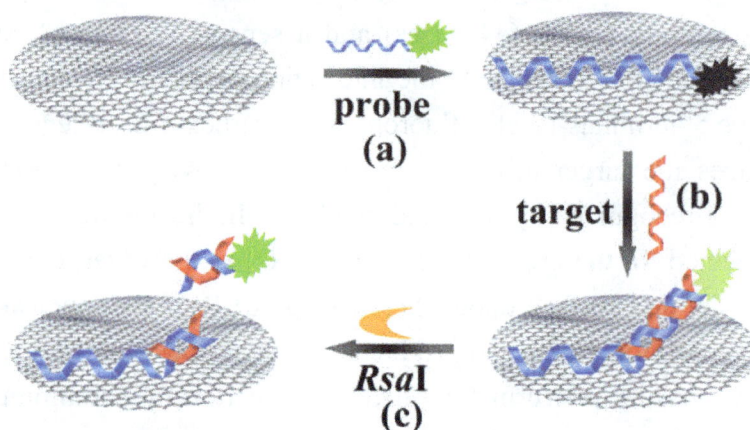

Figure 10. Illustration of miRNA assay based on coupling the fluorescence quenching of GO with site-specific cleavage of an endonuclease. The designed P1, which is labeled at its 5′-terminus with a fluorophore, exhibits partial complementarity to the target (T1). The adsorption of P1 on the GO surface effectively quenches the fluorophore (step a). The hybridization of P1 with T1 leads to a non-GO absorbed duplex region and a single-stranded domain that is associated with the GO surface (step b). In the presence of the endonuclease, the 5′-terminus of the duplex (containing the RsaI-recognized tetranucleotide sequence 5′-GTAC-3′) is digested, resulting in the release of the fluorophore to the solution and recovery of the fluorescence signal (step c). The recovered fluorescence signal depends on the target concentration in solution. Reprinted with permission from [60]. Copyright 2013 American Chemical Society.

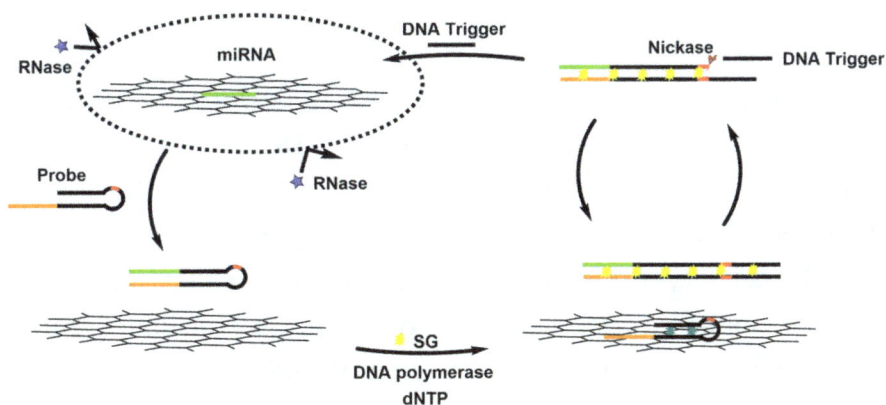

Figure 11. Illustration of the Graphene Fluorescence Switch-Based Cooperative Amplification for Target miRNAs. Reprinted with permission from [64]. Copyright 2014 American Chemical Society.

Although GO has been utilized not only for the detection of miRNAs *in vitro* as a fluorescence quencher of dye-labeled DNA probes but also for the delivery of nucleic acid integrated molecular beacon with help of cationic polymers [65,66], quantitative monitoring of intracellular miRNAs in living cells still remains an important challenge. Recently, Ryoo *et al.* developed a nanosized GO (NCO)-based sensor for quantitative monitoring of target miRNAs expression levels in living cells (Figure 12) [67]. The binding of NCO with fluorophore-labeled peptide nucleic acid (PNA) probe induced fluorescence quenching. The presence of target miRNAs led to the recovery of the fluorescence PNA. In this work, PNA acting as a probe for miRNAs sensing offers many advantages such as high sequence specificity, high loading capacity on the NCO surface compared to DNA and resistance against nuclease-mediated degradation. The sensor allowed the detection of specific target miRNAs with the detection limit as low as ~1 pM and the simultaneous monitoring of three different miRNAs in a living cell.

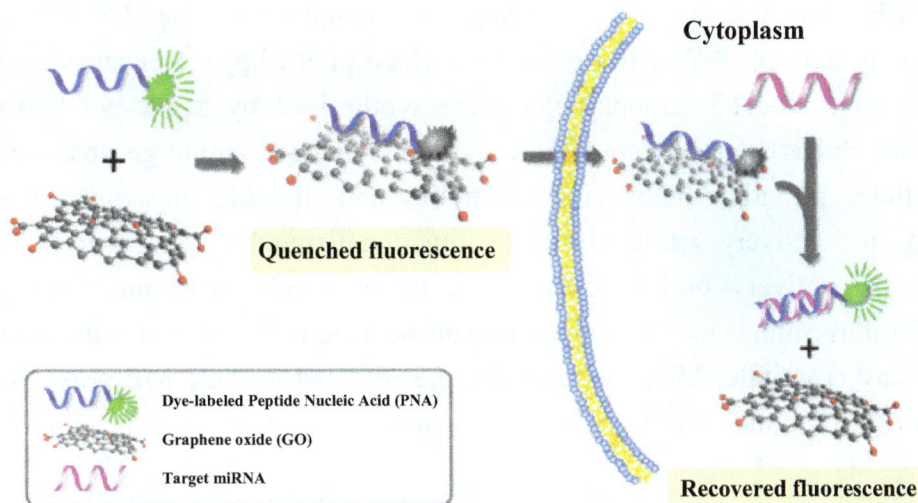

Figure 12. Scheme of strategy for a miRNA sensor based on NGO and PNA. Reprinted with permission from [67]. Copyright 2013 American Chemical Society.

5. Silicon

Recent significant progress in optical imaging techniques has offered the opportunities of noninvasive and repeated real-time analysis of miRNAs in living cells [68]. Heavy metal inorganic nanoparticles have shown demonstrated use in spherical nucleic acid systems. However, the potential toxicity and biodegradability issues of metal nanoparticles remain a concern [4,69,70]. Silicon is well-established to be biocompatible, biodegradable, and earth-abundant, and can exhibit photoluminescence [71]. Recently, multifunctional silicon-based nanomaterials have been used for imaging intracellular target miRNAs. For example, Zhang *et al.* reported the detection of miRNAs in the cells positive to lung cancer using fluorescent silicon-based nanoshells as a molecular imaging agent [72]. These nanoshells were composed of silica spheres with encapsulated tris(2′, 2′-bipyridyl)dichlororuthenium(II) hexahydrate ($Ru(bpy)_3^{2+}$) complexes as cores and thin silver layers as shells. Dong *et al.* designed a multifunctional SnO_2 nanoprobe (mf-SnO_2) that contains a cell-targeting moity (folic acid, FA) for cell-specific delivery and a conjugated gene probe (molecular beacon, MB) for imaging intracellular target miRNAs (Figure 13) [73]. In this method, MB to detect target miRNAs is conjugated by a disulfide linkage, which is sensitive to pH values. Cleavage of the disulfide linkage between the gene probe and the nanoparticle enhances the efficiency of intracellular delivery. The stable aqueous suspension of SnO_2 NPs were obtained by noncovalently functionalizing the SnO_2 NPs with 1,2-distearoyl-*sn*-glycero-3-phosphoethanolamine-N-[amino(polyethylene glycol)2000] (DL-PEG$_{2000}$) conjugated to FA (DL-PEG$_{2000}$-FA) or sulfosuccinimidyl-6-(3′ (2-pyridyldithio) propionamido)hexanoate (DL-PEG$_{2000}$-SPDP). These moieties were adsorbed on the nanoparticles by van der Waals and hydrophobic interactions.

Multicolor fluorescent bioimaging by single-wavelength excitation has been proved to be a powerful tool for simultaneous monitoring of multiple targets in cells. Recently, Li *et al.* designed a multifunctional target-cell-specific fluorescence SiO_2 nanoprobe for target-cell-specific delivery, cancer cells and intracellular miRNAs imaging, and cancer cell growth inhibition (Figure 14) [74]. The nanoprobe (FS-AS/MB) was prepared by simultaneously coupling of the AS1411 aptamer and miRNA-21 molecular beacon (miR-21-MB) onto the surface of $Ru(bpy)_3^{2+}$-encapsulated silica (FS) nanoparticles. In the work, the FS nanoparticles were synthesized by a one-pot two-step reverse microemulsion method; Polyethylene glycol (PEG) and functionalized amine groups were conjugated to these FS nanoparticles, providing better biocompatibility and allowing subsequent bioconjugation, respectively. Cell-specific delivery was achieved by functionalizing FS nanoparticles with AS1411 aptamer, leading to the delivery of MB only inside targeted cells with nucleolin protein. The simultaneous cell and intracellular miRNAs imaging can be accomplished under the same excitation wavelength without any cross-talk. Most importantly, the released miR-21-MB from the nanoprobe can hybridize with miRNA-21 and inhibit cell growth *in vitro*.

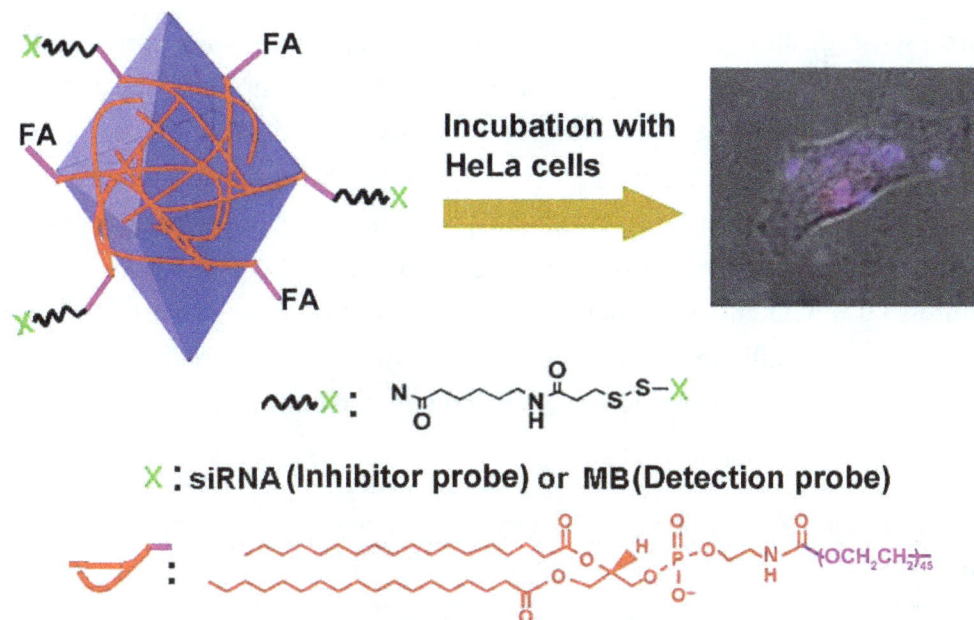

Figure 13. Schematic representation of mf-SnO2 nanoprobe for target-specific-cell imaging and intracellular detection of miRNA. FA = folic acid, MB = molecular beacon. Reprinted with permission from [73]. Copyright 2012 John Wiley and Sons.

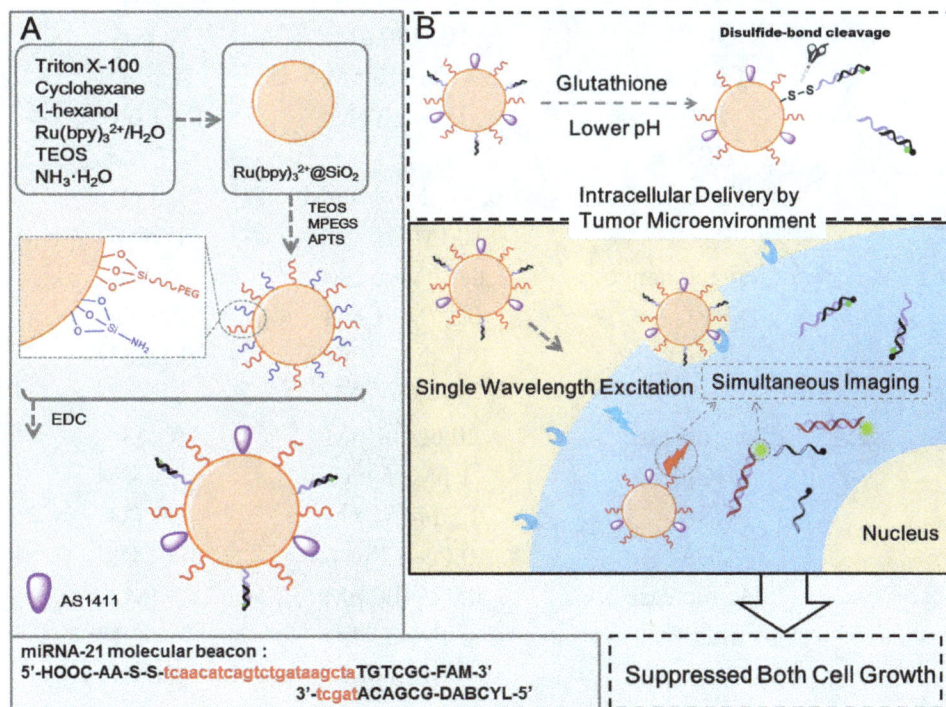

Figure 14. Schematic of the synthesis of FS-AS/MB and strategy of cancer-targeting theranostics using FS-AS/MB. Reprinted with permission from [74]. Copyright 2014 American Chemical Society.

6. Conclusions

MiRNAs, playing important functions in a number of developmental and physiological processes, have been regarded as promising biomarkers and therapeutic targets in cancer treatment.

Nanomaterials-based sensing strategies offer numerous advantages over traditional molecular diagnostic, such as signal amplification, improved sensitivity and simplicity, as well as versatile sensing scheme that can be tailored to a desired target. Recently, considerable efforts have been made to enhance the sensitivity for miRNAs detection by utilizing the unique chemical and physical properties of nanostructures. This work reviewed the progress in the development of fluorimetric methodologies for miRNAs detection based on functional nanoscaffolds of novel nanomaterials, such as metal nanostructures, QDs, GO and SnO_2 nanoparticles. Their analytical performances were shown in Table 1. Although there are still limitations for their practical use as regular methods in clinical diagnostic and prognostic, the advances in nanoscience and nanotechnology promise a better future for the sensing industries.

Table 1. Comparison of the performances of nanomaterials-based fluorimetric miRNAs biosensors.

Nanomaterials	Signal Amplification	Detection ranges	Detection limits	References
AgNCs	-	0~1.5 μM	<0.25 μM	[18]
AgNCs	-	5~125 nM	1.7 nM	[22]
AgNCs	HCR	1.56~400 nM	0.78 nM	[24]
AgNCs	Target recycling	0.5~50 nM	0.16 nM	[23]
AgNCs	TAIEA	10 aM~1 nM	2 aM	[21]
CuNCs	RCR	10~400 nM	10 pM	[29]
CuNCs	TAIEA	1 pM~10 nM	1 pM	[28]
AuNPs	-	0.05~50 pM	0.01 pM	[33]
AuNPs	DSN	-	<25 pM	[34]
CdSe/ZnS	DSN	0.53 pM~3.9 nM	0.28 pM	[47]
CdTe/CdS	-	10 fM~10 nM	10 fM	[45]
CdSe nanocrystals	Cation-Exchange	0.1 pM~5 μM	35 fM	[49]
488QDs	PG-RCA	0.1 fM~1 nM	50.9 aM	[46]
605QDs	EXPAR	0.1 aM–10 fM	0.1 aM	[44]
GO	-	50~400 nM	-	[56]
GO	Endonuclease	20 pM~1 nM	9 pM	[63]
GO	HCR	1 pM~5 nM	-	[59]
GO	DSN	0.5 pM~1 nM	160 fM	[61]
GO	CEA	0.06~12 pM	10.8 fM	[64]
GO	Endonuclease	0.02~100 pM	3 fM	[60]
GO	ISDPR	5 fM~5 pM	2.1 fM	[58]
Nano GO	-	0~100 nM	2 nM	[57]
Nano GO	-	0~1 μM	1 pM	[67]
SiO_2	-	0.5~40 nM	0.18 nM	[74]

Acknowledgments

This work was support by the Fund of Department of Science and Technology Department in Henan (KJT142102310462), the Youth Research Foundation in Pingdingshan University (20120015) and the High-Level Personnel Fund in Pingdingshan University (1011014/G).

Author Contributions

Ming La is responsible for the search of references and wrote the review paper. Lin Liu rewrote and revised the paper. Bin-Bin Zhou oversaw the entire research of the group.

Conflicts of Interest

The authors declare no conflict of interest.

References

1. Grasso, M.; Piscopo, P.; Confaloni, A.; Denti, M.A. Circulating miRNAs as biomarkers for neurodegenerative disorders. *Molecules* **2014**, *19*, 6891–6910.
2. Kong, Y.W.; Ferland-McCollough, D.; Jackson, T.J.; Bushell, M. microRNAs in cancer management. *Lancet Oncol.* **2013**, *13*, e249–258.
3. Ajit, S.K. Circulating microRNAs as biomarkers, therapeutic targets, and signaling molecules. *Sensors* **2012**, *12*, 3359–3369.
4. Dong, H.; Lei, J.; Ding, L.; Wen, Y.; Ju, H.; Zhang, X. MicroRNA: Function, detection, and bioanalysis. *Chem. Rev.* **2013**, *113*, 6207–6233.
5. Chen, C.-D.; La, M.; Zhou, B.-B. Strategies for designing of electrochemical microRNA genesensors based on the difference in the structure of RNA and DNA. *Int. J. Electrochem. Sci.* **2014**, *9*, 7228–7238.
6. Jamali, A.A.; Pourhassan-Moghaddam, M.; Dolatabadi, J.E.N.; Omidi, Y. Nanomaterials on the road to microRNA detection with optical and electrochemical nanobiosensors. *TrAC Trends Anal. Chem.* **2014**, *55*, 24–42.
7. Tian, K.; He, Z.; Wang, Y.; Chen, S.-J.; Gu, L.-Q. Designing a polycationic probe for simultaneous enrichment and detection of microRNAs in a nanopore. *ACS Nano* **2013**, *7*, 3962–3969.
8. Wang, Y.; Zheng, D.; Tan, Q.; Wang, M.X.; Gu, L.Q. Nanopore–based detection of circulating microRNAs in lung cancer patients. *Nat. Nanotechnol.* **2011**, *6*, 668–674.
9. Wanunu, M.; Dadosh, T.; Ray, V.; Jin, J.; McReynolds, L.; Drndić, M. Rapid electronic detection of probe-specific microRNAs using thin nanopore sensors. *Nat. Nanotechnol.* **2010**, *5*, 807–814.
10. Zhang, X.; Wang, Y.; Fricke, B.L.; Gu, L.-Q. Programming nanopore ion flow for encoded multiplex microRNA detection. *ACS Nano* **2014**, *8*, 3444–3450.
11. Gu, L.Q.; Wanunu, M.; Wang, M.X.; McReynolds, L.; Wang, Y. Detection of miRNAs with a nanopore single-molecule counter. *Expert Rev. Mol. Diagn.* **2012**, *12*, 573–584.
12. Xia, N.; Zhang, L. Nanomaterials-based sensing strategies for electrochemical detection of microRNAs. *Materials* **2014**, *7*, 5366–5384.
13. Guo, W.W.; Yuan, J.P.; Dong, Q.Z.; Wang, E.K. Highly sequence-dependent formation of fluorescent silver nanoclusters in hybridized DNA duplexes for single nucleotide mutation identification. *J. Am. Chem. Soc.* **2010**, *132*, 932–934.
14. Su, Y.T.; Lan, G.Y.; Chen, W.Y.; Chang, H.T. Detection of copper ions through recovery of the fluorescence of DNA-templated copper/silver nanoclusters in the presence of mercaptopropionic acid. *Anal. Chem.* **2010**, *82*, 8566–8572.

15. Yu, J.H.; Choi, S.; Dickson, R.M. Shuttle-based fluorogenic silver-cluster biolabels. *Angew. Chem. Int. Ed.* **2009**, *48*, 318–320.

16. Zhang, M.; Liu, Y.-Q.; Yu, C.-Y.; Yin, B.-C.; Ye, B.-C. Multiplexed detection of microRNAs by tuning DNA-scaffolded silver nanoclusters. *Analyst* **2013**, *138*, 4812–4817.

17. Shah, P.; Rørvig-Lund, A.; Chaabane, S.B.; Thulstrup, P.W.; Kjaergaard, H.G.; Fron, E.; Hofkens, J.; Yang, S.W.; Vosch, T. Design aspects of bright red emissive silver nanoclusters/DNA probes for microRNA detection. *ACS Nano* **2012**, *6*, 8803–8814.

18. Yang, S.W.; Vosch, T. Rapid detection of microRNA by a silver nanocluster DNA probe. *Anal. Chem.* **2011**, *83*, 6935–6939.

19. Shah, P.; Cho, S.K.; Thulstrup, P.W.; Bhang, Y.-J.; Ahn, J.C.; Choi, S.W.; Rørvig-Lund, A.; Yang, S.W. Effect of salts, solvents and buffer on miRNA detection using DNA silver nanocluster (DNA/AgNCs) probes. *Nanotechnology* **2014**, *25*, 045101, doi:10.1088/0957-4484/25/4/045101.

20. Shah, P.; Thulstrup, P.W.; Cho, S.K.; Bhang, Y.-J.; Ahn, J.C.; Choi, S.W.; Bjerrum, M.J.; Yang, S.W. In-solution multiplex miRNA detection using DNA-templated silver nanocluster probes. *Analyst* **2014**, *139*, 2158–2166.

21. Liu, Y.-Q.; Zhang, M.; Yin, B.-C.; Ye, B.-C. Attomolar ultrasensitive microRNA detection by DNA-scaffolded silver-nanocluster probe based on isothermal amplification. *Anal. Chem.* **2012**, *84*, 5165–5169.

22. Xia, X.; Ha, Y.; Hu, S.; Wang, J. Hairpin DNA probe with 5′-TCC/CCC-3′ overhangsfor the creation of silver nanoclusters and miRNA assay. Biosens. *Bioelectron.* **2014**, *51*, 36–39.

23. Dong, H.; Hao, K.; Tian, Y.; Jin, S.; Lu, H.; Zhou, S.-F.; Zhang, X. Label-free and ultrasensitive microRNA detection based on novel molecular beacon binding readout and target recycling amplification. *Biosens. Bioelectron.* **2014**, *53*, 377–383.

24. Qiu, X.; Wang, P.; Cao, Z. Hybridization chain reaction modulated DNA-hosted silver nanoclusters for fluorescent identification of single nucleotide polymorphisms in the let-7 miRNA family. *Biosens. Bioelectron.* **2014**, *60*, 351–357.

25. Rotaru, A.; Dutta, S.; Jentzsch, E.; Gothelf, K.; Mokhir, A. Selective dsDNA-templated formation of copper nanoparticles in solution. *Angew. Chem. Int. Ed.* **2010**, *49*, 5665–5667.

26. Qing, Z.H.; He, X.X.; He, D.G.; Wang, K.M.; Xu, F.Z.; Qing, T.P.; Yang, X. Poly(thymine)-templated selective formation of fluorescent copper nanoparticles. *Angew. Chem. Int. Ed.* **2013**, *52*, 9719–9722.

27. Jia, X.; Li, J.; Han, L.; Ren, J.; Yang, X.; Wang, E. DNA-hosted copper nanoclusters for fluorescent identification of single nucleotide polymorphisms. *ACS Nano* **2012**, *6*, 3311–3317.

28. Wang, X.-P.; Yin, B.-C.; Ye, B.-C. A novel fluorescence probe of dsDNA-templated copper nanoclusters for quantitative detection of microRNAs. *RSC Adv.* **2013**, *3*, 8633–8636.

29. Xu, F.; Shi, H.; He, X.; Wang, K.; He, D.; Guo, Q.; Qing, Z.; Yan, L.; Ye, X.; Li, D.; Tang, J. Concatemeric dsDNA-templated copper nanoparticles strategy with improved sensitivity and stability based on rolling circle replication and its application in microRNA detection. *Anal. Chem.* **2014**, *86*, 6976–6982.

30. Xia, N.; Zhang, L.; Wang, G.; Feng, Q.; Liu, L. Label-free and sensitive strategy for microRNAs detection based on the formation of boronate ester bonds and the dual-amplification of gold nanoparticles. *Biosens. Bioelectron.* **2013**, *47*, 461–466.

31. Balcioglu, M.; Rana, M.; Robertson, N.; Yigit, M.V. DNA-length-dependent quenching of fluorescently labeled iron oxide nanoparticles with gold, graphene oxide and MoS₂ nanostructures. *ACS Appl. Mater. Interfaces* **2014**, *6*, 12100–12110.

32. Swierczewska, M.; Lee, S.; Chen, X. The design and application of fluorophore-gold nanoparticle activatable probes. *Phys. Chem. Chem. Phys.* **2011**, *13*, 9929–9941.

33. Tu, Y.; Wu, P.; Zhang, H.; Cai, C. Fluorescence quenching of gold nanoparticles integrating with a conformation–switched hairpin oligonucleotide probe for microRNA detection. *Chem. Commun.* **2012**, *48*, 10718–10720.

34. Degliangeli, F.; Kshirsagar, P.; Brunetti, V.; Pompa, P.P.; Fiammengo, R. Absolute and direct microRNA quantification using DNA–gold nanoparticle probes. *J. Am. Chem. Soc.* **2014**, *136*, 2264–2267.

35. Liu, L.; Gao, Y.; Liu, H.; Xia, N. An ultrasensitive electrochemical miRNAs sensor based on miRNAs-initiated cleavage of DNA by duplex-specific nuclease and signal amplification of enzyme plus redox cycling reaction. *Sens. Actuators B Chem.* **2015**, *208*, 137–142.

36. Rosa, J.; Conde, J.; de la Fuente, J.M.; Lima, J.C.; Baptista, P.V. Gold-nanobeacons for real-time monitoring of RNA synthesis. *Biosens. Bioelectron.* **2012**, *36*, 161–167.

37. Conde, J.; Rosa, J.; de la Fuente, J.M.; Baptista, P.V. Gold-nanobeacons for simultaneous gene specific silencing and intracellular tracking of the silencing events. *Biomaterials* **2013**, *34*, 2516–2523.

38. Liu, L.F.; Miao, Q.Q.; Liang, G.L. Quantum dots as multifunctional materials for tumor imaging and therapy. *Materials* **2013**, *6*, 483–499.

39. Härmä, H.; Pihlasalo, S.; Cywinski, P.J.; Mikkonen, P.; Hammann, T.; Lohmannsröben, H.-G.; Hänninen, P. Protein quantification using resonance energy transfer between donor nanoparticles and acceptor quantum dots. *Anal. Chem.* **2013**, *85*, 2921–2926.

40. Liang, R.-Q.; Li, W.; Li, Y.; Tan, C.-Y.; Li, J.-X.; Jin, Y.-X.; Ruan, K.-C. An oligonucleotide microarray for microRNA expression analysis based on labeling RNA with quantum dot and nanogold probe. *Nucleic Acids Res.* **2005**, *33*, e17.

41. Yuan, Y.; Zhang, J.; Liang, G.L.; Yang, X.R. Rapid fluorescent detection of neurogenin3 by CdTe quantum dot aggregation. *Analyst* **2012**, *137*, 1775–1778.

42. Long, Y.; Zhang, L.-F.; Zhang, Y.; Zhang, C.-Y. Single quantum dot based nanosensor for renin assay. *Anal. Chem.* **2012**, *84*, 8846–8852.

43. Sapsford, K.E.; Granek, J.; Deschamps, J.R.; Boeneman, K.; Blanco-Canosa, J.B.; Dawson, P.E.; Susumu, K.; Stewart, M.H.; Medintz, I.L. Monitoring botulinum neurotoxin a activity with peptide-functionalized quantum dot resonance energy transfer sensors. *ACS Nano* **2011**, *5*, 2687–2699.

44. Zhang, Y.; Zhang, C.-Y. Sensitive detection of microRNA with isothermal amplification and a single-quantum-dot-based nanosensor. *Anal. Chem.* **2012**, *84*, 224–231.

45. Su, S.; Fan, J.; Xue, B.; Yuwen, L.; Liu, X.; Pan, D.; Fan, C.; Wang, L. DNA-conjugated quantum dot nanoprobe for high-sensitivity fluorescent detection of DNA and microRNA. *ACS Appl. Mater. Interfaces* **2014**, *6*, 1152–1157.

46. Zeng, Y.-P.; Zhu, G.; Yang, X.-Y.; Cao, J.; Jing, Z.-L.; Zhang, C.-Y. A quantum dot-based microRNA nanosensor for point mutation assays. *Chem. Commun.* **2014**, *50*, 7160–7162.

47. Jou, A.F.; Lu, C.-H.; Ou, Y.-C.; Wang, S.-S.; Hsu, S.-L.; Willner, I.; Ho, J.A. Diagnosing the miR-141 prostate cancer biomarker using nucleic acid-functionalized CdSe/ZnS QDs and telomerase. *Chem. Sci.* **2015**, *6*, 659–665.

48. Kucur, E.; Boldt, F.M.; Cavaliere-Jaricot, S.; Ziegler, J.; Nann, T. Quantitative analysis of cadmium selenide nanocrystal concentration by comparative techniques. *Anal. Chem.* **2007**, *79*, 8987–8993.

49. Li, J.; Schachermeyer, S.; Wang, Y.; Yin, Y.; Zhong, W. Detection of microRNA by fluorescence amplification based on cation-exchange in nanocrystals. *Anal. Chem.* **2009**, *81*, 9723–9729.

50. Liu, J. Adsorption of DNA onto gold nanoparticles and graphene oxide: surface science and applications. *Phys. Chem. Chem. Phys.* **2012**, *14*, 10485–10496.

51. Manohar, S.; Mantz, A.R.; Bancro, K.E.; Hui, C.Y.; Jagota, A.; Vezenov, D.V. Peeling single-stranded DNA from graphite surface to determine oligonucleotide binding energy by force spectroscopy. *Nano Lett.* **2008**, *8*, 4365–4372.

52. Lu, C.H.; Yang, H.H.; Zhu, C.L.; Chen, X.; Chen, G.N. A graphene platform for sensing biomolecules. *Angew. Chem. Int. Ed.* **2009**, *48*, 4785–4787.

53. He, S.; Song, B.; Li, D.; Zhu, C.; Qi, W.; Wen, Y.; Wang, L.; Song, S.; Fang, H.; Fan, C. A graphene nanoprobe for rapid, sensitive, and multicolor fluorescent DNA analysis. *Adv. Funct. Mater.* **2010**, *20*, 453–459.

54. Chang, H.; Tang, L.; Wang, Y.; Jiang, J.; Li, J. Graphene fluorescence resonance energy transfer aptasensor for the thrombin detection. *Anal. Chem.* **2010**, *82*, 2341–2346.

55. Jang, H.; Kim, Y.K.; Kwon, H.M.; Yeo, W.S.; Kim, D.E.; Min, D.H. A grapheme-based platform for the assay by helicase. *Angew. Chem. Int. Ed.* **2010**, *49*, 5703–5707.

56. Lu, Z.; Zhang, L.; Deng, Y.; Li, S.; He, N. Graphene oxide for rapid microRNA detection. *Nanoscale* **2012**, *4*, 5840–5842.

57. Hizir, M.S.; Balcioglu, M.; Rana, M.; Robertson, N.M.; Yigit, M.V. Simultaneous detection of circulating oncomiRs from body fluids for prostate cancer staging using nanographene oxide. *ACS Appl. Mater. Interfaces* **2014**, *6*, 14772–14778.

58. Dong, H.; Zhang, J.; Ju, H.; Lu, H.; Wang, S.; Jin, S.; Hao, K.; Du, H.; Zhang, X. Highly sensitive multiple microRNA detection based on fluorescence quenching of graphene oxide and isothermal strand-displacement polymerase reaction. *Anal. Chem.* **2012**, *84*, 4587–4593.

59. Yang, L.; Liu, C.; Ren, W.; Li, Z. Graphene surface-anchored fluorescence sensor for sensitive detection of microRNA coupled with enzyme-free signal amplification of hybridization chain reaction. *ACS Appl. Mater. Interfaces* **2012**, *4*, 6450–6453.

60. Tu, Y.; Li, W.; Wu, P.; Zhang, H.; Cai, C. Fluorescence quenching of graphene oxide integrating with the site-specific cleavage of the endonuclease for sensitive and selective microRNA detection. *Anal. Chem.* **2013**, *85*, 2536–2542.

61. Guo, S.; Yang, F.; Zhang, Y.; Ning, Y.; Yao, Q.; Zhang, G.-J. Amplified fluorescence sensing of miRNA by combination of graphene oxide with duplexspecific nuclease. *Anal. Methods* **2014**, *6*, 3598–3603.

62. Tang, Z.; Wu, H.; Cort, J.R.; Buchko, G.W.; Zhang, Y.; Shao, Y.; Aksay, I.A.; Liu, J.; Lin, Y. Constraint of DNA on functionalized graphene improves its biostability and specificity. *Small* **2010**, *6*, 1205–1209.

63. Cui, L.; Lin, X.; Lin, N.; Song, Y.; Zhu, Z.; Chen, X.; Yang, C.J. Graphene oxide-protected DNA probes for multiplex microRNA analysis in complex biological samples based on a cyclic enzymatic amplification method. *Chem. Commun.* **2012**, *48*, 194–196.

64. Liu, H.; Li, L.; Wang, Q.; Duan, L.; Tang, B. Graphene fluorescence switch-based cooperative amplification: A sensitive and accurate method to detection microRNA. *Anal. Chem.* **2014**, *86*, 5487–5493.

65. Paul, A.; Hasan, A.; Kindi, H.A.; Gaharwar, A.K.; Rao, V.T.S.; Nikkhah, M.; Shin, S.R.; Krafft, D.; Dokmeci, M.R.; Shum-Tim, D.; *et al.* Injectable graphene oxide/hydrogel-based angiogenic gene delivery system for vasculogenesis and cardiac repair. *ACS Nano* **2014**, *8*, 8050–8062.

66. Kim, H.; Kim, W.J. Photothermally controlled gene delivery by reduced graphene oxide-polyethylenimine nanocomposite. *Small* **2014**, *10*, 117–126.

67. Ryoo, S.-R.; Lee, J.; Yeo, J.; Na, H.-K.; Kim, Y.-K.; Jang, H.; Lee, J.H.; Han, S.W.; Lee, Y.; Kim, V.N.; *et al.* Quantitative and multiplexed microRNA sensing in living cells based on peptide nucleic acid and nano graphene oxide (PANGO). *ACS Nano* **2013**, *7*, 5882–5891.

68. Hernandez, R.; Orbay, H.; Cai, W. Molecular imaging strategies for in Vivo tracking of microRNAs: A comprehensive review. *Curr. Med. Chem.* **2013**, *20*, 3594–3603.

69. Zhang, P.; Cheng, F.; Zhou, R.; Cao, J.; Li, J.; Burda, C.; Min, Q.; Zhu, J.-J. DNA-hybrid-gated nultifunctional mesoporous silica nanocarriers for dual-targeted and microRNA-responsive controlled drug delivery. *Angew. Chem. Int. Ed.* **2014**, *53*, 2371–2375.

70. Rivera-Gil, P.; de Aberasturi, D.J.; Wulf, V.; Pelaz, B.; Del Pino, P.; Zhao, Y.Y.; de la Fuente, J.M.; de Larramendi, I.R.; Rojo, T.; Liang, X.J.; *et al.* The challenge to relate the physicochemical properties of colloidal nanoparticles to their cytotoxicity. *Acc. Chem. Res.* **2013**, *46*, 743–749.

71. Su, X.; Kuang, L.; Battle, C.; Shaner, T.; Mitchell, B.S.; Fink, M.J.; Jayawickramarajah, J. Mild two-step method to construct DNA-conjugated silicon nanoparticles: Scaffolds for the detection of microRNA-21. *Bioconjug. Chem.* **2014**, *25*, 1739–1743.

72. Zhang, J.; Fu, Y.; Mei, Y.; Jiang, F.; Lakowicz, J.R. Fluorescent metal nanoshell probe to detect single miRNA in lung cancer cell. *Anal. Chem.* **2010**, *82*, 4464–4471.

73. Dong, H.; Lei, J.; Ju, H.; Zhi, F.; Wang, H.; Guo, W.; Zhu, Z.; Yan, F. Target-cell-specific delivery, imaging, and detection of intracellular microRNA with a multifunctional SnO_2 nanoprobe. *Angew. Chem.* **2012**, *124*, 4685–4690.

74. Li, H.; Mu, Y.; Lu, J.; Wei, W.; Wan, Y.; Liu, S. Target-cell-specific fluorescence silica nanoprobes for imaging and theranostics of cancer cells. *Anal. Chem.* **2014**, *86*, 3602–3609.

8

Electrochemical Investigation of the Corrosion of Different Microstructural Phases of X65 Pipeline Steel under Saturated Carbon Dioxide Conditions

Yuanfeng Yang * and Robert Akid

School of Materials, the University of Manchester, Manchester M13 9PL, UK;
E-Mail: Robert.akid@manchester.ac.uk

* Author to whom correspondence should be addressed; E-Mail: yuanfeng.yang@manchester.ac.uk;

Academic Editor: Richard Thackray

Abstract: The aim of this research was to investigate the influence of metallurgy on the corrosion behaviour of separate weld zone (WZ) and parent plate (PP) regions of X65 pipeline steel in a solution of deionised water saturated with CO_2, at two different temperatures (55 °C and 80 °C) and at initial pH~4.0. In addition, a non-electrochemical immersion experiment was also performed at 80 °C in CO_2, on a sample portion of X65 pipeline containing part of a weld section, together with adjacent heat affected zones (HAZ) and parent material. Electrochemical impedance spectroscopy (EIS) was used to evaluate the corrosion behaviour of the separate weld and parent plate samples. This study seeks to understand the significance of the different microstructures within the different zones of the welded X65 pipe in CO_2 environments on corrosion performance; with particular attention given to the formation of surface scales; and their composition/significance. The results obtained from grazing incidence X-ray diffraction (GIXRD) measurements suggest that, post immersion, the parent plate substrate is scale free, with only features arising from ferrite (α-Fe) and cementite (Fe_3C) apparent. In contrast, at 80 °C, GIXRD from the weld zone substrate, and weld zone/heat affected zone of the non-electrochemical sample indicates the presence of siderite ($FeCO_3$) and chukanovite ($Fe_2CO_3(OH)_2$) phases. Scanning Electron Microscopy (SEM) on this surface confirmed the presence of characteristic discrete cube-shaped crystallites of siderite together with plate-like clusters of chukanovite.

Keywords: X65 pipeline steel; corrosion; carbon dioxide; microstructure; weld; SEM; GIXRD; chukanovite; siderite

1. Introduction

Dissolved carbon dioxide (CO_2) in oilfield brines, which accompany extracted oil and gas, causes the internal "sweet" corrosion of steel pipelines. In addition to inducing uniform corrosion, there have been numerous cases of serious localised CO_2 attack—particularly evident in scale or deposit regions [1]. Even though CO_2 corrosion product scales are known to often substantially reduce steel corrosion by offering a physical/diffusion barrier on the steel surface, and thereby inhibiting further corrosion [2,3], localised attack can proceed in regions where the surface scales do not offer sufficient protection [4]. Understanding the formation, chemistry and role of CO_2-induced corrosion scales and surface films in oilfield environments thus remains of considerable interest.

Concerning CO_2 oilfield corrosion, one specific topic of interest in recent years has been the preferential weld corrosion of carbon steels [5–8]. Olsen and co-workers suggest that weld failures are a combination of the corrosion susceptibility of the weld and the film forming properties of the CO_2 saturated environment [9]. In considering CO_2 corrosion in weld regions, studies have shown that there is a strong correlation between susceptibility to corrosion and weld composition. For example, the use of nickel containing weld consumables during the joining process can lead to preferential corrosion along the weld segment in sweet brines, with significantly less corrosion in welds that were produced using filler metals having no deliberate alloying [5,6,10]. CO_2 corrosion scales are suspected to form in the initiation stages of weld corrosion [11,12]. Crolet and co-workers highlighted that the nature of the initial phases can determine whether or not the subsequent layer would be protective [13]. Furthermore, corrosion attack may proceed preferentially along the heat-affected zones (HAZ) if the weld region has been protectively covered by scale [9].

To date, there have only been a few studies reporting CO_2 corrosion/scaling phenomena on low carbon steels in lower pH CO_2 environments—Particularly with the emphasis on comparing parent-plate material (PP) with weld zones (WZ) [3,6,9,14,15]. On this basis, we report here on the results of electrochemical measurements made on individual parent plate and weld regions of X65 pipeline steel at temperatures of 55 °C and 80 °C. In addition, we report on a non-electrochemical immersion experiment performed on a single sample consisting of weld portion together with HAZ and PP regions, at a temperature of 80 °C, in unstirred CO_2 saturated deionised water at an initial pH~4.0. The dissolved oxygen level of the test solution was maintained below 10 parts per billion by volume (ppbv) in order to simulate those near-anaerobic conditions usually found within pipelines. Surface corrosion products (scales) have been investigated using scanning electron microscopy (SEM) and grazing incidence X-ray diffraction (GIXRD) techniques.

2. Experimental Section

2.1. Materials

The original portion of X65 pipeline utilized for the initial set of experiments is shown in Figure 1a, which is illustrated in the "as received" condition. The sample was taken from a commercial pipeline riser, and had been quenched and tempered according to standard industry practice. During production, the pipeline had been welded using gas metal arc welding (GMAW) with a nickel-alloyed carbon steel filler wire. The composition of X65 pipelines steel (UNI EN 10204) parent plate and weld zone filler wire were obtained by x-ray fluorescence spectroscopy and are given in Table 1. Note the difference in nickel content between the PP steel and the composition of the WZ filler. For SEM examination using backscattered electron (BSE) imaging, a water jet cutting system was used to initially cut a region from the original pipeline portion that included part of the weld zone, heat affected zone and parent plate. This sample was polished down to 1/4 µm using diamond paste, then electropolished in 8% perchloric-acetic reagent, at 60 V for 5 s. Figure 1b shows an optical micrograph of the prepared surface of this sample, prior to SEM examination, showing clearly the 3 zones (WZ, HAZ and PP) within the welded pipeline.

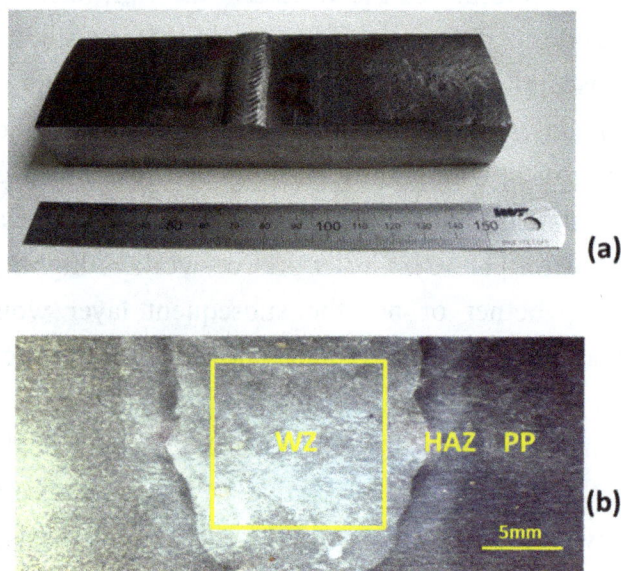

Figure 1. (a) Photograph of original portion of X65 pipeline in the "as received" condition; (b) Optical micrograph of electropolished cross section of small region of sample shown in (a), illustrating part of the Weld Zone (WZ), Heat Affected Zone (HAZ), and Parent Plate (PP). Sample electropolished in 8% perchloric-acetic reagent, at 60 V for 5 s. Yellow box highlighted in weld zone illustrates dimensions of sample that was selected to use in subsequent electrochemical testing.

Table 1. X65 Steel and weld consumable composition (wt%).

Material	C	Mn	Ni	Cr	Mo	Si	Al	Cu	V	P	S	Fe
Parent Plate	0.08	1.08	0.037	0.07	0.13	0.28	0.037	0.16	0.06	0.01	0.001	Balance
Weld Zone filler	0.07	1.46	0.91	0.01	0.01	0.67	0.003	0.11	0.001	0.007	0.009	Balance

Figure 2 shows BSE images obtained using SEM of the X65 parent and weld regions, clearly illustrating the variation in microstructure between the weld zone and parent plate. The microstructure of the parent plate consists primarily of polygonal ferrite grains (Figure 2a). The microstructure of the weld zone consists of finer grains of bainite (A), acicular ferrite (B), grain boundary ferrite (C), and polygonal ferrite (D) (Figure 2b).

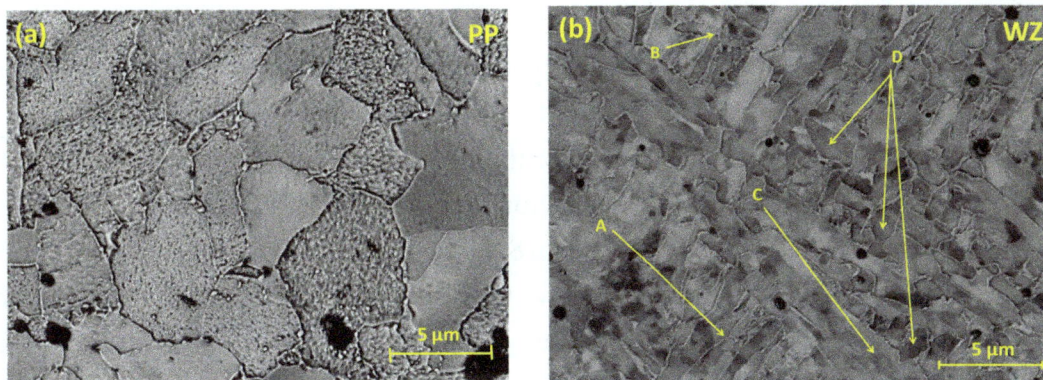

Figure 2. SEM micrographs in BSE mode, of electropolished parent plate and weld regions showing differences in grain morphology: (**a**) Parent Plate (PP) and (**b**) Weld Zone (WZ).

For the electrochemical immersion experiments, samples from both the weld zone and parent plate regions, having dimensions 1.0 cm × 1.0 cm × 1.0 cm, were also cut using a high pressure water jet cutting system. The yellow box highlighted in Figure 1b illustrates the location of the weld zone sample, which was cut from the sample discussed above, whilst the parent plate sample was cut from a site at least 5.0 cm from the weld region, to obviate any effects of heating during welding of the pipeline. The 1.0 cm^2 area samples (WZ and PP) were spot-welded using Cu–Sn wire, to give an electrical connection, and cold mounted in epoxy resin. Exposed sample surfaces were then re-prepared by polishing down to a 5-μm finish using 4000 grade abrasive paper, cleaned with acetone and stored under a nitrogen atmosphere within a glove box before immersion in the CO_2 saturated solution.

An additional sample was also prepared from the original portion of X65 pipeline (shown in Figure 1a), using a high pressure water jet cutting facility, but consisting of a single "multi-zone" sample of weld together with HAZ and PP. This specimen is shown in Figure 3.

Figure 3. Photograph of single "multi-zone" sample consisting of weld portion together with HAZ and PP before immersion.

2.2. Experimental Procedure

The individually prepared 1.0 cm^2 area separate samples from both WZ and PP were used for the electrochemical corrosion experiments. In addition, a set of non-electrochemical experiments, consisting only of immersion tests, was also carried out using the "multi-zone" sample. All experiments, including the "multi-zone" sample, were performed in a jacketed glass electrochemical cell containing 1 L CO_2—Saturated deionised water, at initial pH~4 (all pH values determined using a calibrated Hanna instruments pH meter) and all within a N_2-purged acrylic glovebox. Note that the conductivity of the deionised water was 3 µs/cm and increased to 280 µs/cm when fully saturated with CO_2.

The electrolyte temperature was maintained by circulating heated water from a water bath, through the surrounding jacket cell. The electrolyte was previously purged with high purity CO_2 gas (BOC, Ashton-under-Lyne, Lancashire, UK, 99.995%) for 18–24 h prior to commencing the experiment, to ensure low dissolved oxygen concentration, which was measured using a commercial electrochemical sensor. Once the oxygen concentration was determined to be less than 10 ppb, the CO_2 gas flow was changed to pass over the solution surface (blanket flow), maintaining a constant pressure of 1 bar. This ensured that the solution remained static (natural convention flow) for the experiment duration; and the CO_2—Water equilibrium was maintained. Any water loss was considered to be less than 5%.

Electrochemical measurements were performed using either an X65 parent plate or weld zone sample as the working electrode (WE) with a platinum-mesh counter electrode (CE) and an Ag/AgCl reference electrode (RE). A polymer gel salt bridge was used to connect the RE to the test solution in order to minimise any contamination from the reference electrode reservoir solution (3.5 M KCl). All experiments were of 72 h duration. Throughout almost all of the immersion period, the WE was maintained at open circuit potential (OCP), except for hourly electrochemical impedance spectroscopy (EIS) measurements, involving sample polarisation of ±10 mV vs. OCP, over a frequency range of 10 kHz–50 mHz. EIS data was acquired using a Solartron Potentiostat 2100 A (Farnborough, Hampshire, UK). After the experiment, the WE samples were stored in a vacuum dessicator to avoid surface oxidation. The samples were then subjected to GIXRD and surface SEM analysis. All of the GIXRD measurements used a CuKα source and were carried out at a 3° incidence angle using a Philips XPERT XRD diffractometer (PANalytical Ltd., Cambridge, UK) in the 2θ range from 5° to 85°. A FEI XL-30 FEGSEM (Hillsboro, OR, USA) was used for all SEM examination throughout this work.

2.3. Interpreting EIS for Obtaining Corrosion Rates of X-65 Parent/Weld Samples

Corrosion rate (i_{corr}) was estimated by extracting the observed R_{ct} from the EIS spectra and applying the B parameter [3,16] in order to calculate corrosion rate. Equation (2) was used to estimate the corrosion rate of the steel over the 72 h immersion period.

For calculating the corrosion rates, we applied Faraday's law and the Stern-Geary relation, using the Stern-Geary coefficient $B = 26$ mV:

$$R_p = \frac{B}{i_{corr}} \tag{1}$$

$$CR \ (\text{mm/y}) = \frac{i_{corr} A_w \times 10 \times 3.15 \times 10^7}{nF\rho} \tag{2}$$

Analysis of the single "multi-zone" sample, consisting of weld portion together with HAZ and PP, immersed at 80 °C, is given in Figure 11a. This is supported by GIXRD and SEM data of the individual regions of "multi-zone" sample, as shown in Figure 11b–j.

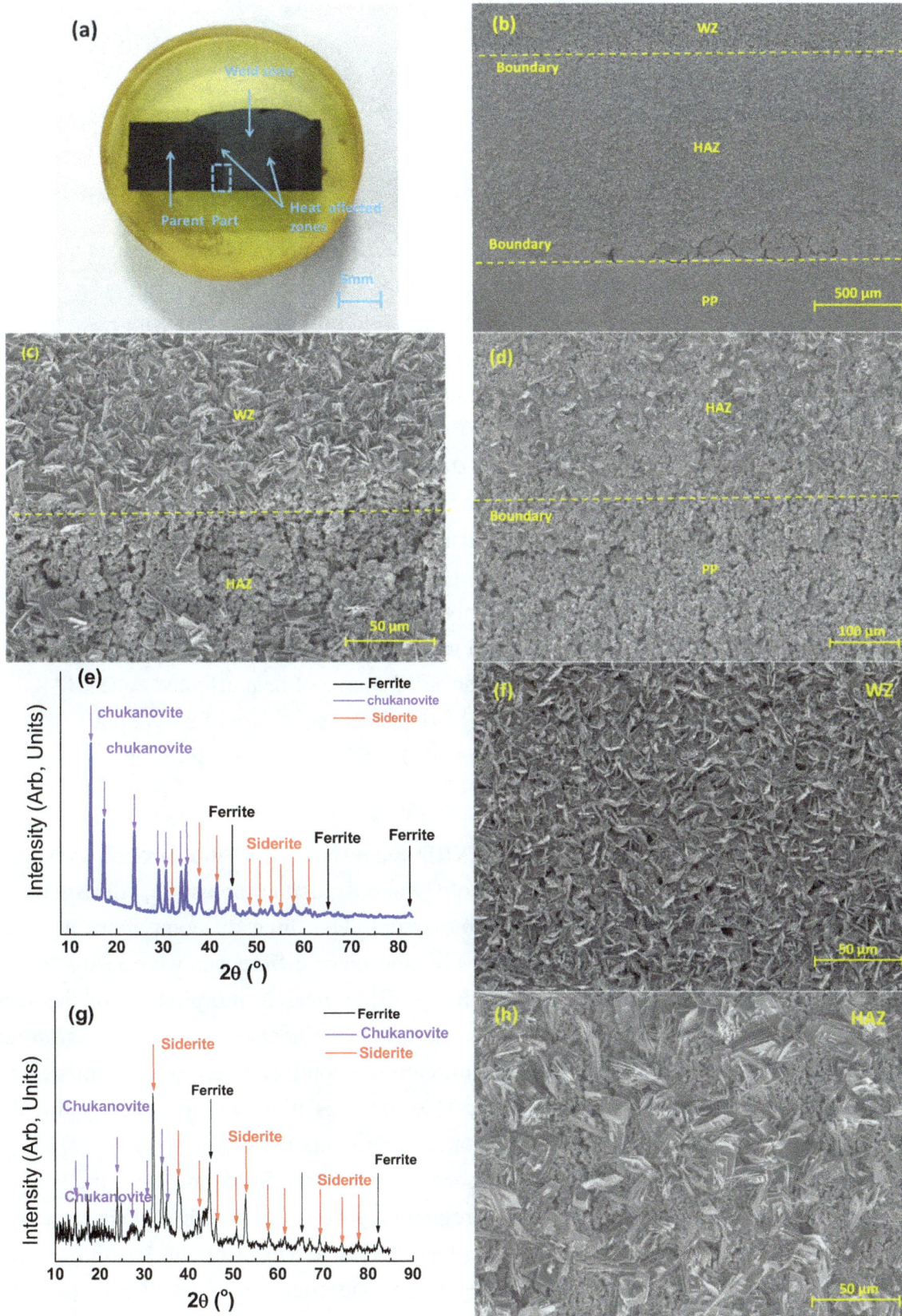

Figure 11. *Cont.*

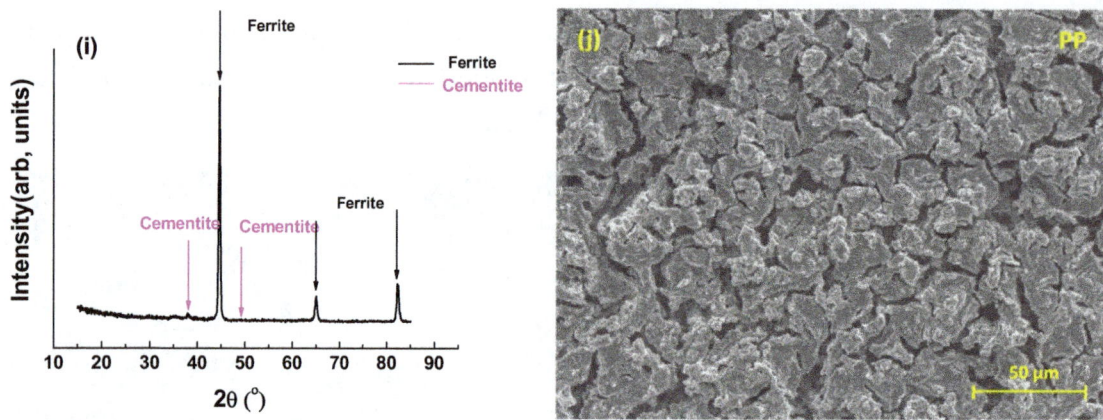

Figure 11. Analysis of "multi-zone" sample after 72 h immersion in CO_2 saturated deionized water at 80 °C, with initial pH~4.0 (Note: after 72 h the pH was 5.2): (**a**) Optical photograph of whole sample after 72 h immersion, the weld portion together with HAZ and PP are clearly visible. Note: Figure 3 illustrates same sample prior to immersion; (**b**) Secondary electron SEM image of area highlighted in blue rectangle in (**a**), Note: all SEM images in this Figure are orientated 90° clockwise; (**c**) Secondary electron SEM image of boundary region between parent plate (PP) and heat affected zone (HAZ), showing clear distinction in the morphology of the surface deposits in the two zones; (**d**) Secondary electron SEM image of the boundary between the heat-affected zone (HAZ) and the weld zone (WZ), also showing a clear distinction in surface morphology; (**e**) GIXRD spectrum from weld zone (WZ) only; (**f**) Secondary electron SEM image of weld zone only, image corresponds to spectrum given in (**e**); (**g**) GIXRD spectrum of heat affected zone (HAZ) only; (**h**) Secondary electron SEM image of heat affected zone (HAZ) only, image corresponds to spectrum given in (**g**); (**i**) GIXRD spectrum of parent plate (PP) region only; (**j**) Secondary electron SEM image of parent plate (PP) region only, image corresponds to spectrum given in (**i**).

Figure 11 shows the results of the SEM and GIXRD analysis of the "multi-zone" sample, and illustrates the significant differences in surface morphology and deposit composition, corresponding to the different sample regions, that occur after 72 h immersion. The three different zones are clearly visible in the optical photograph presented in Figure 11a, and these differences were visible with the naked eye, even after 7 to 8 h immersion. After 72 h, the differences in morphology of the surface deposits on the three zones were clearly discernable in the low magnification secondary electron image shown in Figure 11b. Figure 11c,d present higher magnification secondary electron SEM images of the WZ to HAZ and HAZ to PP interfaces respectively. These images illustrate very clearly the abrupt changes in surface deposit morphology that occur at both zone boundaries.

The individual GIXRD spectra and corresponding secondary electron SEM images of the WZ, HAZ and PP zones are given in Figure 11e–j. For the WZ region (Figure 11e), the GIXRD spectrum of the surface revealed the presence of chukanovite ($Fe_2CO_3(OH)_2$), siderite ($FeCO_3$) and ferrite (α-Fe). The intensity of the chukanovite peaks relative to the ferrite and siderite peaks was clearly far greater, indicating that there might have been a significant surface coverage and presence of this phase. SEM analysis confirmed this particular hypothesis (Figure 11f) showing evidence of typical plate-like chukanovite crystals. On the HAZ region (Figure 11g), the GIXRD spectrum of the surface deposit revealed the presence of chukanovite ($Fe_2CO_3(OH)_2$), siderite ($FeCO_3$) and ferrite (α-Fe), which corresponds to the characteristic surface deposit morphologies observed in the corresponding SEM image (Figure 11h), with evidence of typical cuboidal shaped siderite surrounded by characteristic clusters of plate-like chukanovite crystals. On the PP (Figure 11i), the GIXRD spectrum of the surface deposit revealed only the presence of cementite and ferrite; this was confirmed by SEM observation (Figure 11j).

It is proposed that an even greater quantity of iron will have dissolved from the "multi-zone" sample consisting of a weld portion together with HAZ and PP than the other individual PP and WZ samples, since significant iron-rich corrosion scales were deposited on WZ. After 72 h, the pH attained by the "multi-zone" sample was 5.2, which was the highest pH values recorded across all of the tests conducted. From reports in the literature, scale formation is not usually expected at pH values below 5.5 [18]. At pH conditions less than this, chukanovite and siderite are not expected to precipitate out of solution. Such scales formation has not often reported on steel surfaces in the open literature, particular in fairly acidic bulk solutions. More commonly, chukanovite is detected as the minority phase, and usually secondary to siderite [14,21,22]. Furthermore, acidic solutions would be considered to enhance the dissolution of chukanovite in the first instance, followed by dissolution of siderite [23,24].

In the last decade, chukanovite has been discussed in great detail in the fields of iron archaeology and long-term nuclear storage, typically being found under anaerobic environments. A study by Saheb on archaeological iron, has shown that in carbonate-rich, anaerobic conditions, $Fe_2CO_3(OH)_2$ forms first on iron metal surfaces followed by $FeCO_3$ [21]. There is general agreement in the literature that this compound forms under slightly alkaline conditions (>pH 6) [25,26]. Later work by Remazeilles has suggested that the preferential formation of $Fe_2CO_3(OH)_2$ depended upon the ratio of Fe^{2+} to OH^- and Fe^{2+} to CO_3^{2-} in his deaerated $FeCl_2$/NaOH/Na_2CO_3 solutions [27]. Subsequently, experimental work by Han reported that, even if the bulk solution pH is low (~4), local pH measurements on actively corroding steel surfaces in CO_2 saturated environments at 80 °C showed that the pH at the interface is more alkaline (~6) [28]. In particular, if the relative molar ratio of Fe^{2+}:OH^- is approximately 1 and Fe^{2+}:CO_3^{2-} is approximately 2, thermodynamically stable chukanovite formation dominates in preference to $Fe(OH)_2$ or $FeCO_3$. A very recent report by Refait suggested that the formation of either chukanovite, siderite or other compounds (carbonated green rust or magnetite) at a steel surface under anaerobic carbonate-rich solutions must be controlled primarily by the aforementioned concentration ratios at the metal/solution interface [29]. In such a case, the nature of the surface film should be determined by the kinetics of iron dissolution.

From our electrochemical observations, over the first 30 h the weld zone at 80 °C demonstrated a higher corrosion rate than the parent plate (see Figure 6b). This initial period of "accelerated corrosion" leads to the production of a high local surface concentration of Fe^{2+}. Therefore, on the weld surface, the observed initial severe corrosion rate perhaps equated to the generation of a high interfacial ratio of Fe^{2+} with respect to CO_3^{2-}. This coupled with a probable equivalence of Fe^{2+} to $OH-$ (from the relatively higher interface pH) may have led to the formation of stable chukanovite.

The observed chukanovite observed on the weld zone may have acted as a pre-cursor to siderite formation but this would also need to be confirmed with longer immersion experiments [14,30]. Finally, it remains in debate whether or not it was the initial high corrosion rate that was critical in developing a chukanovite-rich surface scale, or whether the weld microstructure and composition played an important role.

4. Conclusions

- 72 h electrochemical corrosion immersion experiments on X65 parent and weld regions were performed in CO_2-saturated deionised water at 55 °C and 80 °C at an initial pH~4. After 72 h immersion, the measured corrosion rates were related to surface deposit characterisation using GIXRD and SEM.

- At both temperatures, the separate weld zone initially corroded at a higher rate than the parent plate.

- At 55 °C, GIXRD offered no evidence of scaling, and no crystallites could be observed from SEM images of both X65 parent and weld regions. At 80 °C, similar observations held for the X65 parent plate. The sample surfaces exhibit uniform corrosion in these environments.

- The X65 weld sample, at 80 °C, corroded more rapidly during the initial 30 h, but then fell to a lower corrosion rate and remained relatively stable for the remaining period of immersion. The surface was covered with chukanovite ($Fe_2CO_3(OH)_2$), along with some siderite ($FeCO_3$), which was confirmed from GIXRD and SEM studies, and supported by published literature.

- A non-electrochemical immersion experiment was performed on a single "multi-zone" sample consisting of a weld portion together with HAZ and PP zones at 80 °C. A simple optical micrograph of the whole sample surface clearly distinguished the three different zones. From GIXRD and SEM results, the surface of the WZ/HAZ showed extensive presence of chukanovite ($Fe_2CO_3(OH)_2$), along with some siderite ($FeCO_3$), whilst very little scaling was observed on the PP sample surface.

- Understanding the impact of weld microstructure and composition on the initial higher rate of corrosion and possible subsequent chukanovite formation, along with the exact role of this corrosion scale, remains a topic of further research.

Acknowledgments

This work was carried out in the BP Materials and Corrosion Laboratory in the Corrosion and Protection Centre, School of Materials at the University of Manchester. The author Yuanfeng Yang wishes to thank Dr. Robert Lindsay and Mr. Gaurav Ravindra Joshi for assistance with the glove box and experimental set up, and useful and constructive discussion of the results. Dr. Chris Wilkins and Mr. Gary Harrison are thanked for their expert contribution with the SEM and XRD analysis.

Author Contributions

Yuanfeng Yang planed and conducted all experiments, examined the experimental results, performed all data analysis and wrote the article draft. Robert Akid checked through and revised the final draft of the article script.

Conflicts of Interest

The authors declare no conflict of interest.

References

1. Cabrini, M.; Hoxha, G.; Kopliku, A.; Lazzari, L. *Prediction of CO_2 Corrosion in Oil and Gas Wells—Analysis of Some Case Histories*; CORROSION/98; NACE International: San Diego, CA, USA, 1998.

2. Dugstad, A. *Mechanism of Protective Film Formation During CO_2 Corrosion Of Carbon Steel*; CORROSION/98; NACE International: San Diego, CA, USA, 1998.

3. Tanupabrungsun, T.; Brown, B.; Nesic, S. *Effect of pH on CO_2 Corrosion of Mild Steel at Elevated Temperatures*; CORROSION/2013; NACE International: Orlando, FL, USA, 2013.

4. Papavinasam, S.; Doiron, A.; Li, J.; Park, D.Y.; Liu, P. *Sour and Sweet Corrosion of Carbon Steel: General or Pitting or Localised Or All Of The Above*; CORROSION/2010; NACE International: San Antonio, TX, USA, 2010.

5. Lee, C.M.; Bond, S.; Woollin, S. *Preferential Weld Corrosion: Effects of Weldment Microstructure and Composition*; CORROSION/2005; NACE International: San Antonio, TX, USA, 2005.

6. Turgoose, S.; Palmer, J.W.; Dicken, G.E. *Preferential Weld Corrosion of 1% Ni Welds: Effects of Solution Conductivity and Corrosion Inhibitors*; CORROSION/2005; NACE International: San Antonio, TX, USA, 2005.

7. Matinez, M.; Alawadhi, K.; Robinson, M.; Nelson, G.; Macdonald, A. *Control of Preferential Weld Corrosion of X65 Pipeline Steel in Flowing Brines Containing Carbon Dioxide*; CORROSION/2011; NACE International: Houston, TX, USA, 2011.

8. Barker, R.; Hu, X.; Neville, A.; Cushnaghan, S. Assessment of preferential weld corrosion of carbon steel pipework in CO_2-saturated flow-induced corrosion environments. *Corrosion* **2013**, *69*, 1132–1143.

9. Olsen, S.; Sundfaer, B.; Enerhaug, J. *Weld Corrosion in C-Steel Pipelines in CO_2 Environments—Comparison between Field and Laboratory Data*; CORROSION/97; NACE International: Houston, TX, USA, 1997.

10. Alawadhi, K.; Robinson, M.J. Preferential weld corrosion of X65 pipeline steel in flowing brines containing carbon dioxide. *Corros. Eng. Sci. Technol.* **2011**, *46*, 318–329.

11. Chan, E.W. Magnetite and Its Galvanic Effect on the Corrosion of Carbon Steel under Carbon Dioxide Environments. Ph.D. Thesis, Curtin University, Bentley, Australia, December 2011.

12. De Marco, R.; Jiang, Z.T.; Pejcic, B.; Poinen, E. An *in situ* synchrotron radiation grazing incidence X-ray diffraction study of carbon dioxide corrosion. *J. Electrochem. Soc.* **2005**, *152*, 389–392.

13. Crolet, J.L.; Wilhelmsen, S.; Olsen, S. *Observations of Multiple Steady States in the CO_2 Corrosion of Carbon Steel*; CORROSION/95; NACE International: Houston, TX, USA, 1995.

14. Tanupabrungsun, T.; Young, D.; Brown, B.; Nesic, S. *Construction and Verification of Pourbaix Diagrams for CO_2 Corrosion of Mild Steel Valid up to 250 °C*; CORROSION/2012; NACE International: Salt Lake City, UT, USA, 2012.

15. Al-Hassan, S.; Mishra, B.; Olson, D.L.; Salama, M.M. Effect of microstructure on corrosion of steels in aqueous solutions containing carbon dioxide. *Corrosion* **1998**, *54*, 480–491.

16. Li, W.; Brown, B.; Young, D.; Nesic, S. Investigation of pseudo-passivation of mild steel in CO_2 corrosion. *Corrosion* **2014**, *70*, 294–302.

17. Chechirlian, S.; Eichner, P.; Keddam, M.; Takenouti, H.; Mazille, H. A specific aspect of impedance measurements in low conductivity media. Artefacts and their interpretations. *Electrochim. Acta* **1990**, *35*, 1125–1131.

18. Nazari, M.H.; Allahkaram, S.R.; Kermani, M.B. The effects of temperature and pH on the characteristics of corrosion product in CO_2 corrosion of grade X70 steel. *Mater. Des.* **2010**, *31*, 3559–3563.

19. Erdos, V.E.; Altorfer, H. Ein dem Malachit ahnliches basisches Eisenkarbonat als Korrosionsprodukt von Stahl. *Werkst. Korros.* **1976**, *27*, 304–312. (In German)

20. Ruhl, A.S.; Kotré, C.; Gernert, U.; Jekel, M. Identification, quantification and localization of secondary minerals in mixed FeO fixed bed reactors. *Chem. Eng. J.* **2011**, *172*, 811–816.

21. Saheb, M.; Neff, D.; Dillmann, P.; Matthiesen, H.; Foy, E. Long term corrosion behavior of low carbon steel in anoxic envioment: Charactisterisation of archaeological artefacts. *Nucl. Mater. J.* **2008**, *379*, 118–123.

22. Ko, M.; Laucock, N.J.; Infgham, B.; William, D.E. *In situ* synchotron X-Ray diffraction studies of CO_2 corrosion of low carbon steel with scale inhibitors ATMPA and PEI at 80 °C. *Corrosion* **2012**, *68*, 1085–1093.

23. Fajardo, V.; Brown, B.; Young, D.; Nesic, S. *Study of the Solubility of Iron Carbonate in the Presence of Acetic Acid Using EQCM*; CORROSION/2013; NACE International: Houston, TX, USA, 2013; Paper No. 2452.

24. Braun, R.D. Solubility of iron(II) carbonate between 30 and 80 degrees. *Talanta* **1991**, *38*, 205–211.

25. Saheb, M. Iron corrosion in an anoxic soil: Comparison between thermodynamic modelling and ferrous archaeological artefacts characterised along with the local *in situ* geochemical conditions. *Appl. Geochem.* **2010**, *25*, 1937–1948.

26. Reffass, M.; Sabot, R.; Savall, C.; Jeannin, M.; Creus, J.; Refait, P. Localised corrosion of carbon steel in $NaHCO_3/NaCl$ electrolytes: Role of Fe(II)-containing compounds. *Corros. Sci.* **2006**, *48*, 709–726.

27. Rémazeilles, C.; Refait, P. Fe(II) hydroxycarbonate $Fe_2(OH)_2CO_3$ (chukanovite) as iron corrosion product: Synthesis and study by Fourier transform infrared spectroscopy. *Polyhedron* **2009**, *28*, 749–756.

28. Han, J.; Brown, B.; Young, D.; Nesic, S. Mesh-capped probe design for direct pH measurements at an actively corroding metal surface. *J. Appl. Electrochem.* **2010**, *40*, 683–690.

29. Refait, P.; Bourdoiseau, J.A.; Jeannin, M.; Nguyen, D.D.; Romaine, A.; Sabot, R. Electrochemical formation of carbonated corrosion products on carbon steel in deaerated solutions. *Electrochim. Acta* **2012**, *79*, 210–217.

30. Denpo, K.; Ogawa, H. Effects of nickel and chromium on corrosion rate of linepipe steel. *Corros. Sci.* **1993**, *35*, 285–288.

9

Raman Microscopic Analysis of Internal Stress in Boron-Doped Diamond

Kevin E. Bennet [1,2], **Kendall H. Lee** [2], **Jonathan R. Tomshine** [2], **Emma M. Sundin** [3], **James N. Kruchowski** [1], **William G. Durrer** [3], **Bianca M. Manciu** [3], **Abbas Kouzani** [4] and **Felicia S. Manciu** [3,*]

[1] Division of Engineering, Mayo Clinic, Rochester, MN 55905, USA;
     E-Mails: Bennet.Kevin@mayo.edu (K.E.B.); Kruchowski.James@mayo.edu (J.N.K.)
[2] Department of Neurologic Surgery, Mayo Clinic, Rochester, MN 55905, USA;
     E-Mails: Lee.Kendall@mayo.edu (K.H.L.); Tomshine.Jonathan@mayo.edu (J.R.T.)
[3] Department of Physics, University of Texas at El Paso, El Paso, TX 79968, USA;
     E-Mails: emsundin@miners.utep.edu (E.M.S.); wdurrer@utep.edu (W.G.D.);
     bianca.manciu@gmail.com (B.M.M.)
[4] School of Engineering, Deakin University, Waurn Ponds, Victoria 3216, Australia;
     E-Mail: abbas.kouzani@deakin.edu.au

* Author to whom correspondence should be addressed; E-Mail: fsmanciu@utep.edu;

Academic Editor: Jung Ho Je

Abstract: Analysis of the induced stress on undoped and boron-doped diamond (BDD) thin films by confocal Raman microscopy is performed in this study to investigate its correlation with sample chemical composition and the substrate used during fabrication. Knowledge of this nature is very important to the issue of long-term stability of BDD coated neurosurgical electrodes that will be used in fast-scan cyclic voltammetry, as potential occurrence of film delaminations and dislocations during their surgical implantation can have unwanted consequences for the reliability of BDD-based biosensing electrodes. To achieve a more uniform deposition of the films on cylindrically-shaped tungsten rods, substrate rotation was employed in a custom-built chemical vapor deposition reactor. In addition to visibly preferential boron incorporation into the diamond lattice and columnar growth, the results also reveal a direct correlation between regions of pure diamond and enhanced stress. Definite stress release throughout entire film thicknesses was found in the current Raman mapping images for higher amounts of boron addition. There is also a possible contribution to the high values of compressive stress from sp^2 type carbon impurities, besides that of the expected lattice mismatch between film and substrate.

Keywords: confocal Raman mapping; induced stress; boron-doped diamond

---

1. Introduction

Boron-doped diamond (BDD) and other diamond-based materials have been subjects of interest in the past years, especially for their potential in electrochemistry and voltammetry applications [1–9]. In addition to overall material quality, of concern when considering the use of electrodes fabricated from BDD coated tungsten rods in neurosurgical implantations into the brain is the unwanted occurrence of film delamination and dislocation. Deep brain stimulation (DBS), which is the planned application of the current work, is a currently Food and Drug Administration (FDA) approved technique employed for treatment of patients with movement disorders such as Parkinson's disease. By coupling DBS with fast-scan cyclic voltammetry (FSCV) technique, for the purpose of biochemical sensing and feedback control of stimulation parameters, the release of neurotransmitters in the brain can be monitored.

Besides the inherent inhomogeneity of any sample, a significant factor affecting the relevant properties of such thin films and, consequently, long-term electrode performance, is the induced stress during or after deposition. Part of this stress could be reduced by choosing an appropriate substrate for diamond deposition. This would entail matching substrate parameters such as surface energy, lattice, structure, and coefficient of thermal expansion with those of diamond. While there is no ideal candidate as a substrate for diamond, tungsten is the element that most closely approaches diamond in terms of surface energy (e.g., at melting temperature the surface energy of tungsten is approximately 3111 ergs/cm^2 and that of diamond is about 3300 ergs/cm^2) and lattice constant (e.g., tungsten's lattice constant is 3.16 Å and diamond's is 3.57 Å, with about 10% lattice mismatch) [10–13]. On the other hand, tungsten has a significantly different structure (*i.e.*, body-centered cubic for tungsten and face-centered cubic for diamond) and a greater coefficient of thermal expansion (*i.e.*, 4.3×10^{-6}/K for tungsten and 1.18×10^{-6}/K for diamond) [10,12]. In this context, the reported increase of about 10%–15% in thermal expansion for doped synthetic diamond over that of undoped diamond ([11–13] and references therein) is expected to positively impact the characteristics of heavily doped BDD films, not only by increasing the desired electrical conductivity, but also by reducing the induced stress.

Other contributing factors that affect the residual stress induced in the grown material and its other properties are the shape of the substrate, the growth temperature, and the feed gas pressures employed during fabrication [7–21]. Previous studies on synthetic diamond showed that the residual stress has two main components: The intrinsic stress and the extrinsic stress [14–18]. Defects within the film (e.g., nondiamond carbon phases) and lattice mismatch between the deposited material and the substrate are the main reasons for intrinsic stress. Extrinsic stress is likely to happen because of the difference between the thermal expansion coefficients of the diamond and substrate used. The latter type of stress, also known as thermal stress, is mainly compressive. Both tensile and compressive stresses were observed in polycrystalline diamond and BDD films grown by chemical vapor deposition (CVD) [19,20]. Partial release of the stress has been demonstrated for undoped, highly crystalline diamond thin films under proper tuning of growth temperature and amounts of feed gases [18]. More complex, interrelated aspects should be considered for stress release in the case of BDD polycrystalline films. Besides boron incorporation, which results in expansion of the diamond lattice [19], grain boundaries, which are the locations of impurity accumulations, could add to stress relaxation due to the elastic strain induced in the grains [16]. Based on confocal Raman mapping analysis, a comparative assessment of fabrication parameters (*i.e.*, amount of feed gases, shape, type, and temperature of the substrate, which are described later in Section 3.1.) and their contributions to the observed induced stress for doped and undoped diamond films grown on tungsten rods is presented and discussed in this work.

Although other techniques have been employed in analyses of residual stress such as X-ray diffraction [20,21] and the curvature technique [16], Raman spectroscopy still remains the most used method for this type of characterization. One reason is that the Raman technique provides knowledge of the composition of the material at a molecular level, as well as an evaluation of internal stress through the diamond Raman peak shifts. While the majority of previous Raman studies of BDD films have been based on spectral analysis, the confocal Raman mapping used in the current investigations allows for qualitative and relative quantitative evaluations of the material under study through direct visualization of the local distribution of film constituents (e.g., pure diamond, boron incorporation and accumulation, and non-diamond, carbon sp^2 impurities). Additionally, the Raman stress mapping images provide an assessment of stress location (*i.e.*, close to the substrate or to the surface) and its progression throughout the film thickness. This information is essential for improving the properties of BDD coated electrodes, and, consequently, their reliability for DBS studies. The results of this work represent an important component of understanding the interdependence of physical and chemical properties of BDD films necessary for development of stable and high quality electrodes.

2. Results and Discussion

Investigations by confocal Raman mapping of material morphology and internal stress in the currently grown samples for undoped, lightly boron-doped, and heavily boron-doped diamond thin films, are presented in Figures 1a–d, 2a–d, and 3a–d, respectively. The morphological measurements were performed by selecting characteristic Raman vibrational lines at 1332 ± 2 cm^{-1} for diamond, around 1200 cm^{-1} and 500 cm^{-1} for boron incorporation and accumulation in the diamond lattice (*i.e.*, paired boron atoms), respectively, and the band centered at 1500 cm^{-1} corresponding to sp^2 carbon impurities [1,4,8,22–24]. For a relative quantitative identification of these constituents, three pseudo-colors (*i.e.*, red for diamond, blue for BDD, and green for carbon impurities) were used. Comparison of the results of Figure 1a with those of Figures 2a and 3a reveals a decrease of sp^2 carbon (green) with addition of boron. This observation is confirmed in Figures 1d, 2d, and 3d, where the integrated Raman spectra of these images are presented. A decrease in the intensity of the feature around 1500 cm^{-1} can be seen in Figures 2d and 3d as compared with the one in Figure 1d. Thus, boron addition not only contributes to the conductivity of the films [9], but also promotes higher quality of the deposited material, by reducing the amount of unwanted carbon impurities and promoting a more rapid crystallization.

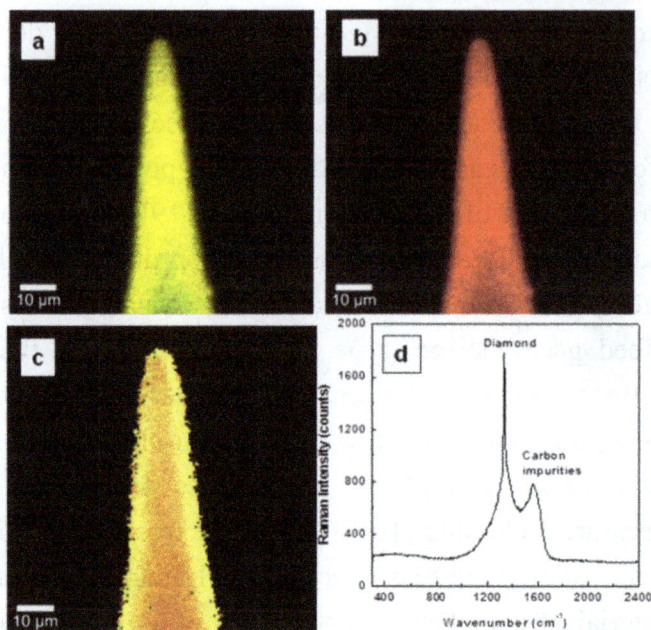

Figure 1. Surface confocal Raman images of undoped diamond film mapping the spatial distribution of: (**a**) diamond and sp^2 carbon, (**b**) diamond only, and (**c**) induced stress. Red pseudo-color is used for diamond and green for sp^2 carbon impurities. While in image (**a**) yellow originates from a combination of red and green, in the stress image (**c**), bright yellow and brown correspond to higher or lower induced stress, respectively. (**d**) integrated Raman spectrum of image (**a**).

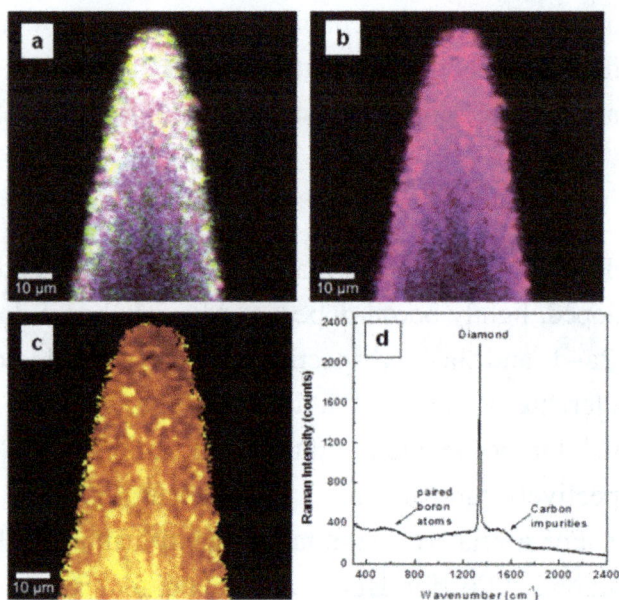

Figure 2. Surface confocal Raman images of lightly doped diamond film mapping the spatial distribution of: (**a**) diamond (red), boron incorporation (blue), and sp^2 carbon impurities (green), (**b**) diamond (red) and boron incorporation (blue), and (**c**) induced stress. Yellow pseudo-color in the stress image corresponds to higher induced stress. White, magenta, and turquoise colors observed in images (**a**) and (**b**) are due to combinations of red with blue and green, of red with blue, and of blue with green, respectively. (**d**) integrated Raman spectrum of image (**a**).

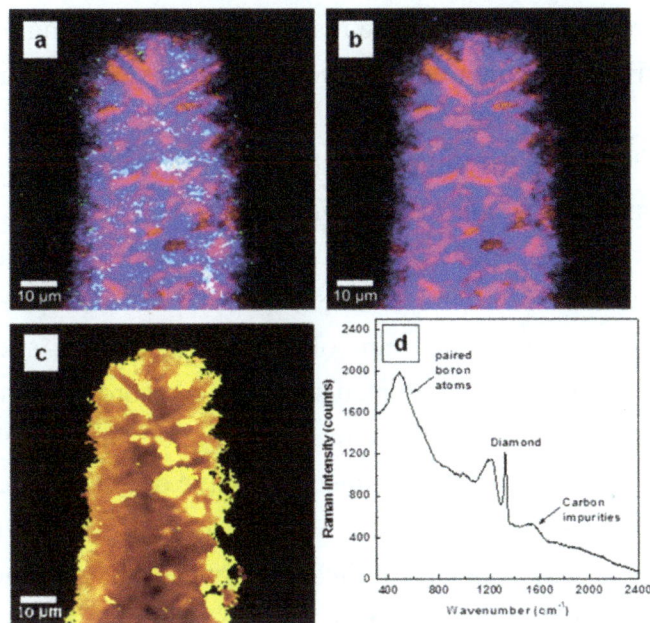

Figure 3. Surface confocal Raman images of heavily doped diamond film mapping the spatial distribution of: **(a)** diamond (red), boron incorporation (blue), and sp^2 carbon impurities (green), **(b)** diamond (red) and boron incorporation (blue), and **(c)** induced stress. Yellow pseudo-color in the stress image corresponds to higher induced stress. White, magenta, and turquoise colors observed in images **(a)** and **(b)** are due to combinations of red with blue and green, of red with blue, and of blue with green, respectively. **(d)** Integrated Raman spectrum of image **(a)**.

For easier examination of film uniformity as related to boron incorporation, the visualization of sp^2 carbon is excluded in Figures 1b, 2b, and 3b. Besides the expected color trend from red (*i.e.*, Figure 1b of undoped material) to blue for a higher boron doping level, in both Raman mapping images presented in Figures 2b and 3b the signature of pure diamond crystallites (red color) is still visible; a phenomenon originating from the known preferential incorporation of boron into the diamond lattice [5–8,22–24].

There is also slightly more uniformity in the incorporation of boron for a lower amount of doping (*i.e.*, dominant magenta color, which is a combination of red and blue). This remark can be explained by the tendency of boron to first substitute for carbon, with the outcome of a magenta pseudo-color, then to incorporate interstitially or even to aggregate for higher doping amounts, resulting in a blue pseudo-color. In this respect, it is worth pointing out that the boron incorporation mechanism for concentrations that impart metallic-like or even superconductive behavior to naturally non-conductive diamond (*i.e.*, more than 10^{20} atoms/cm^3) is still not completely established [11]. Some of the reasons for these uncertainties are the difficulty of accurately determining boron concentration and the low magnitudes of the diamond lattice parameters and thermal expansions. While the tetrahedral arrangement of carbon in the diamond structure is directly related to its low lattice expansion parameter, it is this arrangement that makes diamond such a remarkable material from the points of view of both mechanical and thermal conductive properties, and that makes it a good prospect for future electronic applications. The aggregation of interstitial boron atoms [8,22–24] and the formation of isolated, interacting boron pairs, which is observable in the current Raman spectra by the presence of the band around 500 cm^{-1}, were reported to be the main causes contributing to the dramatic increase of the thermal expansion coefficient of diamond (by several, or dozens, of times larger than that of pure diamond) in heavily doped BDD films [11].

The evaluation of the induced stress in the samples is shown in Figures 1c, 2c, and 3c. Surface confocal Raman mapping was performed to acquire these images, which account for the shift of the characteristic optical-phonon vibration of diamond at 1332 cm^{-1}. More of the bright yellow pseudo-color corresponds to more induced stress. The *Advanced Fitting Tool* of the *WITec Project Plus* software was employed to fit each spectrum of the entire Raman spectral data set with a Lorentzian function (*i.e.*, at every image pixel a Raman spectrum is recorded). It has been shown that this vibrational line shifts around 3 cm^{-1}/1 GPa when the material is under stress [19,25–27]. Also, a peak shift to higher or lower frequency was associated with compressive or tensile stress, respectively [25].

A compressive stress is found, with 4.5 ± 2 to 12 ± 2 cm^{-1} (*i.e.*, 1.5 to 4 GPa) positive shifts and with higher values for the undoped samples than for the doped ones. This statement is supported by the overall lighter, towards-yellow pseudo-color seen in the image presented in Figure 1c and in parts of Figures 2c and 3c. Different causes can contribute to this behavior. For example, a closer, comparative look at these images regarding only the presence of pure diamond crystallites (red pseudo-color in Figures 2b and 3b) demonstrates that more stress (yellow pseudo-color) directly associates with them. This effect is obvious for the heavily doped sample (see Figure 3b,c), as boron addition promotes faster growing rates and formation of larger crystallites. Complementarily, if the decrease in stress is sought, good visual agreement is evident between regions of brown false color and the ones corresponding to higher amounts of boron doping (*i.e.*, blue regions). Thus, as discussed above, the main argument for the stress release could be based on the strong increase in the thermal expansion of diamond with boron addition. Due to carbon impurities, there might be some additional induced surface stress, which is most visible in Figure 1c and in some parts of Figure 3c. The contribution of such impurities to the surface stress in the case of heavily doped samples can be explained by the known phenomenon of their accumulation at the boundaries of BDD crystallites [8,20–23]. The presence of only nanocrystallites in the undoped sample is thus also likely to play a role in the formation of the relatively high amount of sp^2 carbon observed in this case. The broadness seen in the integrated spectrum presented in Figure 1d at the base of the Raman vibrational line characteristic of diamond supports this affirmation.

Another important source of the induced stress in the material is the lattice mismatch between tungsten and diamond, which should occur more near the interface between the two materials. To asses this interface effect, we performed side-wall confocal Raman mapping, and the results are presented in Figures 4a–f and 5a–f. Even though no evidence of tungsten carbide formation was found, Figures 4a,d and 5a,d show a strong accumulation of carbon impurities close to the surface of the tungsten rods. Again, a lower amount of trimethylborane (TMB) feed gas used during deposition results in a more uniform incorporation of boron throughout the entire film thickness (a relatively uniform magenta color is observed in Figure 4d). For a considerably higher amount of TMB, preferential boron incorporation, with an obvious radial, columnar growth, is occurring (see Figure 5b). Comparison between the images for chemical composition of the films and the induced stress (see Figures 4a,c,d,f and 5a,c) reveals an obvious correlation between the stress and the excessive presence of carbon impurities at the interface with tungsten. Shifts of the diamond vibrational line with values as high as 20 ± 1 cm^{-1}

(6.7 GPa) were obtained for these bright yellow regions. We do not exclude the possibility that the lattice mismatch has some contribution to such high values of compressive stress.

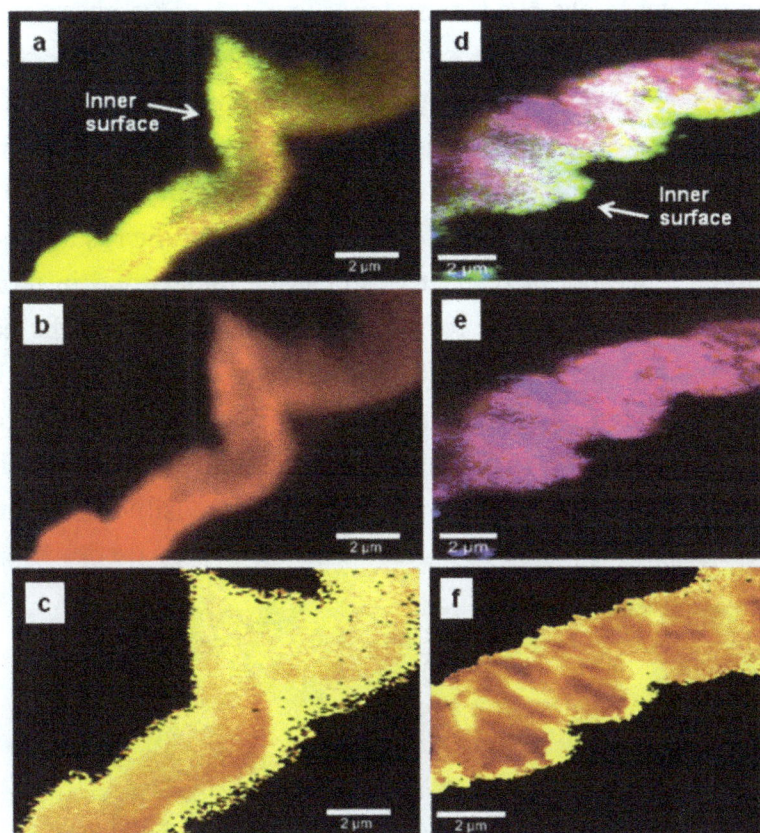

Figure 4. Side-wall confocal Raman images of undoped diamond thin film mapping the spatial distribution of (**a**) diamond (red) and sp^2 carbon (green), (**b**) diamond only, and (**c**) induced stress. Side-wall confocal Raman images of lightly boron-doped thin film showing: (**d**) all the material constituents, (**e**) diamond and boron incorporation, and (**f**) induced stress. Red, blue, and green pseudo-colors are used in image (**d**) for diamond, boron, and carbon impurities, respectively. While in image (**a**) yellow originates from superposition of red and green, in the stress images (**c**) and (**f**), bright yellow and brown correspond to higher or lower induced stress, respectively.

There is also an indication of a radial, channeled release of the stress towards the film surface, where much smaller stress values were found (between 1.5 and 4 GPa). Thus, the relaxation of the grain boundaries at the film surface could also have an influence. As the growth favors the directions with lower strain energy density, a clear analogy between the regions where the boron presence is more visible (blue pseudo-color) and the ones with less stress (dark brown pseudo-color) is observed, especially in Figure 5b,c. No evidence of defects or dislocations that would degrade material quality or create other types of internal stresses is detected.

In order to further investigate our previous assumption that the origin of the stress release is mainly due to enhancement of the thermal expansion parameter of diamond with boron incorporation, we performed side-wall confocal Raman analysis of a two-layer BDD thin film with each layer grown under the same conditions but with different amounts of TMB. These images, which are presented in Figure 5d–f, besides proving the capability of confocal Raman mapping in detecting differences in doping, also give new insights into this matter and its relation to the growth process. For instance, the location of the strongest presence of carbon impurities at the interface with tungsten (see Figure 5d) is different from that of the most intense observed stress (see Figure 5f). This observation implies that the mismatch between the diamond and the substrate is still the dominant factor in the induced stress. More importantly, the noticeable color change corresponding to the induced stress in the two thin films, combined with almost no evidence of carbon impurity accumulation at the interface between them, validates the importance of boron doping for stress release. The nearly uniform distribution of the stress in the entire volume of the undoped diamond films (*i.e.*, at their surfaces and across their thicknesses, as seen in Figures 1c, 4c and 5f, respectively) can be understood in terms of an effective increase in the number of atoms at the surface for the reduced dimensions of the nanocrystallites and absence of a preferential growth direction. The high number of atoms at the surface is likely to affect the grain boundary relaxation effect, while the temperature at the beginning of the growth process will influence the presence of sp^2 carbon, as it might not be high enough to provide sufficient energy for diamond crystallization. Indeed, stabilization of crystal formation is visible in Figure 5d,e for thicker films in comparison with the scenario revealed in Figure 4a,b.

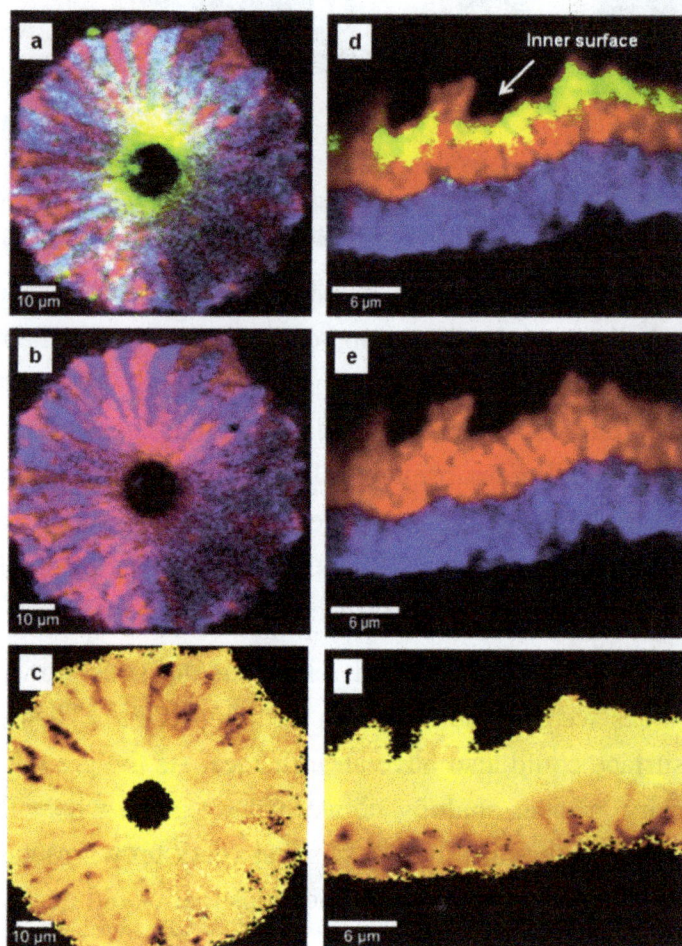

Figure 5. Side-wall confocal Raman mapping images of: (a)–(c) a heavily boron-doped film and its corresponding stress mapping analysis, and (d)–(f) two consecutive layers grown without and with boron addition, together with the associated stress map. Red, blue, and green pseudo-colors are used for diamond, boron, and carbon impurities, respectively. Yellow pseudo-color in the stress images corresponds to higher induced stress.

3. Experimental Section

3.1. Fabrication of Boron-Doped Diamond Thin Films

The BDD films were grown in a hot filament chemical vapor deposition (CVD) reactor using a mixture of CH_4/H_2 gases (nominally 99% H_2 and 1% CH_4) at a total chamber pressure of 20 Torr. The current samples were grown on tungsten rods that were initially electrochemically etched by immersion in 1 M NaOH, then abraded by sonication for 30 min in diamond powder (100 nm, Engis, 105 West Hintz Road, Wheeling, IL, USA)/isopropyl alcohol grit slurry, and finally cleaned by rinsing in deionized water.

For more uniform deposition of the BDD films on the cylindrical shape of the electrodes, the substrate rotation mode of this custom-built reactor (at the Mayo Clinic) was employed. The tungsten rods were mounted parallel to the central axis of the filament coil. A Proportional Integral Derivative (PID) software-based control loop allowing the power to be ramped up to 450 W was used for achieving temperatures of 2300 °C for the filament and about 800 °C for the substrate. The stabilities of these two different temperatures were monitored during film deposition with a Spectrodyne DFP 2000 optical pyrometer and an Omega Engineering type K thermocouple, respectively. Two different amounts of trimethylborane (TMB) flow gas (1000 ppm in hydrogen, Voltaix Products, Branchburg, NJ, USA), namely, chamber concentrations of 10 ppm (2 sccm TMB/hydrogen, 2 sccm methane, and 196 sccm hydrogen) and 100 ppm (20 sccm TMB/hydrogen, 2 sccm methane, and 178 sccm hydrogen), were released into the chamber for obtaining lightly doped and heavily doped thin films, respectively. Undoped samples were also prepared without addition of TMB, under identical temperature growth conditions and using 2 sccm methane and 198 sccm hydrogen.

3.2. Raman System

The confocal Raman measurements were acquired at ambient conditions, in a backscattering geometry, with an *alpha 300R WITec* system (WITec GmbH, Ulm, Germany) using the 532 nm excitation of a frequency-doubled Nd:YAG laser. While 100X/numerical aperture (NA) of 0.90 and 50X/NA of 0.75 objective lenses were used for side-wall measurements of BDD films, to reduce the optical effects of the inherent curvature of the tungsten rods, a 20X/NA of 0.40 objective lens was used for acquiring such data on the surfaces of the films. An array of 150 × 150 Raman spectra were recorded for all Raman images with an integration time of 50 ms/spectrum. All the surface Raman mapping images were acquired with 85 μm × 85 μm scan sizes. Appropriate scan size dimensions, dependent on film thickness, were considered for side-wall maps, while maintaining a similar scale-bar in these images. Also, for these measurements, the samples were cross-sectionally cleaved along their diameters.

Although the spot size of the laser, and implicitly the optical resolution, is diffraction limited, the image quality was improved by using an oversampling procedure. This procedure, applicable for more than 20,000 pixels per image, consists of using increments smaller than the lateral spectral resolution, which varies between 255 and 575 for the currently used excitation of 532 nm and different objective lenses. On the other hand, in confocal Raman microscopy, only a relative determination of concentrations is possible for multi-constituent solid sample materials, as the pseudo-color of a pixel is graded relative to the intensity of the constituent detected at that location (in every pixel a Raman spectrum is recorded).

Besides the *WITec Control* software, which controls the piezoelectric stage for sample scanning and is normally employed for data acquisition, the *Advanced Fitting Tool* of the *WITec Project Plus* software was used for stress data analysis. In this context, small shifts can be detected, as fitting the peaks allows the maxima to be determined with a precision better than 1 cm^{-1}, although the sampling resolution can be lower.

4. Conclusions

In this study we present a detailed analysis by confocal Raman microscopy of the induced stress on a series of undoped and boron-doped diamond thin films grown by chemical vapor deposition on tungsten rods. For consistent film deposition on the cylindrical shapes of these substrates, the custom-built reactor's capability of tungsten rod rotation was employed.

The current results demonstrate a relatively uniform presence of the stress throughout the entire pure diamond film and its radial release that becomes dependent on the sample chemical composition with boron addition. An obviously stronger stress release for a higher amount of boron doping is revealed by some of the surface Raman mapping images (see, and compare, Figures 1c, 2c, and 3c), especially by the side-wall image presented in Figure 5f, where two consecutively grown layers, without and with boron doping, are mapped. Besides an anticipated enhanced compressive stress observed close to the interface between the film and the tungsten substrate, which is due mainly to the lattice mismatch between the two materials, there is also a potential contribution from sp^2 carbon impurities to the high compressive stress values obtained. The high correlation of the presence of pure diamond with the strongest intensity of induced stress, together with the observed stress release with doping, make us conclude that boron addition not only contributes to the electrical conductivity of the material necessary for biosensing applications, but also to its improvement towards desired overall quality. Unwanted occurrence of film delaminations and dislocations during neurosurgical implantation of BDD-based electrodes can negatively affect their reliability in biosensing.

Acknowledgments

This work was supported by the NIH U01 NS090455-01 award, the NIH R01 NS075013 award, The Grainger Foundation, and by a research agreement between the University of Texas at El Paso and the Mayo Clinic.

Author Contributions

Kevin E. Bennet contributed the diamond electrode concept and design of the deposition system. Kevin E. Bennet and Jonathan R. Tomshine developed sample preparation with James N. Kruchowski and Abbas Kouzani. Jonathan R. Tomshine, Kevin E. Bennet and Felicia S. Manciu developed the deposition parameters and active material production. Felicia S. Manciu, William G. Durrer, Emma M. Sundin, and Bianca M. Manciu contributed Raman measurements and data analysis. Kendall H. Lee contributed biological application imperative and data evaluation. All authors contributed significant effort to manuscript preparation.

Conflicts of Interest

The authors declare no conflict of interest.

References

1. Kraft, A. Doped diamond: A compact review on a new, versatile electrode material. *Int. Electrochem. Sci.* **2007**, *2*, 355–385.
2. Shang, F.; Zhou, L.; Mahmoud, K.A.; Hrapovic, S.; Liu, Y.; Moynihan, H.A.; Glennon, J.D.; Luong, J.H.T. Selective nanomolar detection of dopamine using a boron-doped diamond electrode modified with an electropolymerized sulfobutylether-β-cyclodextrin-doped poly(*N*-acetyltyramine) and polypyrrole composite film. *Anal. Chem.* **2009**, *81*, 4089–4098.
3. Nebel, C.E.; Shin, D.; Rezek, B.; Tokuda, N.; Uetsuka, H.; Watanabe, H. Diamond and biology. *J. R. Soc. Interface* **2007**, *4*, 439–461.
4. Soh, K.L.; Kang, W.P.; Davidson, J.L.; Wong, Y.M.; Wisitsoraat, A.; Swain, G.; Cliffel, D.E. CVD diamond anisoptropic film as electrode for electrochemical sensing. *Sens. Actuators B* **2003**, *91*, 39–45.
5. Ramamurti, R.; Becker, M.; Schuelke, T.; Grotjohn, T.; Reinhard, D.; Asmussen, J. Synthesis of boron-doped homoepitaxial single crystal diamond by microwave plasma chemical vapor deposition. *Diam. Relat. Mater.* **2008**, *17*, 1320–1323.
6. Lagrange, J.P.; Deneuville, A.; Gheeraert, E. Activation energy in low compensated homoepitaxial boron-doped diamond films. *Diam. Relat. Mater.* **1998**, *7*, 1390–1393.
7. Taylor, A.; Fekete, L.; Hubík, P.; Jäger, A.; Janíček, P.; Mortet, V.; Mistrík, J.; Vacík, J. Large area deposition of boron doped nano-crystalline diamond films at low temperatures using microwave plasma enhanced chemical vapour deposition with linear antenna delivery. *Diam. Relat. Mater.* **2014**, *47*, 27–34.
8. Bennet, K.E.; Lee, K.H.; Kruchowski, J.N.; Chang, S.Y.; Marsh, M.P.; van Orsow, A.A.; Paez, A.; Manciu, F.S. Development of conductive boron-doped diamond electrode: A microscopic, spectroscopic, and voltammetric study. *Materials* **2013**, *6*, 5726–5741.
9. Manciu, F.S.; Manciu, M.; Durrer, W.G.; Salazar, J.G.; Lee, K.H.; Bennet, K.E. A Drude model analysis of conductivity and free carriers in boron-doped diamond films and investigations of their internal stress and strain. *J. Mater. Sci.* **2014**, *49*, 5782–5789.
10. Kumar, R.; Grenga, H.E. Surface energy anisotropy of tungsten. *Surf. Sci.* **1976**, *59*, 612–618.
11. Brazhkin, V.V.; Ekimov, E.A.; Lyapin, A.G.; Popova, S.V.; Rakhmanina, A.V.; Stishov, S.M.; Lebedev, V.M.; Katayama, Y.; Kato, K. Lattice parameters and thermal expansion of superconducting boron-doped diamonds. *Phys. Rev. B* **2006**, *74*, 140502(R).
12. Han, S.; Pan, L.S.; Kania, D.R. *Diamond: Electronic Properties and Applications*; Kluwer Academic Publishers: Norwell, MA, USA, 1995; pp. 241–278.
13. Saotome, T.; Ohashi, K.; Sato, T.; Maeta, H.; Haruna, K.; Ono, F. Thermal expansion of a boron-doped diamond single crystal at low temperatures. *J. Phys. Condens. Matter* **1998**, *10*, 1267–1272.

14. Baglio, J.A.; Farnsworth, B.C.; Hankin, S.; Hamill, G.; O'Neil, D. Studies of stress related issues in microwave CVD diamond on <100> silicon substrates. *Thin Solid Films* **1992**, *212*, 180–185.

15. Knight, D.S.; White, W.B. Characterization of diamond films by Raman spectroscopy. *J. Mater. Res.* **1989**, *4*, 385–393.

16. Windischmann, H.; Epps, G.F.; Cong, Y.; Collins, R.W. Intrinsic stress in diamond films prepared by microwave plasma CVD. *J. Appl. Phys.* **1991**, *69*, 2231–2237.

17. Fan, Q.H.; Gracio, J.; Pereira, E. Residual stress in chemical vapor deposited diamond films. *Diam. Relat. Mater.* **2000**, *9*, 1739–1743.

18. Chiou, Y.H.; Hwang, C.T.; Han, M.Y.; Jou, J.H.; Chang, Y.S.; Shih, H.C. Internal stress of chemical vapor deposition diamond film on silicon. *Thin Solid Films* **1994**, *253*, 119–124.

19. Wang, W.L.; Polo, M.C.; Sanchez, G.; Cifre, J.; Esteve, J. Internal stress and strain in heavily boron-doped diamond films grown by microwave plasma and hot filament chemical vapor deposition. *J. Appl. Phys.* **1996**, *80*, 1846–1850.

20. Rats, D.; Bimbault, L.; Vandenbulcke, L.; Herbin, R.; Badawi, K.F. Crystalline quality and residual stresses in diamond layers by Raman and X-ray diffraction analyses. *J. Appl. Phys.* **1995**, *78*, 4994–5001.

21. Hempel, M.; Härting, M. Characterization of CVD grown diamond and its residual stress state. *Diam. Relat. Mater.* **1999**, *8*, 1555–1559.

22. Massarani, B.; Bourgoin, J.C.; Chrenko, R.M. Hopping conduction in semiconducting diamond. *Phys. Rev. B* **1978**, *17*, 1758–1769.

23. May, P.W.; Ludlow, W.J.; Hannaway, M.; Heard, P.J.; Smith, J.A.; Rosser, K.N. Raman and conductivity studies of boron-doped microcrystalline diamond, facetted nanocrystalline diamond and cauliflower diamond films. *Diam. Relat. Mater.* **2008**, *17*, 105–107.

24. Bernard, M.; Baron, C.; Deneuville, A. About the origin of the low wave number structures of the Raman spectra of heavily boron doped diamond films. *Diam. Relat. Mater.* **2004**, *13*, 896–899.

25. Boppart, H.; van Straaten, J.; Silvera, I.F. Raman spectra of diamond at high pressures. *Phys. Rev. B* **1985**, *32*, 1423–1426.

26. Yoshikawa, M.; Katagiri, G.; Ishida, H.; Ishitani, A.; Ono, M.; Matsumara, K. Characterization of crystalline quality of diamond films by Raman spectroscopy. *Appl. Phys. Lett.* **1989**, *55*, 2608–2610.

27. Hanfland, M.; Syassen, K.; Fahy, S.; Louie, S.G.; Cohen, M.L. Pressure dependence of the first-order Raman mode in diamond. *Phys. Rev. B* **1985**, *31*, 6896(R).

10

Effect of Experimental Parameters on Morphological, Mechanical and Hydrophobic Properties of Electrospun Polystyrene Fibers

Siqi Huan [1], Guoxiang Liu [1], Guangping Han [1,*], Wanli Cheng [1,*], Zongying Fu [1], Qinglin Wu [2] and Qingwen Wang [1]

1   College of Material Science and Engineering, Northeast Forestry University, Harbin 150040, China; E-Mails: huangsiqi888@hotmail.com (S.H.); gxliunefu@hotmail.com (G.L.); wonwinfu@hotmail.com (Z.F.); qwwang@nefu.edu.cn (Q.W.)

2   School of Renewable Natural Resources, Louisiana State University Agricultural Center, Baton Rouge, LA 70803, USA; E-Mail: qwu@agcenter.lsu.edu

*   Authors to whom correspondence should be addressed;
    E-Mails: guangping.han@nefu.edu.cn (G.H.); nefucwl@nefu.edu.cn (W.C.);

Academic Editor: Jung Ho Je

Abstract: Polystyrene (PS) dissolved in a mixture of *N, N*-dimethylformamide (DMF) and/or tetrahydrofuran (THF) was electrospun to prepare fibers with sub-micron diameters. The effects of electrospinning parameters, including solvent combinations, polymer concentrations, applied voltage on fiber morphology, as well as tensile and hydrophobic properties of the fiber mats were investigated. Scanning electron microscope (SEM) images of electrospun fibers (23% w/v PS solution with applied voltage of 15 kV) showed that a new type of fiber with double-strand morphology was formed when the mass ratio of DMF and THF was 50/50 and 25/75. The tensile strength of the PS fiber film was 1.5 MPa, indicating strong reinforcement from double-strand fibers. Bead-free fibers were obtained by electrospinning 40% (w/v) PS/DMF solution at an applied voltage of 15 kV. Notably, when the ratio of DMF and THF was 100/0, the maximum contact angle (CA) value of the electrospun PS films produced at 15 kV was 148°.

Keywords: electrospinning; polystyrene nanofiber; morphology; tensile strength; hydrophobicity

---

1. Introduction

One-dimensional (1D) nanostructures in the form of fibers, wires, rods, belts, tubes, and rings have attracted interest due to their novel properties and intriguing applications in many areas. Significant advances in the fabrication of 1D nanostructures with well-controlled morphology and chemical composition have been made using a number of manufacturing procedures [1], including drawing, template synthesis, phase separation, self-assembly, and electrospinning, which can be adopted to prepare nanofibers based on polymers, metals, ceramics, and glass. Moreover, the assembly of nanofibers into two-dimensional (2D) and three-dimensional (3D) nanostructures has also been applied to various practical applications [2]. Among these methods, electrospinning, as one of the most versatile techniques for fabricating ultrafine non-woven polymer fibers, has been paid considerable attention in recent decades due to its simplicity, high efficiency, and low cost [3–5]. The obtained mats composed of electrospun nanofibers have several advantageous characteristics such as small diameters, neat morphology, and an interconnected network structure [6]. These outstanding properties make the electrospun mats ideal candidates for different applications, e.g., filtration [7], wound dressing [8], and protective clothing [9]. In particular, polymer mats with superhydrophobicity manufactured by electrospinning have been widely investigated in the field of self-cleaning materials and high-performance coatings [10,11].

Polystyrene (PS) is a class of low cost, promising thermoplastic polymers. The inherent hydrophobicity of PS has also been extensively explored and utilized to create superhydrophobic materials with tailored performances [12]. Nevertheless, the applications of PS films produced with traditional methods are limited in many aspects due to their brittleness. Moreover, in order to improve the film-forming ability of PS, film-forming aids are commonly added during the film manufacturing process, which can lead to volatile organic compound (VOC) emission which pollutes the environment and is harmful to human health. Therefore, the desirability to improve the shortcomings of PS materials is becoming increasingly important and various approaches have been implemented through physical and chemical modifications [13–16]. However, such approaches are either limited to creating PS materials with specific components, or which require a complex fabricating process. In fact, no simple PS modified formulation is commercially available that can offer the potential to overcome all the drawbacks of PS materials. Therefore, the development of simple, environmental friendly methods to prepare high-performance PS materials is critical.

Electrospinning potentially has great utility as a novel manufacturing technique to enhance the properties of PS materials. Numerous research works have been published to study the fabrication and properties of electrospun PS mats [17–19]. Zhan et al. [20] manufactured electrospun PS mats with good tensile properties and different morphologies. Uyar et al. [21] considered the effect of solvent conductivity and produced reproducible uniform electrospun nanofibers for optimal electrospinning conditions. Their results showed that solutions with a higher conductivity yielded bead-free fibers from lower polymer concentrations [21]. In addition, another potential advantage for PS materials manufactured by electrospinning is to realize multi-functionalities by tuning the surface structure of nanofibrous mats, which opens a new avenue to create superhydrophobic PS materials [22].

During the process of electrospinning, the liquid drop elongates with the increasing electric field and distorts into a conical shape when the repulsive force is equal to the surface tension of the liquid.

The fluid extension first occurs uniformly as straight flow lines, and then undergoes a vigorous whipping motion caused by electrohydrodynamic instabilities. As the solvent in the jet solution evaporates, the polymer fibers are collected onto a grounded substrate to form a non-woven mat or network. Based on previous reports [23], beads, necklaces, as well as ribbon-like and branched jet can be formed during the process of electrospinning PS with different experimental parameters. These parameters include spinning solution properties (e.g., viscosity, conductivity, and surface tension), governing variables (e.g., hydrostatic pressure in the capillary tube, electric potential at the capillary tip, and the distance between the tip and the collecting screen), together with ambient parameters (e.g., temperature, humidity, and air velocity in the electrospinning chamber) [24]. However, the conditions for the electrospinning process of PS solutions may be affected mutually, leading to uncontrollable PS fiber morphologies. Therefore, it is necessary to explore the effect of electrospinning parameters on the morphologies of PS fibers, offering potential to controllably produce electrospun PS fibers. Furthermore, it is well known that the properties of electrospun fibers are largely dependent upon the morphologies and surface structures of fibers [25]. Therefore, a fundamental understanding of how to regulate the morphology of electrospun PS fibers is also a prerequisite for manufacturing electrospun PS nanofibrous mats with high performances and novel functionalities. Additionally, determining the link between electrospinning parameters and fiber morphologies will further allow for sophisticated design and precise control of polymeric mats to meet specific application needs.

In the present work, PS nanofibrous mats were manufactured by electrospinning PS solutions (Figure 1). *N,N*-dimethylformamide (DMF) and tetrahydrofuran (THF) were used as solvents for preparing PS solutions with different solvent combinations and PS concentrations. The physiochemical properties (e.g., viscosity, surface tension, and conductivity) of these solutions were characterized. The effects of solvents, solution concentrations, and applied voltage on the morphological appearance of the obtained PS beads and 1D fibers were investigated using scanning electron microscope (SEM). The contact angle and tensile property were characterized to explore further application of the electrospun PS nanofibers as superhydrophobic nanofibrous mats. Most importantly, a new type of controllable parallel double-strand PS fibers with high hydrophobicity was successfully produced using a mixed solvent of DMF and THF at the ratio of 50/50 and 25/75 at a solution concentration of 23% (w/v). To our knowledge, the fabrication and control of special morphological PS fibers, especially for the parallel double-strain structure in this paper, have not been fully investigated.

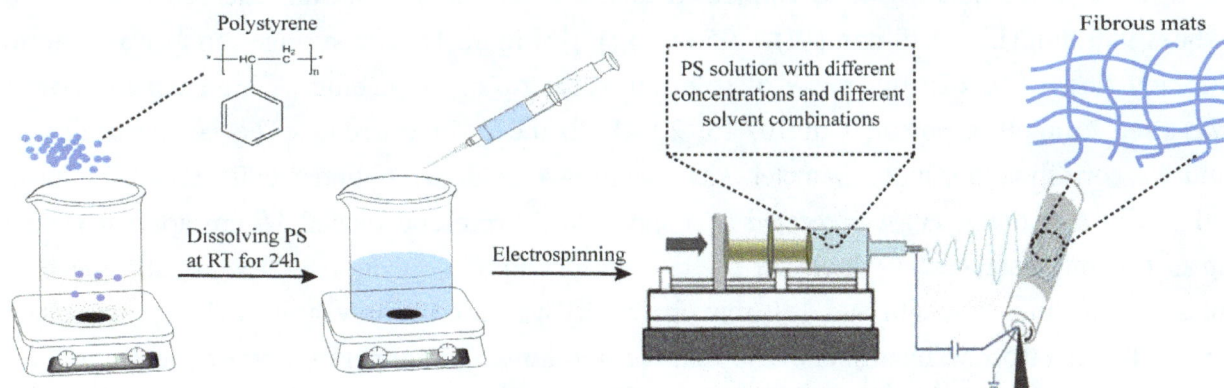

Figure 1. Schematic illustration of electrospinning process of polystyrene (PS) solutions and manufactured PS nanofibrous mats.

2. Experimental

2.1. Materials

PS particles (melt flow rate = 4.0 g/10 min) with average molecular weight of 260,000 were purchased from Aladdin (Shanghai, China). The solvents used were a mixture of DMF (AR grade, Kermel Co., Tianjin, China) and THF (AR grade, Kermel Co., Tianjin, China). These materials were used as received without further purification.

2.2. Preparation of As-Spun Solutions

Different types of solvents for electrospinning were prepared by using a mixture of DMF and THF in the weight ratios of 100/0, 75/25, 50/50, 25/75 and 0/100 of 23% (w/v). Polymer solutions in DMF with 10, 23, 27, 32, 40% (w/v) concentrations were made simultaneously. PS particles were dissolved in the given solvents at room temperature (25 ± 2 °C) with 12 h stirring to obtain homogeneous solutions.

2.3. Characterization of Electrospinning Solutions

Viscosity, surface tension and conductivity of the PS solutions were characterized at room temperature by using a digital rotational viscometer (SNB-1, Heng Ping Co., Shanghai, China), surface tension meter (JK 98B, Shanghai, China), and conductivity meter (DDSJ-318, Lei Ci Co., Shanghai, China), respectively.

Viscoelastic properties of PS solutions were carried out using an AR2000ex (TA Instruments) controlled strain rheometer equipped with a 40 mm Teflon plate-and-plate geometry. The gap was fixed at 1 mm. Sample edges were covered with paraffin oil to prevent evaporation during measurements. The mechanical spectrum, storage modulus (G') and loss modulus (G"), were obtained at room temperature in the range of 0.1–100 Hz. Three replicates prepared at each composition were measured.

2.4. Electrospinning Set-up

The electrospinning apparatus (Yong Kang Le Ye Co., Beijing, China) is composed of a 5-mL syringe (Zhi Yu Co., Shanghai, China) connected to a syringe pump that was encased in a vented Plexiglas box [950 mm (L) × 820 mm (W) × 950 mm (H)] (Figure 1). The syringe pump was used to supply a steady flow of 0.0625 mL·min^{-1} of solution to the tip of the needle. A high-voltage power supply was used to apply a potential of 10, 15, 20 kV to the syringe needle. The needle used was size 22 and the corresponding inner diameter was 0.4 mm. A grounded cylinder collector (340 mm in length and 108 mm in diameter) was rotated at a speed of 80 rpm and placed 15 cm apart from the needle tip to test the distances at which the fibers were dry upon collection. A rectangular piece of aluminum foil (240 mm in length and 150 mm in width) was used to cover the cylinder to collect nonwovens of the electrospun nanofibers. The relative humidity and temperature were measured by a hygrothermograph placed inside the electrospinning chamber. They were kept at constant values of 13% and 25 °C, respectively.

2.5. Characterization of Electrospun PS Fibers

Scanning electron microscope (SEM, QUANTA-200, FEI, Hillsboro, OR, USA) was used to obtain microphotographs of the non-woven nanofibers formed after electrospinning. The nanofibrous mats were collected on the aluminum foil. The mats were cut into small pieces, and then coated with a layer of gold-palladium before being observed with SEM at an accelerating voltage of 12.5 kV. The diameter and distribution of the electrospun nanofibers were analyzed from the SEM images by using Nanometer software (Fudan University, China). At least ten beads were measured for the aspect ratio (major axis: bead length along the fibers; minor axis: bead length perpendicular to the fiber) and 100 fibers were measured to obtain the average fiber diameter.

The wettability of the PS surface was characterized on a contact angle meter (OCA20, Dataphysics, Bad Vilbel, Germany) at ambient conditions. Contact angle (CA) was measured using a sessile drop method. A droplet of 5 µL volume was used. The CA values of the right side and the left side of the water droplet were both measured and averaged. All the CA data were an average of five measurements at different locations on the surface.

Dynamic mechanical properties of the electrospun PS mats were measured in tensile mode using a dynamic mechanical analyzer (DMA, TA Instruments Q800). The measurements were performed at a constant frequency of 1 Hz and strain amplitude of 0.01%, for temperature range of room temperature to 150 °C, using a heating rate of 5 °C/min and a gap between jaws of 10 mm. Three samples were used to characterize each material.

Mechanical properties of the electrospun mats were determined from the stress-strain curves from tensile tests. The tensile test of the samples was carried out on a Model 3365universal testing machine (Instron Co., Norwood, MA, USA) with a tensile rate of 50 mm·min^{-1} according to ASTM D 882-09 at room temperature and 30% humidity. The size of each sample was of 15 mm length and 5 mm width. Three replicates of the mats prepared at each composition were measured.

3. Results and Discussion

3.1. Fiber Morphology

3.1.1. Effect of Solvent Types

In the electrospinning process of a polymer solution, the solvent is one of the main contributors for solution properties, e.g., conductivity, surface tension, and viscosity [26]. Hence, the morphology of electrospun fibers can be altered by changing the solvent composition to control the surface tension and viscosity of solutions at constant solution concentration. The properties of the prepared solvents and solutions are summarized in Tables 1 and 2, respectively.

Table 1. Basic properties of solvents [a].

Solvent	Viscosity (mPa·s)	Surface tension (mN·m^{-1})	Conductivity (F·m^{-1})	Boiling point (°C)
DMF	0.80	35.20	0.55	152.8
DMF/THF (75/25)	0.73	31.34	0.47	—
DMF/THF (50/50)	0.67	30.80	0.43	—
DMF/THF (25/75)	0.61	28.60	0.35	—
THF	0.53	26.40	—	66

[a] Value reported for 25 °C. The ratio represented the weight ratio of dimethylformamide (DMF) to tetrahydrofuran (THF).

Table 2. Characteristic properties of tested solutions [a].

Solution	Concentration (w/v, %)	Viscosity (mPa·s)	Surface tension (mN·m^{-1})	Conductivity (F·m^{-1})
PS/DMF	10	25	37.54	0.45
PS/DMF/THF (100/0)	23	301	37.87	0.42
PS/DMF	27	715	37.90	0.39
PS/DMF	32	1080	38.51	0.34
PS/DMF	40	1210	40.21	0.30
PS/DMF/THF (75/25)	23	321	33.92	0.29
PS/DMF/THF (50/50)	23	343	32.28	0.28
PS/DMF/THF (25/75)	23	357	30.71	0.14
PS/DMF/THF (0/100)	23	362	28.97	—

[a] Value reported for 25 °C. The ratio represented the weight ratio of DMF to THF.

Figure 2 presents the fiber morphology variation with an increasing amount of THF in the mixture of DMF/THF. Figure 2a shows the bead-on-string structure with different fiber sizes and aspect ratios of beads formed with 23% (w/v) PS/DMF. In Figure 2b, when THF was added to the solvent at 25%, bead-on-string fibers were gradually tuned to a collapsed bead surface. The formation of beads could be related to insufficient resistance of the electrospinning solutions to resist electrical force stretching caused by low viscosity, high surface tension of the solutions and high boiling point of the solvent [24,27]. Table 1 shows that the solvent DMF has a higher boiling point of 152.8 °C than THF (66 °C), and Table 2 shows that the solution of 23% (w/v) PS/DMF/THF yielding bead-on-string morphology had the lowest viscosity of 301 mPa·s and the highest surface tension of 37.87 mN·m^{-1} compared to the other four solutions. From Figure 2c, it can be seen that the formed PS fibers are uniform without beads. The magnified SEM image in Figure 2c clearly shows that parallel double-strand fibers with a rough surface were manufactured in the solutions having an equal amount of DMF and THF. In Figure 2d, as the ratio of THF increased to 75%, smooth parallel double-strand fibers were clearly produced. It is attributed to the fact that with increasing the addition amount of THF, a lower surface tension and boiling point, as well as a higher viscosity of polymer solutions could be obtained, leading to significant resistance to sustain the elongation of polymer chains. Therefore, the collapsed bead-on-string morphology was transformed to homogeneous fibers without beads. Additionally, for the formation of parallel double-strand fibers, a possible mechanism is proposed. It is known that increasing the amount of THF in the solution can decrease the charge density due to its lower conductivity than that of DMF, leading to a trend of stable whipping and larger fiber diameter during electrospinning. It is also known that stable whipping tends to form well-organized fibers. When the solution with increased THF amount sprays out from the nozzle in a stable manner, the low surface tension and high viscosity of the solution provides less resistance to form continuous orderly fibers. Moreover, different evaporation rates of THF and DMF could result in different accumulation rates of fibers [28], namely producing a spatial fluctuation fiber structure. As a result, PS fibers with larger parallel double-strand morphology can be manufactured with proper ratios of DMF to THF, manifesting a controllable manipulation of PS fiber morphology by regulating the properties of electrospinning solutions.

In order to obtain bead-free fibers, reducing surface tension might be an appropriate approach. However, it should be applied with caution. It is not necessary that a lower surface tension of solvent will always be more suitable for electrospinning. Interestingly, it was observed in Figure 2e that flat ribbon fibers with a wide range of diameters were produced with the solvent THF due to its lower surface tension compared to the other solvents. The fiber diameter had a wide range from 3.68–19.46 μm. This result is ascribed to the higher viscoelastic force of the PS/THF solution (362 mPa·s) and low boiling point of THF (66 °C). Since jet splitting is difficult when the viscoelastic force is too large, the average fiber diameters had a wide range of values.

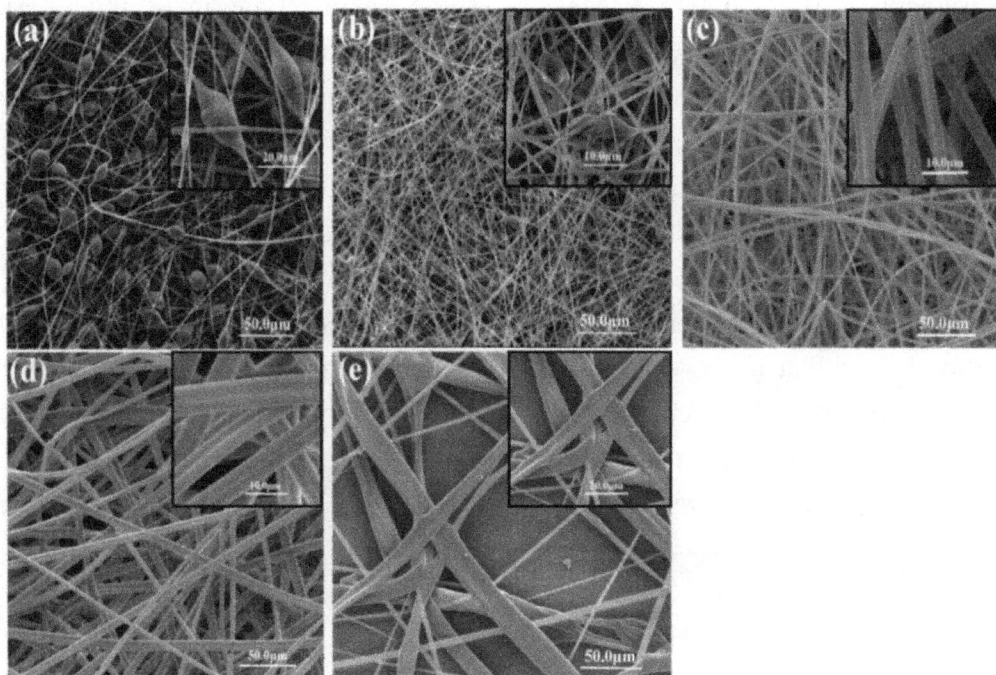

Figure 2. Scanning electron micrographs of electrospun PS fibers on stationary Al foil of 80 rpm, 23 (w/v)% solution concentration and applied voltage of 15 kV with different solvent combinations (**a**) DMF/THF = 100/0; (**b**) DMF/THF = 75/25; (**c**) DMF/THF = 50/50; (**d**) DMF/THF = 25/75; (**e**) DMF/THF = 0/100.

Figure 3 shows the average diameter of electrospun PS fibers prepared with different solvent combinations. The addition of 25% THF reduced the average fiber diameter from 1.14 to 0.64 μm. However, the average fiber diameter markedly increased from 0.64 to 3.92 μm with further increase of the amount of THF due to its lower surface tension of 30.71 mN·m^{-1} and higher viscosity of 357 mPa·s.

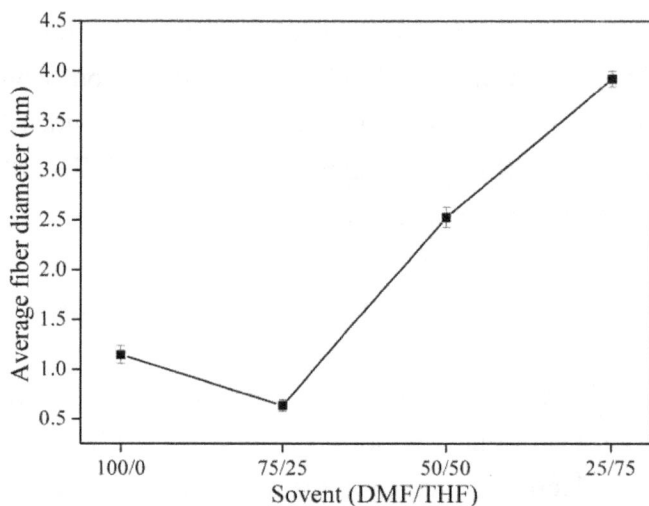

Figure 3. Average fiber diameter of electrospun PS fibers with different solvent combinations.

3.1.2. Effect of Solution Concentration

Solution concentration is one of the most important parameters for electrospinning and mainly affects solution viscosity. It is known that viscosity is the characterization of the intermolecular interactions in polymer solutions. The intermolecular interaction in a polymer-solvent system is either attractive or repulsive, depending on the type of solvent. Therefore, solution viscosity has a large effect on electrospinning. According to the results in Table 2, an increase in the solution concentration increased the viscosity value of the solution from 25 to 1210 mPa·s, whereas it did not significantly affect the value of the surface tension and conductivity. Figure 4 shows G' and G'' of the spinning solutions with different concentrations at an angler frequency of 2.5 Hz. It can be seen that G' and G' both increased with increasing solution concentration but the difference between G' and G' was larger at higher concentration than that at lower concentration, indicating that the elastic property of PS solution was more significant at high concentrations. This is attributed to the fact that at higher concentration, the number of PS chains per unit volume of solvent is much more than that at lower concentration, resulting in enhanced interaction and entanglement of PS chains to resist deformation.

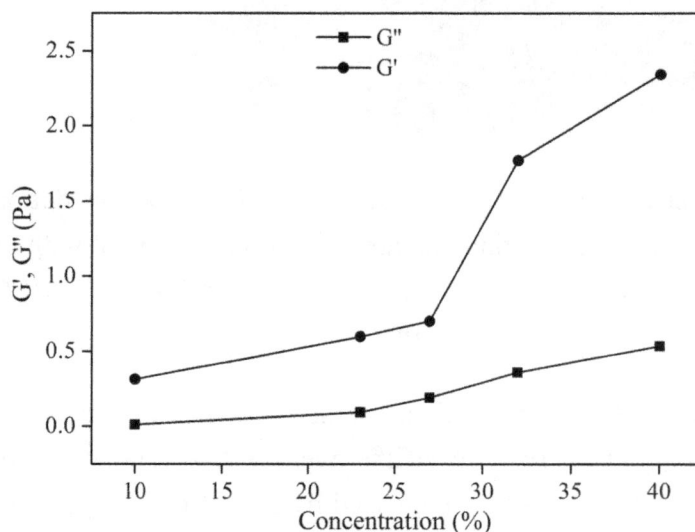

Figure 4. Viscoelasticity (G', G'') of PS solutions with different concentrations.

Figure 5 exhibits SEM photographs of electrospun PS fiber prepared from PS/DMF solutions at concentrations of 10%, 23%, 27%, 32%, and 40% (w/v). The amount of beads decreased with increased polymer concentration, and eventually formed straight fibers. It was observed in Figure 5a for the diluted solution of 10% (w/v) that irregular beads were formed because the very low viscosity did not suffice to sustain the elongation of the liquid jet, and therefore the thin jet of solution left the nozzle instantly and shrunk to droplets. Moreover, at 10% (w/v), PS solution exhibited typical viscoelastic property, which means that elasticity is too small to provide resistance to sustain the elongation resulting from electrostatic force, leading to bead morphology. With further increase to concentrated solutions of 23%–32% (w/v), ultrafine fibers were formed through beads as shown in Figure 5b–d. This is because the high viscosity and polymer content cause the solution jet to elongate and solidify quickly. More importantly, the difference between G′ and G″ was larger at 23%–32%, leading to the fact that the elasticity gradually turned to be a key role in the PS solution's rheological behavior (shown in Figure 4), and the larger elasticity could offer more resistance to sustain elongation during the initial electrospinning stage. However, the inherent viscous characteristic could also contribute to the unrecoverable deformation of PS at these concentrations, partially counteracting the elastic effect. Thus, the enhanced resistance against electrostatic force can produce fibers with bead-on-string structure. As the concentration increases to 40%, the combination of viscosity and viscoelastic property of the PS solution was large enough to favor fiber formation, shown in Figure 5e.

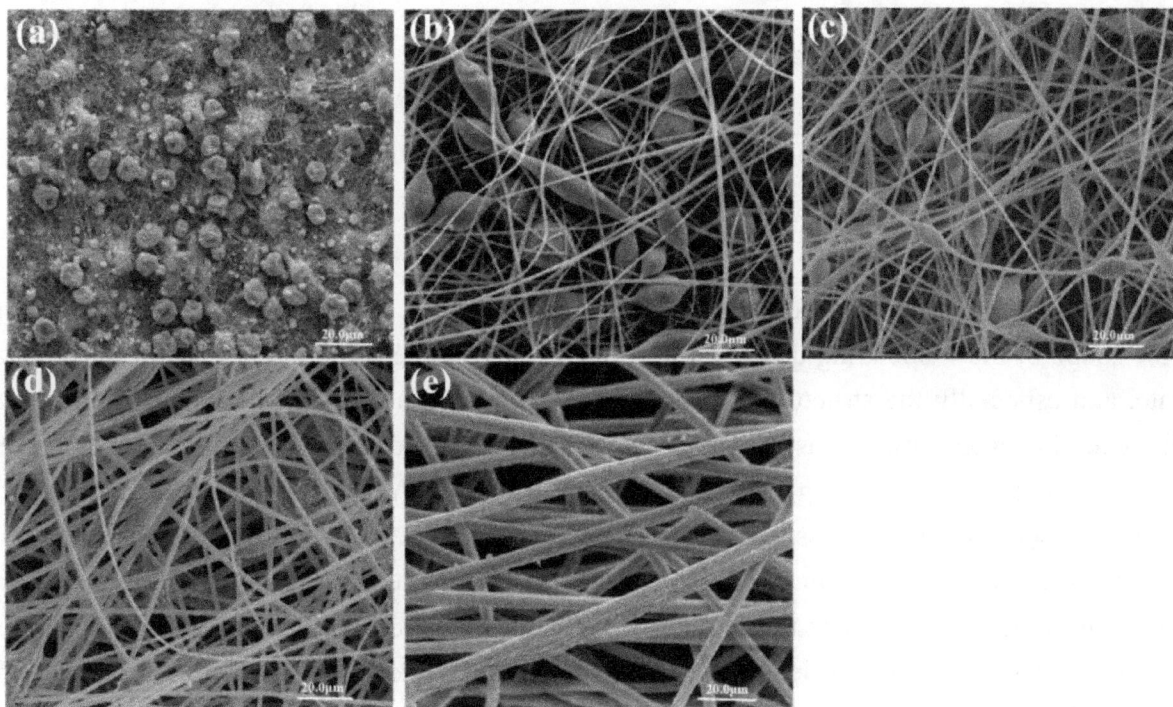

Figure 5. Scanning electron micrographs of electrospun PS fibers on stationary Al foil of 80 rpm and applied voltage of 15 kV with different concentrations of PS/DMF solutions **(a)** 10% **(b)** 23% **(c)** 27% **(d)** 32% **(e)** 40% (w/v).

Figure 6 shows the aspect ratio of beads and fiber diameters as a function of the polymer concentration. The aspect ratio and fiber diameter increased proportionally with the increase of polymer concentration. In order to explain the effect of solution concentration on the diameters or widths of the as-spun fibers, analysis of all forces acting on a small segment of a charged jet is necessary. Jarusuwannapoom *et al.* [18] summarized six types of forces to be considered: gravitational force, electrostatic force, Coulombic force, viscoelastic force, surface tension, and drag force. Among these forces, only the Coulombic, the viscoelastic, and the surface tension forces are responsible for the formation of beads as well as the shrinking of the charged jet during its flight to the grounded target [18]. As to the polymer solution, solvent molecules tend to aggregate and form a spherical shape due to the effect of surface tension at lower concentration which has more solvent fraction. The viscosity increase with the concentration of polymer illustrates the stronger effect on the interaction between polymeric chains and solvent. Consequently, solvent molecules tend to make the entangled molecular chains separate to reduce the tendency for aggregation and shrinkage.

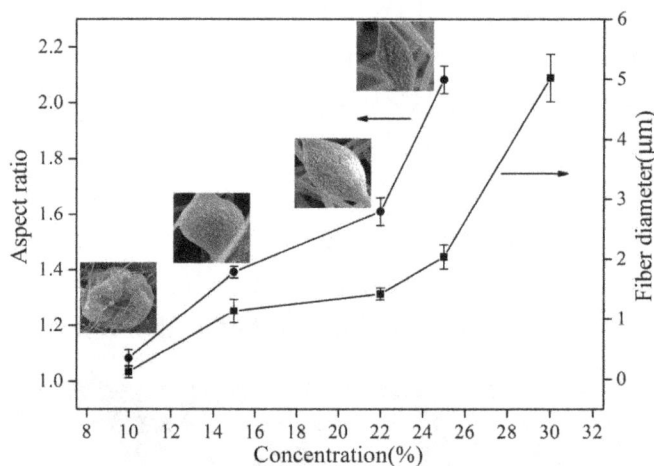

Figure 6. Aspect ratio of beads and fiber diameter of PS fibers with different concentrations.

3.1.3. Effect of Spinning Voltage

As the applied voltage directly affects the dynamics of the liquid flow, the changes in the voltage reflect on the shape of the suspending droplets out of the nozzle of the spinneret, on its surface charge, dripping rate, and especially the structural morphology (*i.e.*, fiber diameter) of the electrospun fibers. The jet diameter becomes smaller as it travels to the ground, attributed to two factors: solvent evaporation and continuous stretching caused by electrical force. Although some researchers have shown that the applied voltage has less effect on the bead formation and fiber diameter than the other parameters studied above [29], certain conclusions can still be reached in this study. Figure 7 shows the resulting bead morphology, aspect ratio and average fiber diameter of PS/DMF electrospun fibers at varying voltage ranges. Three applied voltages of 10, 15 and 20 kV were individually applied for electrospinning 23% (w/v) PS/DMF solutions while keeping all other parameters constant, setting the collection distance to15 cm, and the rotational speed at 80 rpm. Figure 7a–c show that as the applied voltage increased, the average fiber diameter increased. The applied voltage of 20 kV in Figure 7c made the polymer solution eject in a more fluid jet and resulted in an average fiber diameter of 1.136 μm, while the voltages of 15 kV and 10 kV yielded average diameters of 0.903 μm and 1.052 μm, respectively. From these results, it is interesting to note that the distribution of the fiber diameters was very broad when the applied voltage was 10 kV, was quite narrow at the applied potentials of 15 kV, and was broad again at the applied potential of 20 kV. Another interesting point is the effect of applied potential on the morphology of the obtained beads. Figure 7 clearly shows that the aspect ratio of beads formed at 15 kV was much higher than those of the beads formed at 10 and 20 kV.

The effect of applied voltage on the fiber diameters can be explained in terms of the relationships between the three major forces (the Coulombic, the viscoelastic and the surface tension), which influence the fiber diameters. At low applied potentials (e.g., 10 kV), the Coulombic force was not high when compared with that of the surface tension. This resulted in as-spun fibers with large diameters and the presence of large beads along the fibers. At a moderate applied potential (e.g., 15 kV), all three forces were well-balanced, resulting in a narrow distribution of the fiber diameters. With further increase in the applied potential (e.g., 20 kV), the Coulombic force was much greater than the viscoelastic force. This might result in a higher possibility of breakage of an over-stretched charged jet during its flight to the target. Moreover, with an increased applied potential, a charged jet travelled to the grounded target much faster. The solvent, therefore, had less time to evaporate. The charged jet retracts and some of the jets were neutralized, which led to bigger but irregular fibers.

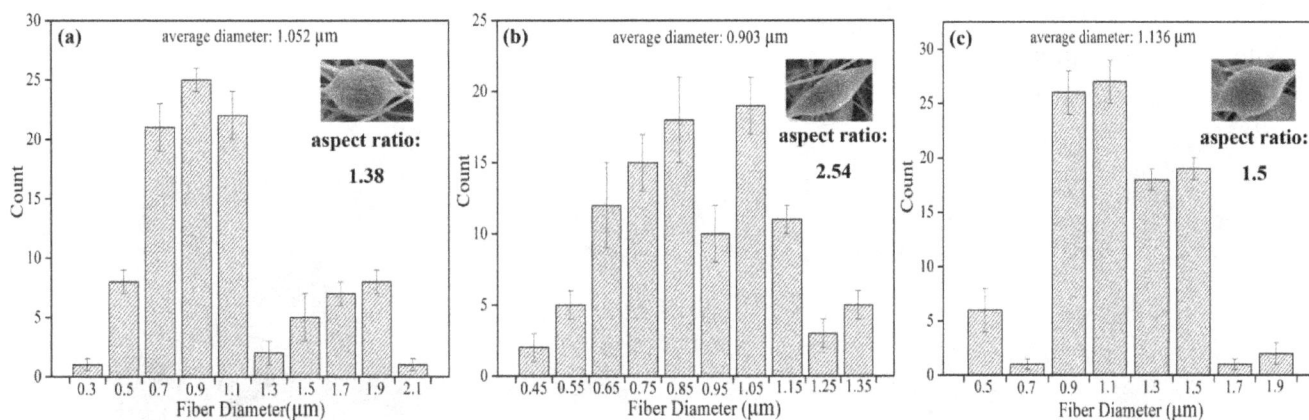

Figure 7. Fiber diameter distributions of electrospun 20 wt% PS fibers with different applied voltages **(a)** 10 kV **(b)** 15 kV **(c)** 20 kV.

3.2. Fiber Mat Wettability

The water contact angle values of the PS films, which were fabricated by solvent evaporation on an aluminum plate from the solution of PS in DMF and THF with a concentration of 23% (w/v), were $106 \pm 1.5°$ and $90 \pm 1.2°$, respectively. It is well known that PS is chemically hydrophobic. Therefore its water contact angle is higher than those of other hydrophilic polymers. However, the value of the water contact angle is not sufficient enough to meet the superhydrophobic requirements for many practical applications. In addition, the surface morphology of electrospun mats can be designed and controlled during the electrospinning process. Therefore, it is reasonable to deduce that taking advantage of the hydrophobic behavior of PS and the special characteristic of the electrospinning technique, electrospun mats with various PS fibers morphologies can be prepared with good hydrophobicity.

The water contact angles of various electrospun PS mats fabricated with different solvents and concentrations are shown in Figure 8. The CA value decreased with increasing amount of THF in the solvent. Moreover, Figure 8b shows that the CA value could be significantly affected by solution concentration and the smallest CA value was found using concentration of 32% (w/v). From Figure 8a,b, it can also be seen that the CA value of 23% (w/v) PS/DMF is as high as 147°, which is much higher than those of the films resulting from solvent evaporation, indicating that the electrospun PS mats are sufficient enough to exhibit superhydrophobicity in practical applications. The actual behavior of a water droplet on the surface of electrospun PS fiber is shown in Figure 8c. To our knowledge, the surface structure of the electrospun mat plays a leading role in determining the hydrophobicity. Moreover, the electrospun fibers with bead, porous and protuberant structure also contribute to the roughness of the electrospun mat. More air can be trapped as the roughness of the PS surface increases, which is beneficial to improve the hydrophobicity of the membranes as the water CA of air is considered to be 180° based on the Cassie equation [30]. Therefore, it is believed that beads have a predominant effect on the resultant hydrophobic behavior of the electrospun mat, and the fibers with wide diameter range are with the limitation of advancing superhydrophobicity. Based upon the illustrated effect of solvents above, with more THF added, the amount of beads decreased gradually and the rough surface of the electrospun products turned smooth. Furthermore, the effect of solution concentration demonstrates that beads morphology can be easily produced when the solution concentration is low. Thus, the CA value gradually decreases with the bead-on-string fiber turning to fibers, which is shown in Figure 8a,b.

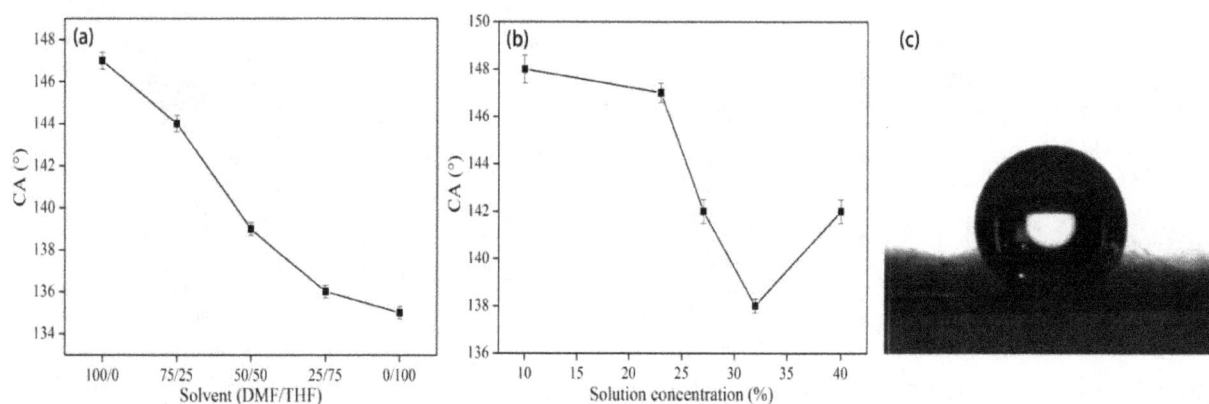

Figure 8. Contact angle of electrospun PS nanofibrous mats with (**a**) different solvent combinations and (**b**) different concentrations (**c**) water droplet on electrospun PS fibers from 23% (w/v) PS/DMF, CA = 147°.

Figure 9 presents the contact angles of PS mats made out of different fibers prepared using different applied voltages. In Figure 9, it is shown that the CA value of mats with their fibers prepared using an applied voltage of 15 kV is up to 147°, while the other two values using 10 kV and 20 kV are 137° and 141°, respectively. During the electrospinning process, the diameter of fibers prepared using an applied voltage of 15 kV is much larger than the other two according to Section 3.1.3, while the CA of 15 kV is also the largest as shown in Figure 9. On other hand, large diameter fibers are not favorable for increasing the roughness to form a better hydrophobic surface. Since the electrospun nanofibers had a small diameter, consequently the water droplet came into contact with a relevant small area of the electrospun mat compared with other materials and resulted in smaller CA values.

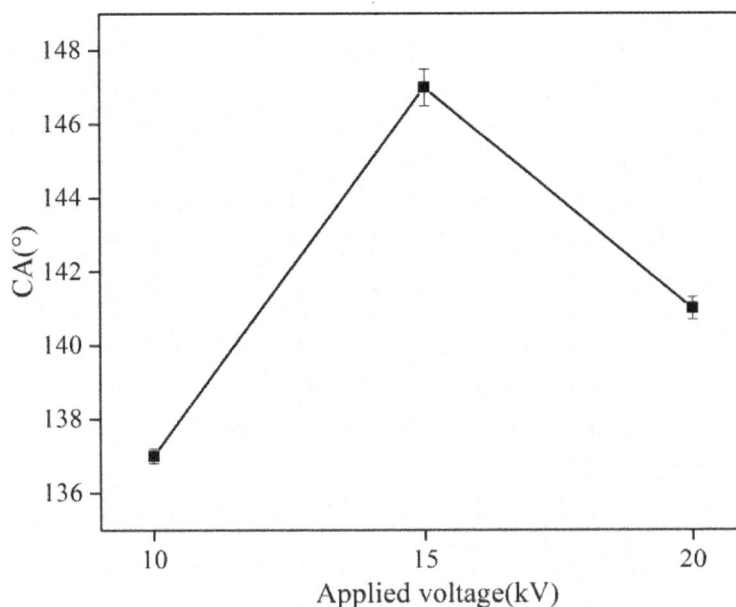

Figure 9. Contact angle of electrospun PS nanofibrous mats with different applied voltages.

3.3. Fiber Mat Dynamic Mechanical Property

The storage modulus (E′) of the electrospun double-strand PS fibrous mats is shown in Figure 10. All the curves represent a typical change in E′ as temperature increases, of an amorphous, high-molecular weight thermoplastic polymer. For temperatures below the glass transition region, the E′ of these mats decreased slightly with temperature because the PS was in the glassy state and the molecular motions were largely restricted to vibration and short-range rotation. The corresponding viscoelastic relaxation phenomenon at the transition region was observed at 100 °C and produced a drastic drop in E′ due to unrecoverable melting deformations of the polymer matrix. DMA curves clearly showed that the double-strand structure could significantly reinforce the E′ of PS fibrous mats, producing the maximum E′ of approximately 5.13 MPa. This reinforcing effect can be attributed to the further restriction in mobility of PS fibers by the rough double-strand structure with chain friction and entanglement.

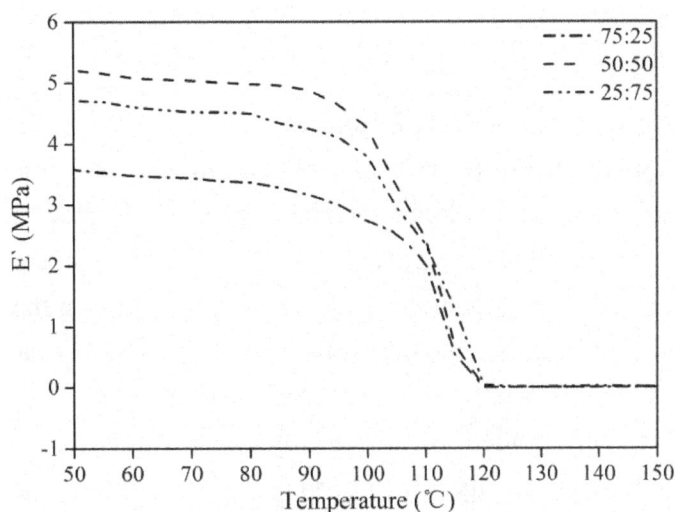

Figure 10. Dynamic mechanical analyzer (DMA) curves of electrospun PS nanofibrous mats.

3.4. Fiber Mat Tensile Strength

The tensile properties of electrospun mats with a double-strand structure are shown in Figure 11. The mechanical property of electrospun mats was enhanced with the presence of double-strand fibers. As bead-on-string fibers appeared using solvent of DMF/THF at a ratio of 75/25 in the prepared mat, the mat demonstrates a weaker tensile strength of less than 0.2 MPa. This behavior is mainly attributed to the unique characteristic of PS. Electrospun PS mats are generally soft and flexible. When the weight ratio of THF increased, double-strand fibers with rough surface formed. The tensile strength of the formed mats was almost 1.5 MPa, which indicated that the presence of double-strand fibers contributed to the reinforcement of electrospun PS mats and overcame the inherent shortcomings of PS mats. This result could be explained by the fact that during the tensile test, the friction effect and close compacting of double-strand fibers with rough surface provided effective neutralization to the stress and partially dissipated the tensile energy, resulting in an increase in stress and a corresponding decrease in strain. However, the tensile strength decreased to 0.4 MPa as THF increased, attributed to the reduced dissipation effect of the smooth fiber surface.

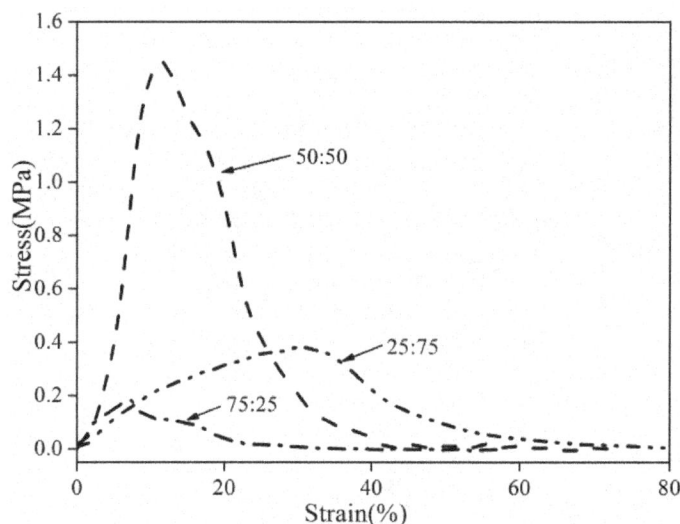

Figure 11. Stress-strain curves of electrospun PS nanofibrous mats.

4. Conclusions

In the present work, the effects of solvents, solution concentrations and applied voltage on the electro-spinnability of PS solutions along with the morphological appearance of the as-spun PS fibers and nanofibrous mat properties, were all characterized. Various surface morphologies, including beads with different sizes and shapes, bead-on-string structure with different aspect ratios of the beads, as well as fibers with different diameters and shapes, were formed by tuning the physical properties of the solvents and PS solutions. High surface tension, low viscosity, and typical viscoelasticity of the polymer solution contribute to the formation of the bead structure. A new kind of fiber with a double-strand structure was formed at moderate viscosity by using a mixed solvent of DMF and THF. It was also found that the increased viscosity and the typical elasticity characteristic of the polymer solution led to a tendency to form fibers. Distribution of fiber diameter was quite broad when the applied voltage was 10 and 20 kV, but narrowed with a voltage of 15 kV. The maximum water CA value of electrospun PS films obtained from DMF at 15 kV reached 148°. The DMA results showed that the double-strand could significantly reinforce the storage modulus of PS nanofibrous mats. The tensile strength of electrospun double-strand fibers with rough surfaces was 1.5 MPa, while the corresponding tensile strength of the mats with smooth surfaces was only 0.4 MPa.

Acknowledgments

This work was financially supported by the State Forestry Bureau 948 project (Grant No. 2013-4-11), the National Natural Science Foundation of China (Grant No. 31470580), and the Fundamental Research Funds for the Central Universities (Grant No. 2572014AB14).

Author Contributions

Siqi Huan and Guangping Han conceived and designed the experiments. Siqi Huan and Guoxiang Liu performed the experiments. Siqi Huan, Wanli Cheng, and Zongying Fu analyzed and discussed the data. Siqi Huan wrote the manuscript, with revisions by Guangping Han, Qinglin Wu, and Qingwen Wang.

Conflicts of Interest

The authors declare no conflict of interest.

References

1. Lu, X.; Wang, C.; Wei, Y. One-dimensional composite nanomaterials: Synthesis by electrospinning and their applications. *Small* **2009**, *5*, 2349–2370.
2. Sun, B.; Long, Y.Z.; Zhang, H.D.; Li, M.M.; Duvail, J.L.; Jiang, X.Y.; Yin, H.L. Advances in three-dimensional nanofibrous macrostructures via electrospinning. *Prog. Polym. Sci.* **2014**, *39*, 862–890.
3. Reneker, D.H.; Yarin, A.L.; Fong, H.; Koombhongse, S. Bending instability of electrically charged liquid jets of polymer solutions in electrospinning. *Polymer* **2008**, *49*, 2387–2425.
4. Greiner, A.; Wendorff, J.H. Electrospinning: A fascinating method for the preparation of ultrathin fibers. *Angew. Chem. Int. Ed. Engl.* **2007**, *46*, 5670–5703.
5. He, D.; Hu, B.; Yao, Q.F.; Wang, K.; Yu, S.H. Large-scale synthesis of flexible free-standing SERS substrates with high sensitivity: Electrospun PVA nanofibers embedded with controlled alignment of silver nanoparticles. *ACS Nano* **2009**, *3*, 3993–4002.
6. Park, S.M.; Kim, D.S. Electrolyte-assisted electrospinning for a self-assembled, free-standing nanofiber membrane on a curved surface. *Adv. Mater.* **2015**, *27*, 1682–1687.
7. Santos, C.; Silva, C.J.; Büttel, Z.; Rodrigo, G.; Sara, B.P.; Paula, T.; Andrea, Z. Preparation and characterization of polysaccharides/PVA blend nanofibrous membranes by electrospinning method. *Carbohyd. Polym.* **2014**, *99*, 584–592.
8. Abdelgawad, A.M.; Hudson, S.M.; Rojas, O.J. Antimicrobial wound dressing nanofiber mats from multicomponent (chitosan/silver-NPs/polyvinyl alcohol) systems. *Carbohyd. Polym.* **2014**, *100*, 166–178.

9. Schreuder-Gibson, H.L.; Gibson, P.K.; Senecal, M.; Sennett, J.; Walker, W.; Yeomans, D.Z.; Tsai, P.P. Protective textile materials based on electrospun nanofilbers. *J. Adv. Mater.* **2002**, *34*, 44–55.

10. Zhu, M.F.; Zuo, W.W.; Yu, H.; Yang, W.; Chen, Y.M. Superhydrophobic surface directly created by electrospinning based on hydrophilic material. *J. Mater. Sci.* **2006**, *41*, 3793–3797.

11. Menini, R.; Farzaneh, M. Production of superhydrophobic polymer fibers with embedded particles using the electrospinning technique. *Polym. Int.* **2008**, *57*, 77–84.

12. Celia, E.; Darmanin, T.; de Givenchy, E.T.; Amigoni, S.; Guittard, F. Recent advances in designing superhydrophobic surfaces. *J. Colloid. Interface Sci.* **2013**, *402*, 1–18.

13. Krishnan, K.A.; Anjana, R.; George, K.E. Effect of alkali-resistant glass fiber on polypropylene/polystyrene blends: Modeling and characterization. *Polym. Compos.* **2014**, doi:10.1002/pc.23193.

14. Ma, M.L.; Hill, R.M.; Lowery, J.L.; Fridrikh, S.V.; Rutledge, G.C. Electrospun poly(Styrene-block-dimethylsiloxane) block copolymer fibers exhibiting superhydrophobicity. *Langmuir* **2005**, *21*, 5549–5554.

15. Hardman, S.J.; Sarih, N.M.; Riggs, H.J.; Thompson, R.L.; Rigby, J.; Bergius, W.N.A.; Hutchings, L.R. Electrospinning superhydrophobic fibers using surface segregating end-functionalized polymer additives. *Macromolecules* **2011**, *44*, 6461–6470.

16. Bai, L.; Gu, J.Y.; Huan, S.Q.; Li, Z.G. Aqueous poly (vinyl acetate)-based core/shell emulsion: Synthesis, morphology, properties and application. *RSC Adv.* **2014**, *4*, 27363–27380.

17. Yoon, Y.L.; Moon, H.S.; Lyoo, W.S.; Lee, T.S.; Park, W.H. Superhydrophobicity of cellulose triacetate fibrous mats produced by electrospinning and plasma treatment. *Carbohyd. Polym.* **2009**, *75*, 246–250.

18. Jarusuwannapoom, T.; Hongrojjanawiwat, W.; Jitjaicham, S.; Wannatong, L.; Nithitanakul, M.; Pattamaprom, C.; Koombhongse, P.; Rangkupan, R.; Supaphol, P. Effect of solvents on electro-spinnability of polystyrene solutions and morphological appearance of resulting electrospun polystyrene fibers. *Eur. Polym. J.* **2005**, *41*, 409–421.

19. Kang, M.; Jung, R.; Kim, H.S.; Jin, H.J. Preparation of superhydrophobic polystyrene membranes by electrospinning. *Colloid. Sur. A* **2008**, *313*, 411–414.

20. Zhan, N.; Li, Y.; Zhang, C.; Song, Y.; Wang, H.; Sun, L.; Yang, Q.; Hong, X. A novel multinozzle electrospinning process for preparing superhydrophobic PS films with controllable bead-on-string/microfiber morphology. *J. Colloid. Interface Sci.* **2010**, *345*, 491–495.

21. Uyar, T.; Besenacher, F. Electrospinning of uniform polystyrene fibers: The effect of solvent conductivity. *Polymer* **2008**, *49*, 5336–5343.

22. Zheng, J.F.; He, A.H.; Li, J.X.; Xu, J.; Han, C.C. Studies on the controlled morphology and wettability of polystyrene surfaces by electrospinning or electrospraying. *Polymer* **2006**, *47*, 7095–7102.

23. Fong, H.; Chun, I.; Reneker, D.H. Beaded nanofibers formed during electrospinning. *Polymer* **1999**, *40*, 4585–4592.

24. Doshi, J.; Reneker, D.H. Electrospinning process and applications of electrospun fibers. *J. Electrostat.* **1995**, *35*, 151–160.

25. Rojas, O.J.; Montero, G.A.; Habibi, Y. Electrospun nanocomposites from polystyrene loaded with cellulose nanowhiskers. *J. Appl. Polym. Sci.* **2009**, *113*, 927–935.

26. Marie, R.L.; Christian, P. Partial disentanglement in continuous polystyrene electrospun fibers. *Macromolecules* **2015**, *48*, 37–42.

27. Fong, H.; Reneker, D.H. Elastomeric nanofibers of styrene–butadiene–styrene triblock copolymer. *J. Polym. Sci. B Polym. Phys.* **1999**, *37*, 3488–3493.

28. Wang, L.F.; Pai, C.L.; Boyce, M.C.; Rutledge, G.C. Wrinkled surface topographies of electrospun polymer fibers. *Appl. Phys. Lett.* **2009**, *94*, 151916.

29. Jacobs, V.; Anandjiwala, R.D.; Maaza, M. The influence of electrospinning parameters on the structural morphology and diameter of electrospun nanofibers. *J. Appl. Polym. Sci.* **2010**, *115*, 3130–3136.

30. Cassie, A.B.D.; Baxter, S. Wettability of porous surfaces. *Trans. Faraday Soc.* **1944**, *40*, 546–551.

11

Semi-Analytic Solution and Stability of a Space Truss Using a High-Order Taylor Series Method

Sudeok Shon [1], Seungjae Lee [1,*], Junhong Ha [2] and Changgeun Cho [3]

[1] School of Architectural Engineering, Korea University of Technology and Education, Cheonan 330-708, Korea; E-Mail: sdshon@koreatech.ac.kr
[2] School of Liberal Arts, Korea University of Technology and Education, Cheonan 330-708, Korea; E-Mail: hjh@koreatech.ac.kr
[3] School of Architecture, Chosun University, Gwangju 501-759, Korea; E-Mail: chocg@chosun.ac.kr

* Author to whom correspondence should be addressed; E-Mail: leeseung@koreatech.ac.kr;

Academic Editor: Richard Thackray

Abstract: This study is to analyse the dynamical instability (or the buckling) of a steel space truss using the accurate solutions obtained by the high-order Taylor series method. One is used to obtain numerical solutions for analysing instability, because it is difficult to find the analytic solution for a geometrical nonlinearity system. However, numerical solutions can yield incorrect analyses in the case of a space truss model with high nonlinearity. So, we use the semi-analytic solutions obtained by the high-order Taylor series to analyse the instability of the nonlinear truss system. Based on the semi-analytic solutions, we investigate the dynamical instability of the truss systems under step, sinusoidal and beating excitations. The analysis results show that the reliable attractors in the phase space can be observed even though various forces are excited. Furthermore, the dynamic buckling levels with periodic sinusoidal and beating excitations are lower, and the responses react sensitively according to the beating and the sinusoidal excitation.

Keywords: steel space truss; Taylor series method; semi-analytical solution; sinusoidal excitation; beating excitation; attractor; dynamic buckling

1. Introduction

A space truss system has been applied to a variety of structural systems ranging from traditional roof trusses, bridge and reticulated spatial structures to deployable structures. Furthermore, this system has much potential for future use because it is composed of discrete steel members and can form large spaces with a relatively small volume. However, a shallow space truss exhibits unstable phenomena, such as dynamic snapping, due to its nonlinearity.

Studies on structural stability have primarily dealt with static or dynamic buckling of continuous systems, such as shells or arches, in the past, and studies on space trusses have focused on the critical buckling below the static load [1–10]. Kassimali and Bidhendi [11] investigated the stability problem based on an Eulerian formulation that considered arbitrarily large displacements, and Tada and Suito [12] examined the static and dynamic post-buckling of discrete compressive members using a vibration model. Kim *et al.* [13] considered damping in their study of dynamic buckling of shallow trusses and explained that the structure was more sensitive to indirect snapping than direct snapping [14,15]. Although there are reduction techniques for dynamic analysis [16,17], there are few studies on analytical approaches or the nonlinear dynamic stability of space trusses. Moreover, an exact or an accurate solution of a nonlinear equation must be attained to overcome this problem, and there is a need to analyse the change and characteristics of the periodic orbit. However, it is difficult to obtain the exact solution or an analytical solution of the nonlinear governing equations of a space truss composed of discrete members, and numerical methods [18] as the Newmark-β method and Runge-Kutta method are widely used.

Recently, an analytical approach for both weakly and strongly nonlinear problems has been introduced [19]. The traditional analytic or semi-analytic methods, such as the Taylor's power series method [20,21], Adomian decomposition method [22,23], the homotopy perturbation method [24–26] and so on, have been developed. While these methods are limited in their application and depend on parameters, the Taylor Series Method (TSM) has a long history and is very reliable in terms of offering an analytical solution. Recently, Barrio [21] reported that one of the advantages of the TSM was its easy formulation with the variable-order and variable step-size method, and explained that it was very useful in the analysis of many dynamic systems requiring an accurate analytical solution. Additionally, it has the advantage of allowing the error limit to be calculated so that the computed solution can be assured, and it has also been reported that a very accurate solution can be attained in a very short time by appropriately adjusting for the number of terms and the error limit [20,27–29].

Moreover, the accurate solution based on an analytical approach is also needed to deal with the inverse problem or the identification of these steel trusses because the governing equation has a large coefficient and the convergence of the 4th-order Runge-Kutta method (RK4) is $O(h^4)$. In addition, because TSM, unlike other methods, directly computes the differential coefficients, it is useful when a high-precision solution is required. Furthermore, it is relatively simple in terms of applying the analytical excitation.

Accordingly, the goal of this study is to apply the TSM to a shallow steel space truss and analyse the nonlinear dynamic response. Governing equations are formulated by considering geometric nonlinearity, and a semi-analytical solution is computed by applying the high-order TSM. This paper is organised as follows. Section 2 discusses the formulation of the nonlinear governing equation and the theoretical analysis technique of TSM. Section 3 obtains a semi-analytical solution of the governing equation of a single-free-node (SFN) model and investigates the dynamic instability of SFN model under the various loads, *i.e.*, step, sinusoidal and beating load. The effect of changing damping coefficients is also considered. Section 4 investigates the unstable behaviour of a double-free-node (DFN) model in consideration with an initial imperfection. Lastly, Section 5 proposes conclusions of this study.

2. A Discrete Nonlinear Dynamic System and the Taylor Series Method (TSM)

2.1. Formulation of Nonlinear Motion Equations of a Space Truss

The dimension for the 3D bar element is defined using a local system (x, y, z) and a global system (X, Y, Z), and the nodal vectors are assumed to be f, d, F, and D with 3 degrees of freedom (DOF) per node. Here, f and d as well as F and D denote the force and displacement vectors of the local and global system, respectively. Each vector in the local and global systems can be transformed by matrix T. The displacement function represented by the nodal vector of the element is defined with a Lagrangian interpolation function, N_i and N_j, as:

$$u = \left[N_i I_3 \ \vdots \ N_j I_3 \right] T^T D \tag{1}$$

Using large deformation theory, the strain–displacement relationship of an elastic material is assumed to be as follows [8]:

$$\epsilon = \frac{du}{dx} + \frac{1}{2} \left\{ \left(\frac{du}{dx} \right)^2 + \left(\frac{dv}{dx} \right)^2 + \left(\frac{dw}{dx} \right)^2 \right\} \tag{2}$$

The following applies the principle of virtual work to obtain the stiffness equations of the element:

$$\delta d^T f = \int \delta \epsilon^T \sigma \, dV \tag{3}$$

When the nodal vector in the global system is substituted and integrated, the following equation is the result:

$$\delta D^T F = A \, E \, l \, (\delta \epsilon^T \epsilon) \tag{4}$$

where A, E and l are an area, elastic modular and a length of each member, respectively. $\delta\epsilon$ and ϵ are expressed in the following equations by obtaining the differential of the displacement functions and substituting it into the above relational equations:

$$\epsilon = UT^T D + \frac{1}{2} D^T K_s D \tag{5}$$

$$\delta\epsilon = UT^T \delta D + D^T K_s \delta D \tag{6}$$

The matrices, U and K_s, in the above equations are defined as:

$$U = \{ N_{i,x} \ 0 \ 0 \ \vdots \ N_{j,x} \ 0 \ 0 \} \tag{7}$$

$$K_s = \frac{1}{l^2} \begin{bmatrix} I_3 & -I_3 \\ -I_3 & I_3 \end{bmatrix} \tag{8}$$

The following stiffness equation is obtained by substituting each term of Equation (4) and solving for the nodal displacement vector using the global system:

$$F = EAl\left\{TU^TUT^TD + \left(K_sDUT^T + \frac{1}{2}TU^TD^TK_s\right)D + \frac{1}{2}K_sDD^TK_sD\right\} \tag{9}$$

Examining the terms on the right hand side of the above equation, the first term is the first-order, the second and third terms are the second-order, and the last term is the third-order term of the unknown displacement vector. Accordingly, the discrete nonlinear dynamic system is simplified and expressed in the equation below and includes the mass and damping from the above Equation (9).

$$M\ddot{D} + C\dot{D} + \{K_1 + K_2(D) + K_3(D^2)\}D = F \tag{10}$$

where M is the lumped mass matrix, C is the damping matrix, and K_1, $K_2(D)$, $K_3(D^2)$ are the stiffness matrices for each degree of the terms in Equation (9).

2.2. High-Order Taylor Series Method (TSM)

Equation (10) can be expressed in the equation bellow and defined as the following to attain the solution for an initial-value problem of the governing equations using the TSM and an approximate power series solution:

$$\ddot{D} = f(t, \dot{D}, D), \quad t \in I, \quad D = D(t) \in \mathcal{R}^n \tag{11}$$

$$D(t_0) = D_0, \dot{D}(t_0) = D_1 \tag{12}$$

where I is an open interval containing t_0, f is a sufficiently smooth function on $I \times R^n \times R^n$. $D(t)$ is an analytic function, which can be expressed by:

$$D(t) = \sum_{n=0}^{\infty} \frac{D^{(n)}(t_0)}{n!}(t - t_0)^n \tag{13}$$

$$D(t) = \sum_{k=0}^{n} \frac{D^{(k)}(t_0)}{k!}(t - t_0)^k + R(n, t, t_0) \tag{14}$$

When an analytical series solution is to be obtained for the nth-degree term, the remainder term, $R(n, t, t_0)$, is expressed as the following, and c^*, which satisfies the equation below, exists in the open interval. Additionally, the error is defined by Equation (16).

$$R(n, t, t_0) = \frac{D^{(n+1)}(c^*)}{(n + 1)!}(t - t_0)^{n+1} \tag{15}$$

$$|R(n, t, t_0)| \leq \frac{1}{(n + 1)!}max_{t_0 \leq t \leq t_0 + t_h}\left|D^{(n+1)}(t)\right|t_h^{n+1}, \quad t_0 \leq t \leq t_0 + t_h \tag{16}$$

TSM defines the solution of Equation (10) as an nth-degree series except for the R-term in Equation (14) and obtains the solution of coefficients of each differential, $D^{(k)}(t_0)$, from Equation (10). The computed solution has the precision of the solution within the error range of Equation (16) in the defined range of $[t_0, t_0 + t_h]$. Here, it is efficient to use order n and step-size t_h in Equation (16) as the parameters for the multi-step solution, and the variable order (VO) and variable step-size (VS) scheme adjusts for n and t_h, respectively. Also, the two parameters, n and t_h, can determine the error limit, and the accuracy of and the time to compute the solution are determined by these two parameters. This study defines the solution of Equation (10) as a finite Taylor series as expressed in Equation (17) and computes the solution in multiple steps.

$$\boldsymbol{D}_i(t) = \sum_{k=0}^{n} \frac{\boldsymbol{D}_i^{(k)}(t_{0i})}{k!}(t - t_{0i})^k , \quad (t_{0i} \leq t \leq t_{0i} + t_h) \quad i = 1,2,\ldots, total\ step \tag{17}$$

where the initial value pertaining to the ith-step can be found by the previous step.

3. A Single-Free-Node (SFN) Steel Space Truss Model

This chapter discusses the dynamic analysis and instability of SFN model under step, sinusoidal and beating excitations. SFN model, shown in Figure 1, is composed of five nodes and four elements, and only the top node is free with the others being clamped. This model is widely used to investigate dynamic snapping. In the case of the model, the governing equation, described by Equation (18), can be easily induced by Equation (10), and the coefficients of the differentials $D^{(k)}(t_0)$ can be obtained from Equation (18). For example, $D^{(k)}(t_0)$ is described by Equation (18a~e) for order $n = 5$. In this equation, the stiffness terms are $K_1 = 4EAH^2/L^3$, $K_2 = 12EAH/2L^3$ and $K_3 = 4EA/2L^3$. Where, E and A are Young's modulus and the cross-sectional area of the elements, respectively. H and L are the height and the half-width of SFN model as shown in Figure 1.

$$m\ddot{D} + c\dot{D} + K_1 D + K_2 D^2 + K_3 D^3 = F_0 \tag{18}$$

$$D^{(0)}(t_0) = D_0 \tag{18a}$$

$$D^{(1)}(t_0) = \dot{D}_0 \tag{18b}$$

$$D^{(2)}(t_0) = -2hw_0\dot{D}_0 - w_0^2 D_0 - \frac{K_2}{m}D_0^2 - \frac{K_3}{m}D_0^3 + \frac{F_0}{m} \tag{18c}$$

$$\begin{aligned} D^{(3)}(t_0) = (2hw_0)^2\dot{D}_0 + 2hw_0^3 D_0 + 2hw_0\frac{K_2}{m}D_0^2 + 2hw_0\frac{K_3}{m}D_0^3 - 2hw_0\frac{F_0}{m} \\ - \frac{K_2}{m}D_0\dot{D}_0 - w_0^2\dot{D}_0 - 3\frac{K_3}{m}D_0^2\dot{D}_0 \end{aligned} \tag{18d}$$

$$\begin{aligned} D^{(4)}(t_0) = -(2hw_0^2)^2 D_0 - w_0^2\frac{F_0}{m} - 6\frac{K_3}{m}D_0\dot{D}_0^2 - 2\frac{K_2}{m}\dot{D}_0^2 + w_0^4 D_0 - (2hw_0)^3\dot{D}_0 \\ + (2hw_0)^2\frac{F_0}{m} + 3\left(\frac{K_3}{m}\right)^2 D_0^5 + 2\left(\frac{K_2}{m}\right)^2 D_0^3 + 4hw_0^3\dot{D}_0 + 3w_0^2\frac{K_2}{m}D_0^2 \\ + 4\,w_0^2\frac{K_3}{m}D_0^3 - (2hw_0)^2\frac{K_2}{m}D_0^2 - (2hw_0)^2\frac{K_3}{m}D_0^3 + 5\frac{K_2}{m}\frac{K_3}{m}D_0^4 - 3\frac{K_3}{m}\frac{F_0}{m}D_0^2 \\ - 2\frac{K_2}{m}\frac{F_0}{m}D_0 + 8hw_0\frac{K_2}{m}D_0\dot{D}_0 + 12hw_0\frac{K_3}{m}D_0^2\dot{D}_0 \end{aligned} \tag{18e}$$

where w_0 is the natural angular frequency, $w_0 = \sqrt{K_1/m}$, with the damping coefficient $h = c/2mw_0$, and initial values $D(t_0) = D_0$ and $\dot{D}(t_0) = \dot{D}_0$. Accordingly, if the Taylor series is expanded at each step using the above coefficients, an approximate analytical solution can be attained, and the number of terms increases as the order increases. This study used a relatively high order n and a smaller step-size t_h. In this paper, the adopted model has the following structural information: the material density (ρ), Young's modulus (E) and the cross-sectional area (A) are 7.85×10^{-3} kg/m^3, 2.06×10^5 MPa and 11.2 cm^2, respectively. The shape parameter μ (= $H/2L$) is defined by the rise (H) to span ($2L = 10$ m) ratio.

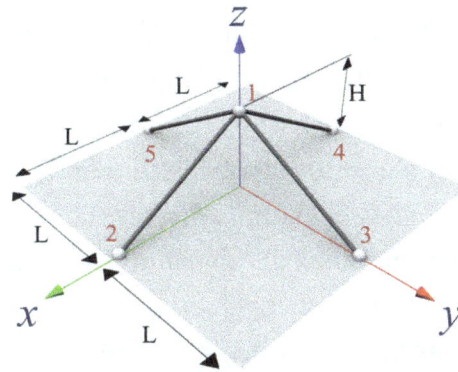

Figure 1. Shape of single-free-node (SFN) model.

3.1. Dynamic Response under a Step Excitation

The analytical solution of the model directly using TSM can be obtained by calculating the coefficients of the differentials, similar to Equation (18). In this case, the order n is 7, the rise-span ratio is $\mu = 0.1$ and damping is not considered. Also, the step excitation F_0 is applied.

As shown in Figure 2, the two cases that use high-order TSM are compared with the result of the 4th order Runge-Kutta method (RK4). One case is the model when $\mu = 0.05$ at $F_0 = 200$ kN (see Figure 2a), and the other model is when $\mu = 0.15$ at $F_0 = 3300$ kN (see Figure 2b).

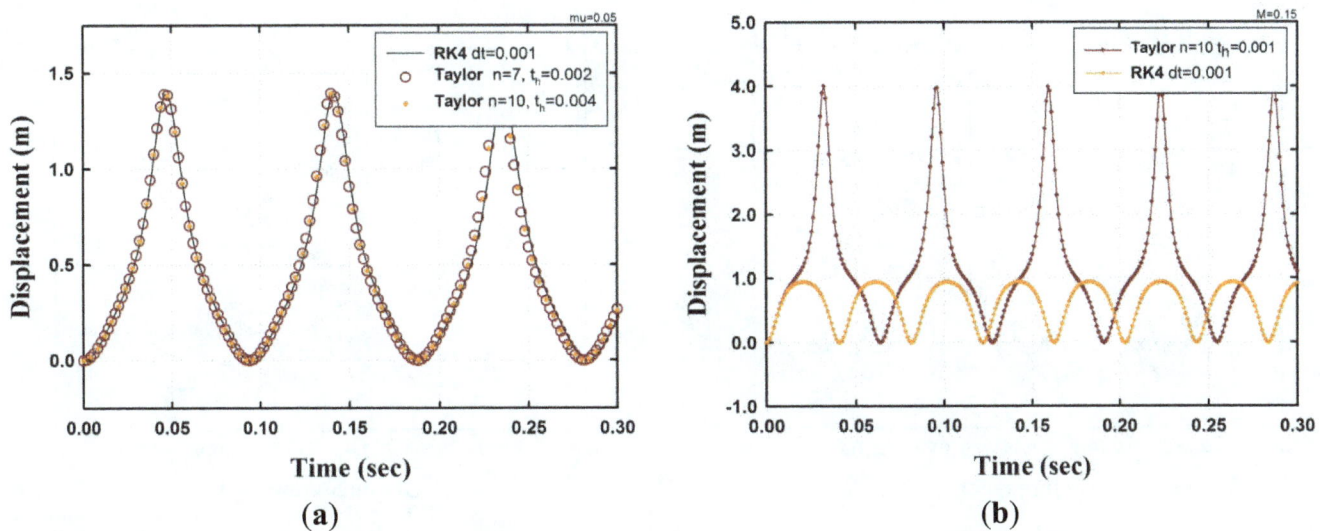

(a)

(b)

Figure 2. Comparing the result from SFN model using TSM with the result from RK4;
(a) Displacement $\mu = 0.05, F_0 = 200$ kN; (b) Displacement $\mu = 0.15, F_0 = 3300$ kN.

In the first case, shown in Figure 2a, the time-step is set to 0.0005 for the RK4, and the order n and step size t_h of the Taylor method are 7 and 10 and 0.002 and 0.004, respectively. As shown in the figure, the results of the Taylor method agree with the result from the RK4. As mentioned earlier, the increasing order can reduce the errors; however, reducing the step-size can also result in accurate results, which means that an accurate solution with a relatively long time-step can be obtained with a high order.

In the second case, shown in Figure 2b, the result shows that to obtain an accurate solution, the area near the dynamic buckling load is more sensitive and difficult than the other state. In this case, the time parameter of both high-order TSM and the RK4 was set at the same time-step, $t_h = 0.001$, and the order is $n = 10$. As shown in Figure 2b, the result of the high-order TSM is different from the result of the RK4 and corresponds to the state of post-buckling, whereas the result of the RK4 is pre-buckling. However, increasing F_0 by only a small amount, the dynamic snap-buckling will also appear in the case of the RK4. Therefore, to obtain a more accurate solution, using the analytical approach with a high order of the differential equations as in TSM would be appropriate.

To investigate the dynamic instability, let us consider another case with $\mu = 0.1$ and $h = 0.0$. The parameters for TSM are set at $n = 10$ and $t_h = 0.0005$.

First, the time history of the model with $F_0 = 900, 1000, 1100$ and 1200 kN is shown in Figure 3. Figure 3a shows that the period of the model gradually increases up to when the load $F_0 = 1000$ kN and then decreases afterwards. The maximum displacement is suddenly amplified at $F_0 = 1100$ kN and does not vary in proportion to F_0 because of the effect of the geometrical nonlinearity. In addition, the rapidly changing maximum displacement as F_0 increases, between 1000 and 1100 kN, is expected as the critical point, and the F_0 at the point refers to the dynamic buckling load under a step excitation [1,14]. Figure 3b shows the trajectory in the phase space to observe the attractor before and after dynamic buckling. In the phase space of the figure, the shape of the trajectory is changing from a limit cycle attractor with a single centre point to a limit cycle attractor with two centres.

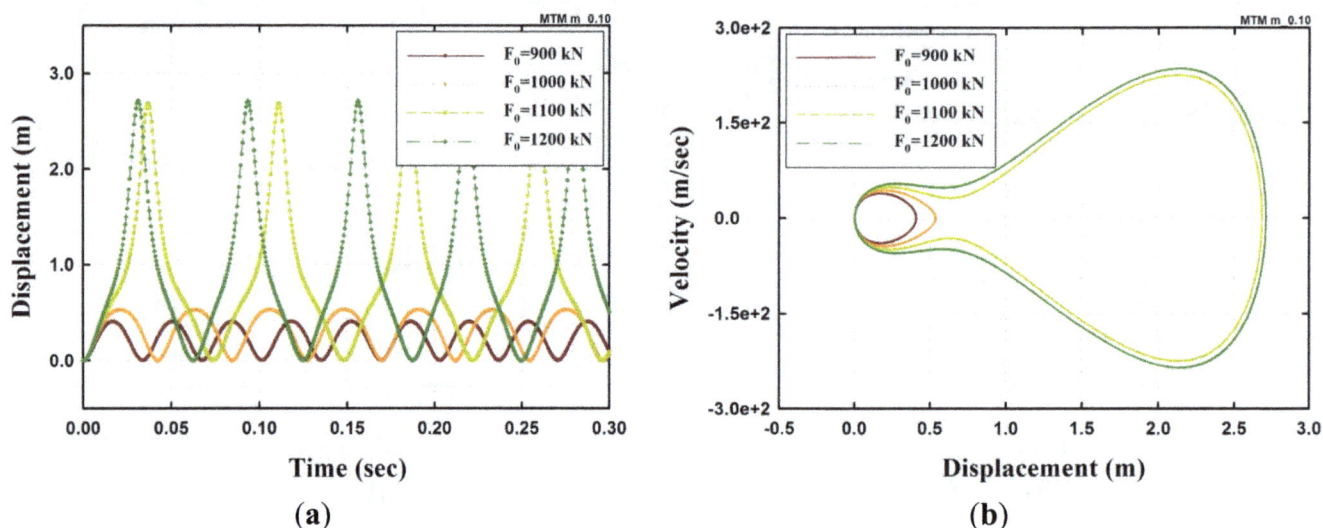

Figure 3. Dynamic analysis results of an undamped SFN model under a step excitation ($\mu = 0.1, h = 0.0$): **(a)** Displacement; **(b)** Trajectory of the phase space.

Next, the results of the model with different damping coefficients h are shown in Figure 4 with $h = 0.01, 0.03$ and 0.05. Here, Figure 4a,b shows the result of a pre-buckling load with $F_0 = 1000$ kN and a post-buckling load with $F_0 = 1200$ kN, respectively. The figures indicate that the curves converge well even with post-buckling. The trajectory in the phase space is shown in Figure 5. In the case $F_0 = 1000$ kN, the trajectory converged to near the centre point at the trajectory, as shown in Figure 5a. The trajectory when $F_0 = 1200$ kN converges on the other centre in the limit cycle attractor, as shown in Figure 5b. However, in this case, it is possible to have two points of the system converge to a limit set, and these fixed point attractors and trajectories in the phase space are sensitive to the initial condition [14].

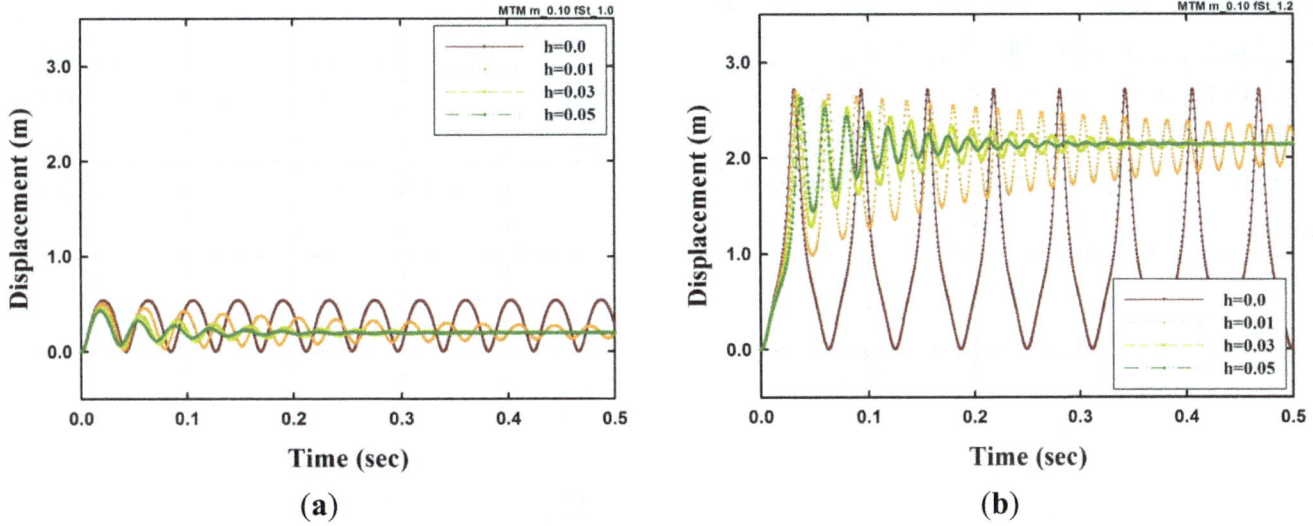

Figure 4. Dynamic analysis results of a damped SFN model under a step excitation ($\mu = 0.1$): **(a)** Displacement $F_0 = 1000$ kN; **(b)** Displacement $F_0 = 1200$ kN.

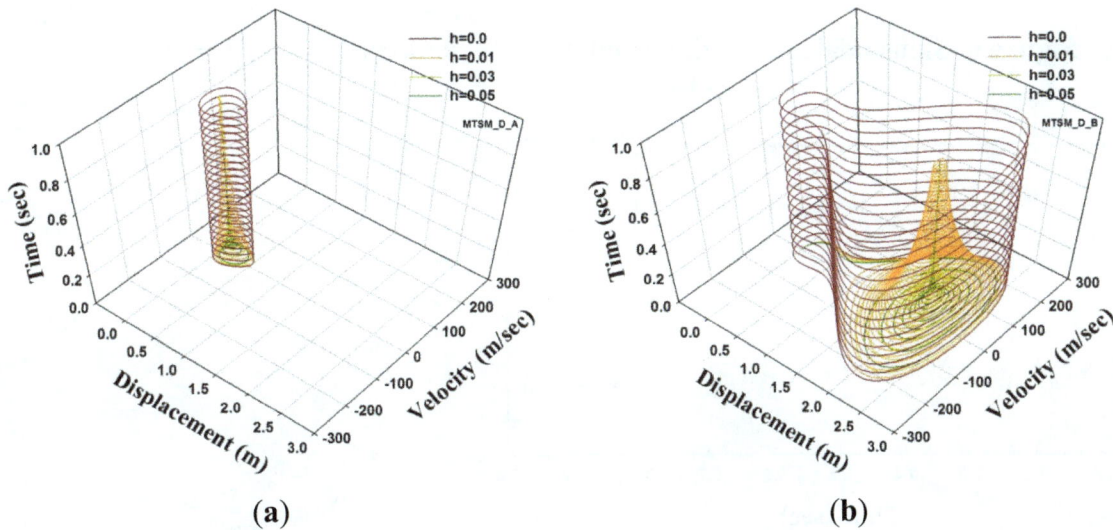

Figure 5. Phase space of a damped SFN model under a step excitation ($\mu = 0.1$): **(a)** Extended phase diagram, $F_0 = 1000$ kN; **(b)** Extended phase diagram, $F_0 = 1200$ kN.

3.2. Dynamic Response under a Periodic Excitation

Periodic excitation, such as a sinusoidal and a beating excitation, is defined in Equations (19) and (20), and is shown in Figure 6. In the equations, w_0 is the natural angular frequency of the analysis model and the periodic parameters α and β are introduced and are applied in the model to investigate the dynamic response under periodic excitations. Let us consider the model with a rise-span ratio of $\mu = 0.1$ and a damping coefficient of $h = 0.0$.

$$F = F_0 \cdot sin(\alpha w_0 t) \tag{19}$$

$$F = F_0 \cdot 0.5\{cos(\alpha w_0 t) - cos(\alpha(1 - \beta)w_0 t)\} \qquad (20)$$

Figure 7 shows the result of the sinusoidal excitation with $F_0 = 402$ kN. The analysis result indicated that when $\alpha = 1.0$, the amplitude of the displacement was the greatest, as shown in the Figure 7a, and dynamic buckling was observed. However, dynamic buckling did not occur when $\alpha = 0.5$ and 1.5. These results are depicted in the phase space of Figure 7b. The strange attractor for $\alpha = 1.0$ was formed, as shown in the figure, and the amplitude of response conspicuously increased. In particular, $\alpha = 1.0$ indicates that the frequency of excitation is the same as the first natural angular frequency of the adopted model.

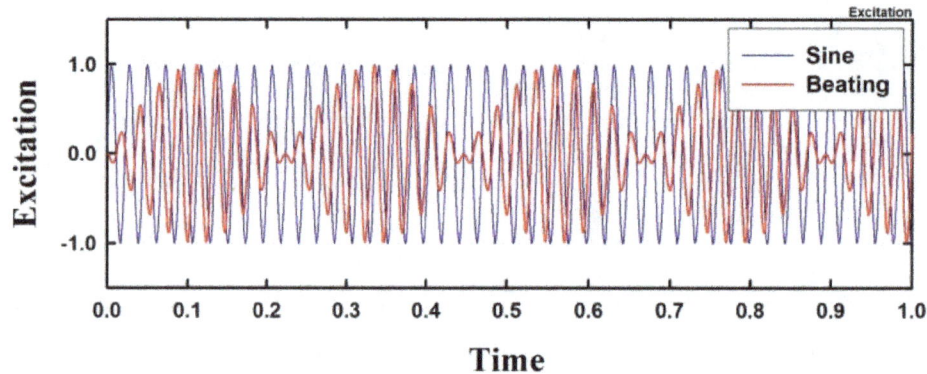

Figure 6. Sinusoidal and beating excitations excitation ($\alpha = 1.0, \beta = 0.1$).

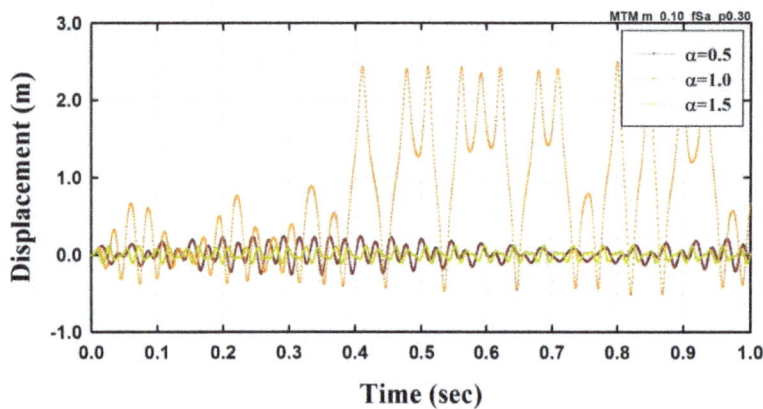

(a)

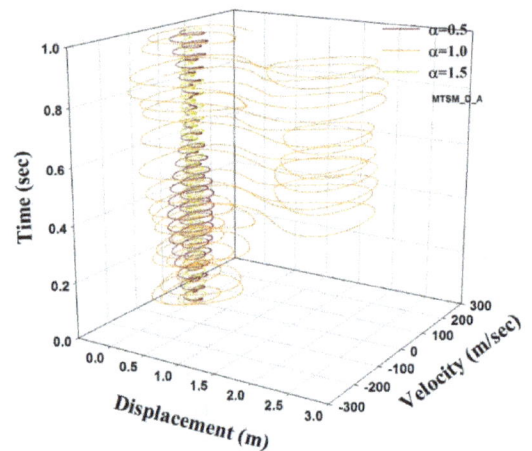

(b)

Figure 7. Dynamic analysis results of SFN model under a sinusoidal excitation ($\mu = 0.1, \alpha = 1.0, F_0 = 402$ kN): **(a)** Displacement **(b)** Trajectory in the phase space.

In the case of the model under a beating excitation, the periodic parameters, $\alpha = 1.0$ and $\beta = 0.1$, are considered. Here, α and β represent two periods of the beating excitation. Figure 8 shows the analysis results when $F_0 = 134$ and 268 kN. The amplitude of the displacement markedly increases when $F_0 = 268$ kN, as shown in Figure 8a. This result can be easily observed from the trajectory of the phase space of Figure 8b, and the change of trajectory is manifested as a strange attractor when $F_0 = 268$ kN.

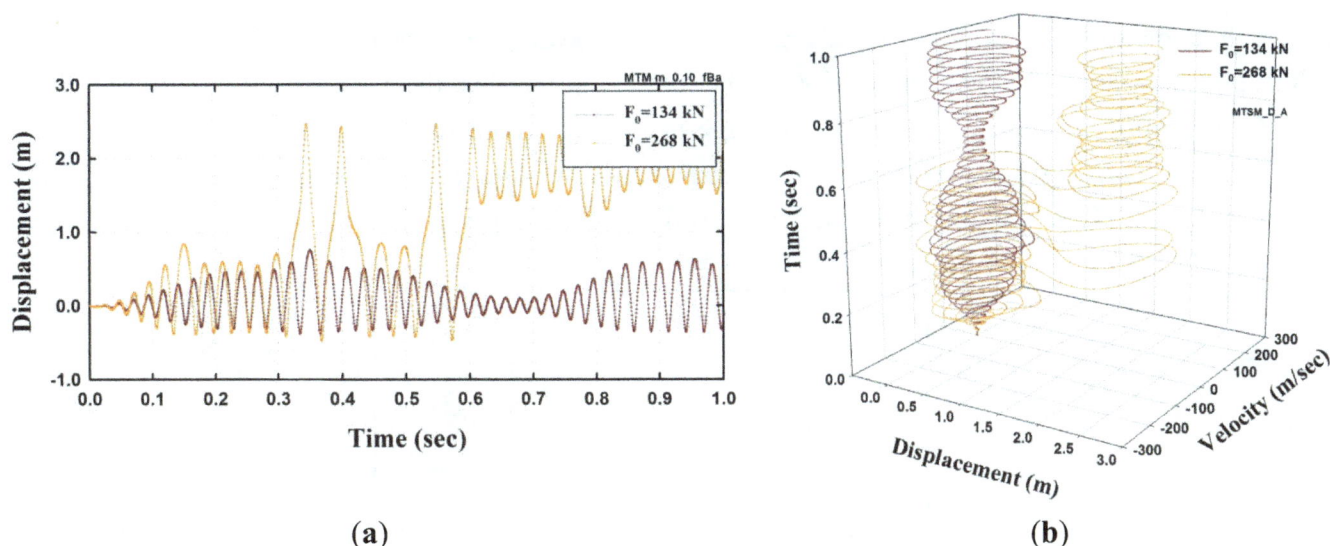

(a)   (b)

Figure 8. Dynamic analysis results of SFN model under a beating excitation ($\mu = 0.1$, $\alpha = 1.0$, $\beta = 0.1$). (**a**) Displacement; (**b**) Trajectory in the phase space.

3.3. Dynamic Instability and Buckling Load

Dynamic instability with different rise-span ratios μ of a shallow SFN model under a step excitation is investigated. Five ratios of μ were used: 0.05, 0.1, 0.15, 0.2 and 0.25. Here, the static critical buckling load level sP_{cr} is shown in Table 1, and the dynamic buckling load was determined using the Budiansky-Roth criterion [1,14].

Table 1. Static buckling load of SFN model (sP_{cr}).

M	0.05	0.1	0.15	0.2	0.25
sP_{cr} (kN)	175	1340	4214	9098	15886

In the case of the undamped model, the dynamic buckling load dP_{cr} is shown in Table 2. The result shows that the dP_{cr} increases as μ increases, and the non-dimensional load dP_{cr}/sP_{cr}, is approximately 77% of the static buckling. Next, the dynamic buckling load of the damped model, $h = 0.05$, is shown in Table 3. The table shows that dP_{cr} varies in proportion to the increase in the damping coefficient (h), and the dynamic buckling was approximately 83% of the static buckling.

Table 2. Dynamic buckling load of SFN model (dP_{cr}) ($h = 0.0$).

M	0.05	0.1	0.15	0.2	0.25
dP_{cr} (kN)	136	1032	3244	7004	12229
dP_{cr}/sP_{cr}	0.7771	0.7703	0.7698	0.7698	0.7698

Table 3. Dynamic buckling load of SFN model (dP_{cr}) ($h = 0.05$).

M	0.05	0.1	0.15	0.2	0.25
dP_{cr} (kN)	146	1106	3479	7511	13113
dP_{cr}/sP_{cr}	0.8343	0.8256	0.8256	0.8255	0.8255

Figure 9 compares the dynamic buckling load of the beating excitation with that of the static buckling and step and sinusoidal excitations. The figures show that the beating excitation and the sinusoidal excitation resulted in a sensitive change in the maximum displacement. Additionally, the lowest buckling load was observed with the beating excitation followed by the sinusoidal excitation, step excitation, and static buckling in an increasing order of buckling load.

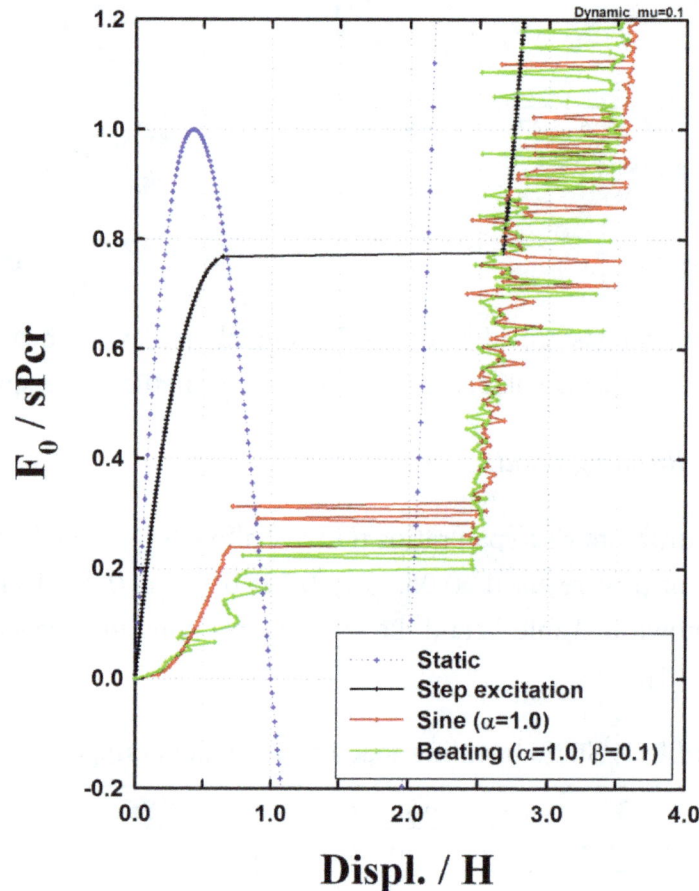

Figure 9. Maximum displacement response under various excitations ($\mu = 0.1$).

4. Double-Free-Nodes (DFN) Steel Space Truss Model

The second exemplar case structure is double-free-nodes (DFN) model as shown in Figure 10 [14,15]. The adopted model is composed of 10 nodes and 11 elements, and node 1 and 2 are free while the others are clamped. This model is more complex than the first one. In this study, the vertical displacements of node 1 and 2 (*i.e.*, D_{Z1} and D_{Z2}) are only considered to simplify the problem. In this case of DFN model, the governing equation, described by Equation (21), can be driven. In the governing equation, $\omega_0 = \sqrt{(5EAH^2)/(m\alpha^3L^3)}$ and $\alpha = L_e/L$. Where, L_e is a length of an inclined element as shown in Figure 10.

$$\ddot{D}_{Z1} + 2\omega_0 h \dot{D}_{Z1} + \omega_0^2 D_{Z1} + \frac{1}{m}(k_2 D_{Z1}^2 + k_3 D_{Z1}^3 + k_4 D_{Z1} D_{Z2}^2 + k_5 D_{Z1}^2 D_{Z2} + k_6 D_{Z2}^3 - F) = 0 \quad (21)$$

The coefficients of Equation (21) are as follow:

$$k_1 = \frac{5EAH^2}{\alpha^3 L^3}, \qquad k_2 = -\frac{7.5EAH}{\alpha^3 L^3}, \qquad k_3 = \frac{0.5EA}{L^3} + \frac{2.5EA}{\alpha^3 L^3},$$

$$k_4 = \frac{1.5EA}{L^3}, \qquad k_5 = -\frac{1.5EA}{L^3}, \qquad k_6 = -\frac{0.5EA}{L^3} \tag{22}$$

where Young's modulus (E), the density (ρ), the cross-sectional area (A) and span parameter (L) are equal to those of SFN model.

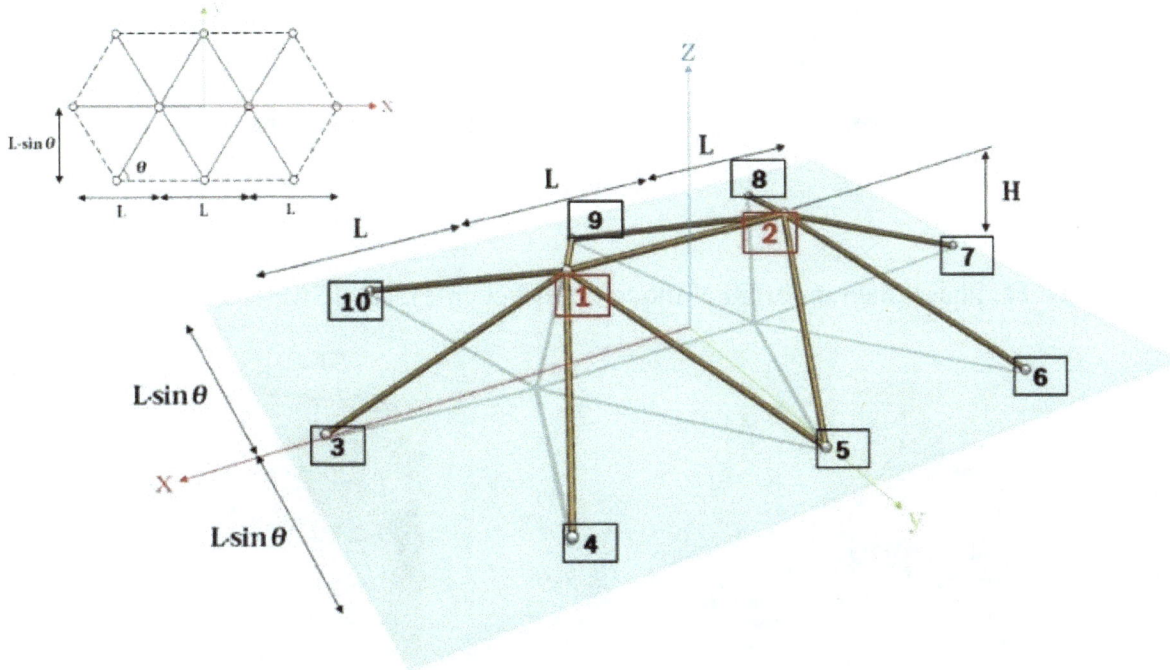

Figure 10. Shape of DFN model.

In this section, the shape parameter $\mu = 0.1$, i.e., (H) = 1 m, and load level $F = 1300\text{kN}$ are considered. To analyze DFN model using TSM, let us consider the parameter $n = 7$ and the step $t^{\text{int}} = 0.001$. For the comparison of the analysis results, the RK4 solution for $t^{\text{int}} = 0.001$ is adopted. To observe a dynamic buckling phenomenon, we consider two different cases; one is a perfect shape and the other is an imperfect one. For the second case, 0.1% of H is applied to account for the initial imperfection. The analysis results using TSM and RK4 are as shown in Figures 11 and 12.

In Figures 11 and 12, we present a comparison between TSM and RK4 for both cases. In the figures, we observe that the attractors of TSM and RK4 agree with each other. For the first case, the result figure shows that a periodic orbit is observed, i.e. limit cycle. But Figure 12a in consideration of initial imperfection shows that a strange attractor is observed and displacement rapidly changes due to the influence of coupling under asymmetric imperfection. This figure indicates that the shallow DFN model is very sensitive to the initial condition. Generally, the DFN model is well known as a shallow space truss dome which is sensitive to the initial condition.

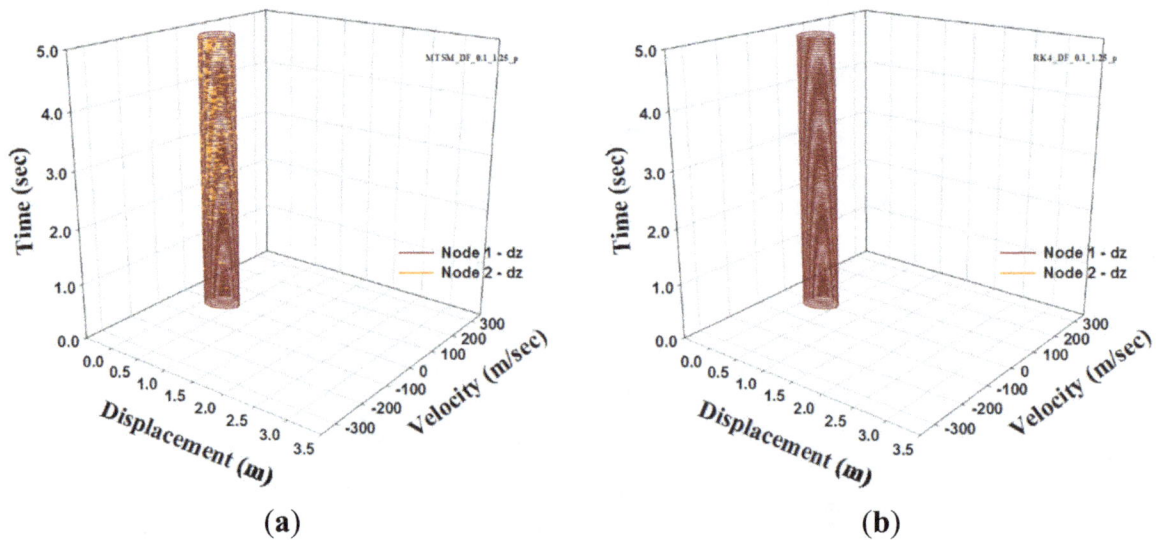

Figure 11. Phase diagram of DFN model. (Perfect case): (**a**) TSM; (**b**) RK4.

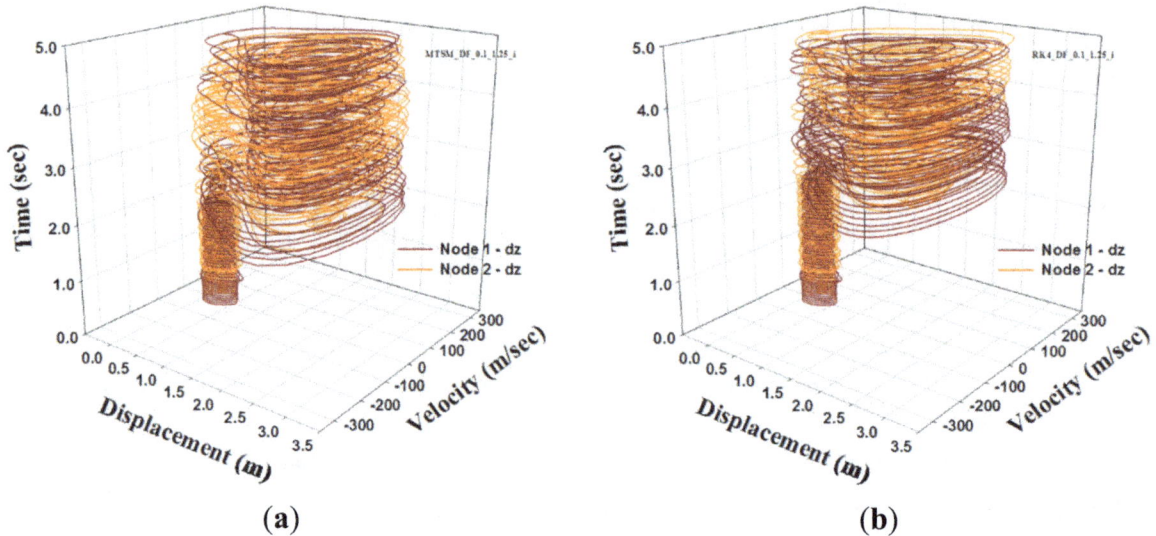

Figure 12. Phase diagram of DFN model. (Imperfect case): (**a**) TSM; (**b**) RK4.

5. Conclusions

This paper described the nonlinear dynamic analysis of a steel space truss using TSM and investigated the dynamic instability under various excitations. The governing equations are formulated by considering geometrical nonlinearity, where an accurate analytical solution with a relatively long time-step could be obtained using the high-order TSM. In the investigation, the nonlinear dynamic response and trajectory of the phase space were analysed with step, sinusoidal and beating excitations. By analysing the response of the adopted models, the attractors and maximum displacement could delineate the characteristics of dynamic snapping, which occurs in a shallow shell under various excitations, and the change to an asymptotically stable state of the model with different levels of damping was well reflected. In investigating the dynamic instability of the SFN model, dynamic buckling occurred at approximately 77% of the static buckling with a step excitation. The dynamic buckling was approximately 83% of the static buckling when damping was considered, and the same result was obtained for all rise-span ratios (μ). The buckling load was lower with the model under periodic excitations compared with that of the model under a step excitation, and the model under a beating excitation reacted more sensitively.

Acknowledgments

This research was supported by Basic Science Research Program through the National Research Foundation of Korea (NRF) funded by the Ministry of Science, ICT & Future Planning (NRF-2014R1A2A1A01004473).

Author Contributions

The background research of this work was carried out under supervision of both Sudeok Shon and Junhong Ha. The paper was written by Sudeok Shon in cooperation with Seungjae Lee, Junhong Ha and Changgeun Cho. The theoretical part of the High-order Taylor series method was written by Junhong Ha. All authors participated in the final corrections of the paper.

Conflicts of Interest

The authors declare no conflict of interest.

References

1. Budiansky, B.; Roth, R.S. Axisymmetric dynamic buckling of clamped shallow spherical shells. In *Collected Papers on Instability of Shells Structures*; NASA TN D 1510; The National Aeronautics and Space Administration (NASA): Washington, DC, USA, 1962; pp. 597–606.
2. Ball, R.E.; Burt, J.A. Dynamic buckling of shallow spherical shells. *J. Appl. Mech.* **1973**, *40*, 411–416.
3. Akkas, N. Bifurcation and snap-through phenomena in asymmetric dynamic analysis of shallow spherical shells. *Comput. Struct.* **1976**, *6*, 241–251.
4. Papadrakakis, M. Post-buckling analysis of spatial structures by vector iteration methods. *Comput. Struct.* **1981**, *14*, 393–402.
5. Hill, C.D.; Blandford, G.E.; Wang, S.T. Post-bucking analysis of steel space trusses. *J. Struct. Eng.* **1989**, *115*, 900–919.
6. Kong, X.; Wang, B.; Hu, J. Dynamic snap buckling of an elastoplastic shallow arch with elastically supported and clamped ends. *Comput. Struct.* **1995**, *55*, 163–166.
7. Blair, K.B.; Krousgrill, C.M.; Farris, T.N. Non-linear dynamic response of shallow arches to harmonic forcing, *J. Sound Vib.* **1996**, *194*, 353–367.
8. Blandford, G.E. Progressive failure analysis of inelastic space truss structures. *Comput. Struct.* **1996**, *58*, 981–990.
9. Sansour, C.; Wriggers, P.; Sansour, J. Nonlinear dynamics of shells: Theory, finite element formulation, and integration schemes. *Nonlinear Dyn.* **1997**, *13*, 279–305.
10. Xu, J.X.; Huang, H.; Zhang, P.Z.; Zhou, J.Q. Dynamic stability of shallow arch with elastic supports-application in the dynamic stability analysis of inner winding of transformer during short circuit. *Int. J. Non-Linear Mech.* **2002**, *37*, 909–920.

11. Kassimali, A.; Bidhendi, E. Stability of trusses under dynamic loads. *Comput. Struct.* **1988**, *29*, 381–392.

12. Tada, M.; Suito, A. Static and dynamic post-buckling behavior of truss structures. *Eng. Struct.* **1998**, *20*, 384–389.

13. Ha, J.H.; Gutman, S.; Shon, S.D.; Lee, S.J. Stability of shallow arches under constant load. *Int. J. Nonlinear Mech.* **2014**, *58*, 120–127.

14. Shon, S.D.; Lee, S.J.; Lee, G.G. Characteristics of bifurcation and buckling load of space truss in consideration of initial imperfection and load mode. *J. Zhejiang Univ. Sci. A* **2013**, *14*, 206–218.

15. Kim, S.D.; Kang, M.M.; Kwun, T.J.; Hangai, Y. Dynamic instability of shell-like shallow trusses considering damping. *Comput. Struct.* **1997**, *64*, 481–489.

16. Slaats, P.M.A.; Jongh, J.; Sauren, A.A.H.J. Model reduction tools for nonlinear structural dynamics. *Comput. Struct.* **1995**, *54*, 1155–1171.

17. Coan, C.H.; Plaut, R.H. Dynamic stability of a lattice dome. *Earthq. Eng. Struct. Dyn.* **1983**, *11*, 269–274.

18. Belytschko, T. A survey of numerical methods and computer programs for dynamic structural analysis. *Nucl. Eng. Des.* **1976**, *37*, 23–34.

19. Sadighi, A.; Ganji, D.D.; Ganjavi, B. Travelling wave solutions of the Sine-Gordon and the coupled Sine-Gordon equations using the homotopy perturbation method. *Sci. Iran. Trans. B Mech. Eng.* **2007**, *16*, 189–195.

20. Barrio, R. Performance of the Taylor series method for ODEs/DAEs. *Appl. Math. Comput.* **2005**, *163*, 525–545.

21. Barrio, R.; Blesa, F.; Lara, M. VSVO Formulation of the Taylor method for the numerical solution of ODEs. *Comput. Math. Appl.* **2005**, *50*, 93–111.

22. Adomian, G.; Rach, R. Generalization of adomian polynomials to functions of several variables. *Comput. Math. Appl.* **1992**, *24*, 11–24.

23. Adomian, G.; Rach, R. Modified adomian polynomials. *Math. Comput. Model.* **1996**, *24*, 39–46.

24. He, J.H. Homotopy perturbation method: A new nonlinear analytical technique. *Appl. Math. Comput.* **2003**, *135*, 73–79.

25. He, J.H. Application of homotopy perturbation method to nonlinear wave equations. *Chaos Solitons Fractals* **2005**, *26*, 695–700.

26. Chowdhury, M.S.H.; Hashim, I.; Momani, S. The multistage homotopy-perturbation method: A powerful scheme for handling the Lorenz system. *Chaos Solutions Fractals* **2009**, *40*, 1929–1937.

27. Barrio, R.; Rodriguez, M.; Abad, A.; Blesa, F. Breaking the limits: the Taylor series method. *Appl. Math. Comput.* **2011**, *217*, 7940–7954.

28. Abad, A.; Barrio, R.; Blesa, F.; Rodriguez, M. TIDES: A Taylor series integrator for differential equations. *ACM Trans. Math. Softw. TOMS* **2012**, *39*, doi:10.1145/2382585.2382590.

29. Rodriguez, M.; Barrio, R. Reducing rounding errors and achieving Brouwer's law with Taylor Series Method. *Appl. Numer. Math.* **2012**, *62*, 1014–1024.

12

Application of Turkevich Method for Gold Nanoparticles Synthesis to Fabrication of SiO$_2$@Au and TiO$_2$@Au Core-Shell Nanostructures

Paulina Dobrowolska [1], Aleksandra Krajewska [1], Magdalena Gajda-Rączka [1], Bartosz Bartosewicz [1], Piotr Nyga [1] and Bartłomiej J. Jankiewicz [1,*]

Institute of Optoelectronics, Military University of Technology, Warsaw 00-908, Poland;
E-Mails: paulina.doborowolska@student.wat.edu.pl (P.D.); aleksandra.krajewska@wat.edu.pl (A.K.);
magdalena.gajda-raczka@wat.edu.pl (M.G.R.); bartosz.bartosewicz@wat.edu.pl (B.B.);
piotr.nyga@wat.edu.pl (P.N.)

* Author to whom correspondence should be addressed; E-Mail: bartlomiej.jankiewicz@wat.edu.pl;

Academic Editor: Gururaj V. Naik

Abstract: The Turkevich synthesis method of Au nanoparticles (AuNPs) was adopted for direct fabrication of SiO$_2$@Au and TiO$_2$@Au core-shell nanostructures. In this method, chloroauric acid was reduced with trisodium citrate in the presence of amine-functionalized silica or titania submicroparticles. Core-shells obtained in this way were compared to structures fabricated by mixing of Turkevich AuNPs with amine-functionalized silica or titania submicroparticles. It was found that by modification of reaction conditions of the first method, such as temperature and concentration of reagents, control over gold coverage on silicon dioxide particles has been achieved. Described method under certain conditions allows fabrication of semicontinuous gold films on the surface of silicon dioxide particles. To the best of our knowledge, this is the first report describing use of Turkevich method to direct fabrication of TiO$_2$@Au core-shell nanostructures.

Keywords: SiO$_2$@Au; TiO$_2$@Au; core-shell nanostructures; Au nanoparticles; Turkevich Method

1. Introduction

In recent years, nanotechnology is one of the fastest growing fields in science and engineering. One of the results of extremely rapid progress in this field is the discovery of the large number of functional materials in which at least one dimension is in the range of 1 to 100 nm. Some of the most prominent examples of these materials are noble metal nanostructures, such as nanoparticles (NPs) [1], nanowires [2,3], nanorods [3–5], nanotriangles [6], nanostars [7] and many others. Various properties and thus also possible applications of noble metal nanostructures, are dependent on their composition, shape and size [8–10]. The noble metals were also used for fabrication of more complex hybrid nanostructures, where they are combined with other materials. This often results in new functionalities as compared to nanostructures made from noble metal only. The core-shell nanostructures consisting of oxide core and noble metal shell is an interesting example of such hybrid nanostructures.

The core-shell nanostructures have received increasing attention since Halas and co-workers [11] reported studies on the synthesis of SiO_2@Au nanoparticles. In this work, a continuous gold nanoshell was grown on the surface of amine-functionalized silica in a two-step process. The amine-functionalized silica particles were decorated with very small Au nanoparticles (1–2 nm in diameter) in the first step. In the next step, the growth of the shell on gold-decorated silica particles was realized by a subsequent reduction of an aged mixture of chloroauric acid and potassium carbonate by a solution of sodium borohydride. The Au nanoparticles on silica surface served as a nucleation sites for the reduction. One of the most important characteristics of the core-shell nanostructures is that their optical properties, localized surface plasmon resonance (LSPR) band positions, can be tuned by their geometry, core radius/shell thickness ratio. The possibility of core-shell nanostructures optical properties tuning has opened the door to many applications such as cancer therapy [12], surface enhanced spectroscopies [13,14] or steam generation for solar autoclave [15].

Following work of Halas [11], several approaches to fabrication of core-shell nanostructures based on the silica core and gold shell have been reported [16–33]. The most commonly used approach to synthesis of core-shell nanostructures based on silica and gold involves a two-step process, deposition of gold seeds on functionalized silica surface, followed by gold shell growth. Research studies based on this approach were mainly focused on the investigations of the influence of silica surface functionalization and reducing agent used for shell growth on the formation of gold shell.

The silica surface functionalization is crucial for deposition of small gold nanoparticles, which serve as seeds for shell growth. The surface of silica was most frequently functionalized with various organosilanes containing functional groups with high affinity to gold, such as amino ($-NH_2$) or mercapto (-SH) group [11,17], or using polymers containing multiple $-NH_2$ groups, such as polyethyleneimine (PEI) [18] or poly(diallyldimethylammonium chloride) (PDADMAC) [19]. In most cases, the organosilanes-containing amino group were used for functionalization because they provided uniform and relatively high surface coverage (up to 30%) with gold nanoparticles, which is very important for obtaining smooth continuous shell.

Most of the studies regarding a two-step process focused on the influence of the gold salts reducing agents on the shell formation. In the first studies on synthesis of silica-gold core-shell nanostructures, sodium borohydride was used as a reducing agent [11], however one of the most often used reducing agent in SiO_2@Au nanostructures synthesis is formaldehyde [20–22]. Due to the toxicity of formaldehyde, researchers have also looked for safer alternatives to it, such as glucose [23] or ascorbic acid [24]. Also, hydroxylamine [25] and CO [26] were used as reducing agents. Use of gaseous CO as reducing agent results in uniform shell formation [26].

Only a few studies regarding direct growth of gold shell on the non-functionalized silica surface have been reported. These studies include the reduction of gold (I) chloride in acetonitrile with ascorbic acid [27], formaldehyde stimulated shell growth on Sn-seeded silica particles [28] and ultrasound assisted deposition of gold nanoparticles on silica particles surface [29]. Similarly, studies on direct growth of gold shell on amino-functionalized silica particle are limited to a few articles. For example, gold salts were directly reduced on functionalized surface of SiO2 particles using formaldehyde [30]. An interesting approach involves direct growth of gold shell on functionalized silica surface by reduction of gold salts with trisodium citrate [31–33]. Trisodium citrate was used by Turkevich [34] and Frens [35] for the synthesis of monodispersed gold nanoparticles and has become probably one of the most often used reducing agents for this purpose. It is not surprising since it is cheap, not hazardous, by varying its concentration one can control the size of synthesized gold nanoparticles, and, most importantly, it acts as both reducing agent and stabilizing agent. In studies reported by Zhang et al. [31], gold salts were reduced directly on the surface of amino-functionalized silica particles (average diameter of 227 nm) and the silica particles decorated with gold particles (average diameter of 14 nm) at 100 °C by rapidly injected trisodium citrate. In both cases rather non-uniform coverage with gold was achieved, however the coverage was higher in case of reduction on silica particles decorated with gold nanoparticles. Storti et al. [32] synthesized SiO2@Au core-shell nanoparticles by gold salts reduction on amino-functionalized silica nanoparticles (average diameter 16 nm) obtained in a one-step process. The formation of complete gold shell in these studies was carried out by slow addition of HAuCl4 solution to silica particles suspension containing trisodium citrate at 100 °C [32]. In studies of Zhang et al. [33], gold shell was grown on the tris(2,2′-bipyridyl)ruthenium(II) chloride embedded amino-functionalized SiO2 particles (diameter ~70nm). In this method, reducing agent was added dropwise to mixture containing functionalized cores and gold salt. Au shells of different thickness were obtained by adjusting the ratio of the SiO2 NPs and gold salt, while maintaining trisodium citrate in excess.

Herein, we report studies on the application of modified Turkevich synthesis method of Au nanoparticles to direct fabrication of SiO2@Au core-shell nanostructures. The effect of reaction temperature and amount of silica particles used in the reaction on deposition of gold nanostructures on the amino-functionalized silica particle was investigated. In addition, this method was for the first time applied to synthesis of TiO2@Au core-shell nanostructures.

2. Results and Discussion

The synthesis of SiO2@Au and TiO2@Au core-shell nanostructures described in this article was carried out according to procedures shown in Figure 1. The commercially available submicrometer silica and titania (rutile) particles, which are used as cores, were functionalized with -NH2 groups. The amino groups served as anchor groups, which bind gold nanoparticles to surface of functionalized SiO2 and TiO2 particles. In the first method used in these studies, amino-functionalized cores were decorated with 16–20 nm AuNPs synthesized using Turkevich method [34]. These particles served as reference structures to nanostructures obtained by direct reduction of gold salts on functionalized core using modification of Turkevich method. The syntheses of SiO₂@Au and TiO₂@Au nanostructures via direct reduction were carried out at different temperatures. In addition, the influence of different amounts of silica cores on the morphology of gold shell was investigated.

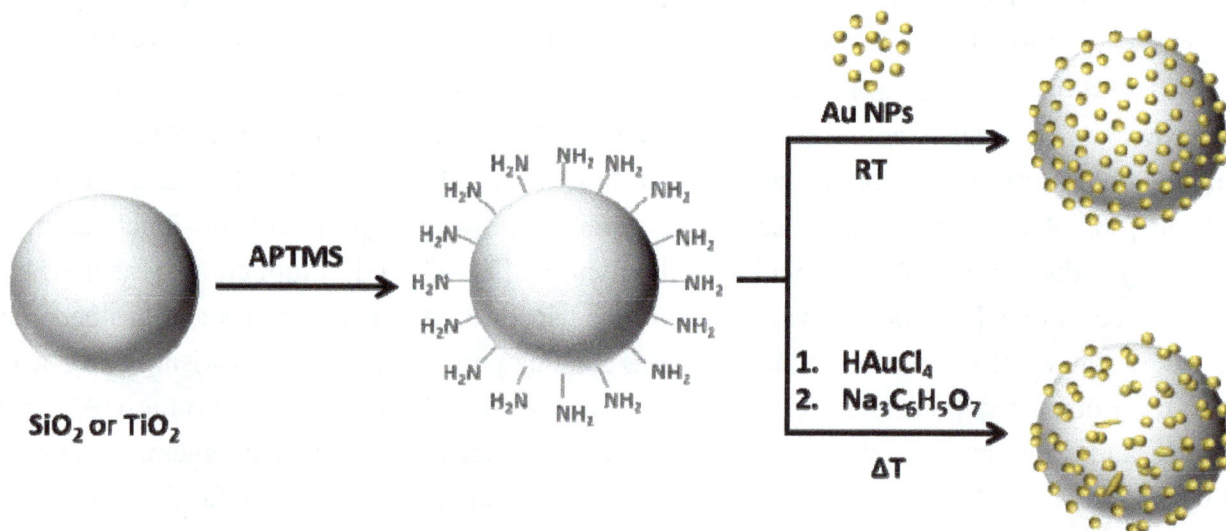

Figure 1. Schematic illustration of SiO₂@Au and TiO₂@Au core-shell nanostructures fabrication via decoration of NH₂-functionalized cores with AuNPs and via direct reduction of gold salts.

2.1. Deposition of AuNPs on Nh2-Functionalized Silicon Dioxide and Titanium Dioxide Cores

The silicon dioxide and titanium dioxide cores shown in Figure 2a,c were functionalized with amino groups and mixed with Turkevich AuNPs to yield structures shown in Figure 2b,d. For both core materials, the deposition yields structures with monodispersed gold nanoparticles evenly distributed on the surface of SiO₂ and TiO₂ cores. Gold nanoparticles are usually well separated on the SiO₂ core surface, however in some areas of core surface clusters containing from two up to several AuNPs are also observed. This phenomenon is more pronounced in the case of TiO₂ cores, where AuNPs are more densely packed. The degree of surface coverage with AuNPs and the way AuNPs are distributed on the core surface are dependent on the surface modifier used for core surface functionalization [36] and can be controlled by amount of AuNPs used for this process.

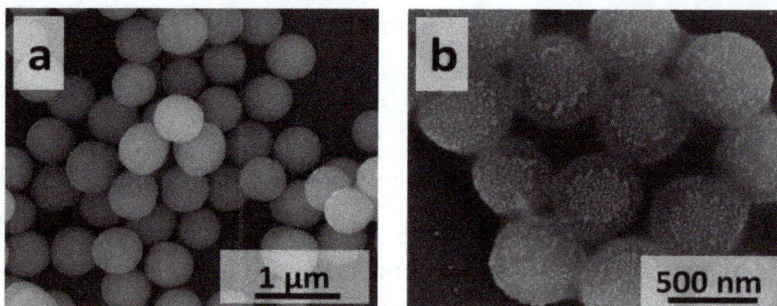

Figure 2. *Cont.*

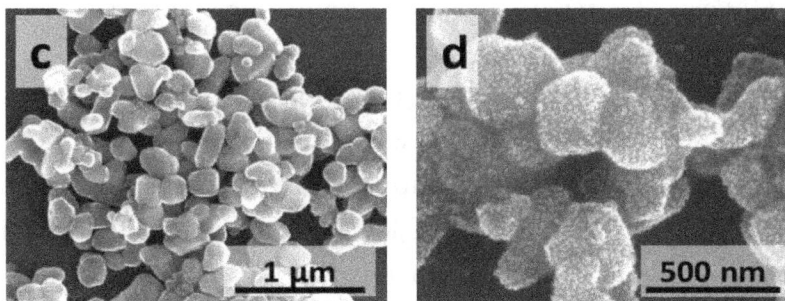

Figure 2. Scanning Electron Microscopy (SEM) images of silicon dioxide and titanium dioxide cores before and after deposition of Turkevich AuNPs: (**a,b**) SiO_2 and (**c,d**) TiO_2.

Different gold coverage on the SiO_2 and TiO_2 cores may be associated with the different content of OH groups on the surface of non-functionalized SiO_2 and TiO_2 cores, which may be up to four times higher for TiO_2 [37–39]. The higher number of OH groups on the surface of TiO_2 particles may result in higher number of NH_2 groups after functionalization step and thus provide more anchoring groups for AuNPs binding.

The UV-Vis spectra of silicon dioxide and titanium dioxide cores before and after deposition of Turkevich AuNPs are shown in Figure 3. Due to separation of AuNPs on the surface of silica cores, UV-Vis spectrum of SiO_2@Au nanostructures closely resembles spectrum of AuNPs (not presented) and has extinction band centered at about 550 nm. The significant red shift of the maximum of absorption (λ_{max} = 670 nm) was observed in case of TiO_2@Au nanostructures. The spectral location of plasmon resonance of single gold nanoparticle is dependent on the refractive index of surrounding medium. Therefore, the red shift observed for TiO_2@Au in the UV-Vis spectrum as compared to SiO_2@Au is likely related to the higher refractive index of TiO_2 (n ~ 2.2–2.6) particles compared to SiO_2 particles (n ~ 1.45). In addition, both red shift and broadening of the spectrum may be associated with dense packing of AuNPs on TiO_2 particles surface resulting in Au nanoparticles plasmon-plasmon coupling.

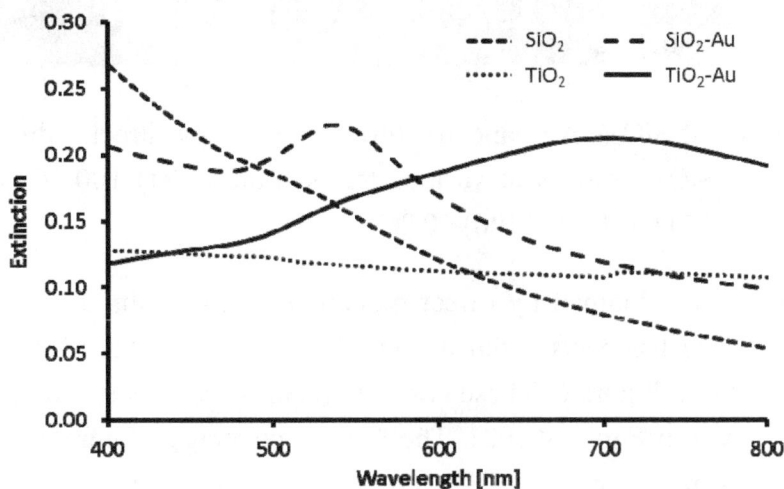

Figure 3. UV-Vis spectra of silicon dioxide and titanium dioxide cores before and after deposition of Turkevich AuNPs.

2.2. Direct Reduction of Gold Salts on NH₂-Functionalized Silicon Dioxide Cores

The SiO₂@Au nanostructures obtained by direct reduction of gold salts on functionalized SiO₂ particles at temperatures in the 50–100 °C range are shown in Figure 4. For all temperatures, in addition to desired core-shell structures, formation of large amount of gold nanoparticles not attached to SiO₂ particles was observed. The aim of lowering the reaction temperature was to slow down gold salts reduction process and therefore to provide more time for gold seeds to bind to NH₂-functionalized silicon dioxide cores. As a result, higher coverage of cores with gold was expected. However, the opposite effect was observed. Denser and more uniform coverage was observed for higher reaction temperatures 80–100 °C. In these temperatures, syntheses were more reproducible. Direct reduction of gold salts on SiO₂ cores at higher temperatures yields SiO₂@Au nanostructures with denser coverage of gold as compared to SiO₂@Au nanostructures obtained by AuNPs deposition. In addition, the gold nanostructures fabricated on core surface by direct reduction are of different shape and sizes, ranging from a few to several nanometers. Despite the fact that variety of gold nanostructures were obtained on SiO₂ cores, their UV-Vis spectra are very similar to the UV-Vis spectrum of silica cores decorated with AuNPs (Figures 3 and 5).

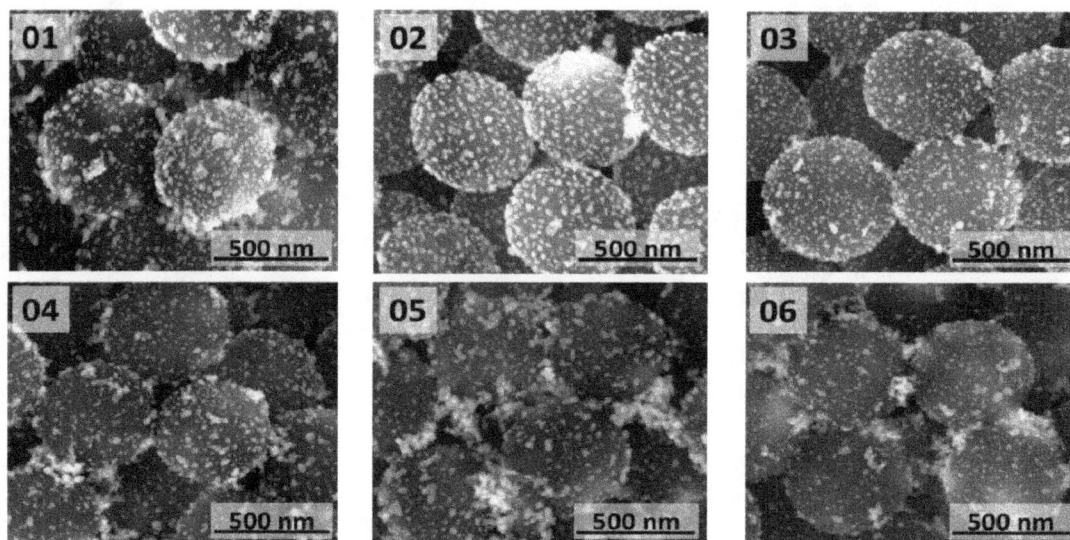

Figure 4. SEM images of SiO₂@Au nanostructures obtained by direct reduction of gold salts on functionalized SiO₂ particles at various temperatures: **(01)** 100 °C; **(02)** 90 °C; **(03)** 80 °C; **(04)** 70 °C; **(05)** 60 °C; and **(06)** 50 °C.

The SiO₂@Au nanostructures obtained by direct reduction of gold salts at 90 °C and 100 °C in reaction mixture containing different amounts of functionalized SiO₂ particles (2.5, 13, 50 and 125 μg per mL of reaction mixture) are shown in Figure 6. These core-shell nanostructures could also be compared to SiO₂@Au nanostructure shown in Figure 4 (02). The highest coverage of core surface was obtained when only 2.5 μg/mL of silica core particles were used (Figure 6 (07)). In these conditions, semicontinuous gold film formed on the silica surface. Unfortunately, due to low number of used silica cores, it was impossible to measure UV-Vis spectra of these structures. Structures with such shell morphology are expected to have high extinction in broad spectral range extending to longer wavelengths [40,41]. Based on SEM images (Figure 6), it is clearly seen that by increasing number of functionalized cores, lower gold coverage is achieved and also less Au particles and larger separation between them are observed. The UV-Vis spectra of (08)–(10) are qualitatively the same as for the (03)–(04) samples, but with less pronounced AuNPs extinction peak.

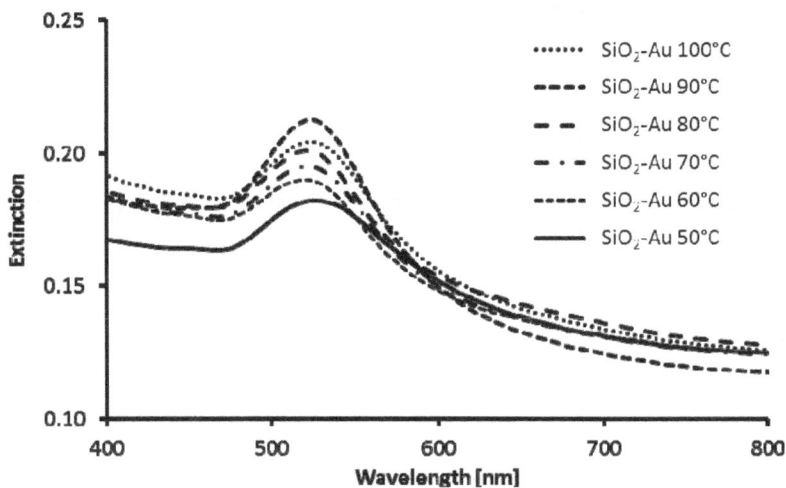

Figure 5. UV-Vis spectra of SiO_2@Au nanostructures obtained by direct reduction of gold salts on functionalized SiO_2 particles at various temperatures: (**01**) 100 °C; (**02**) 90 °C; (**03**) 80 °C; (**04**) 70 °C; (**05**) 60 °C; and (**06**) 50 °C.

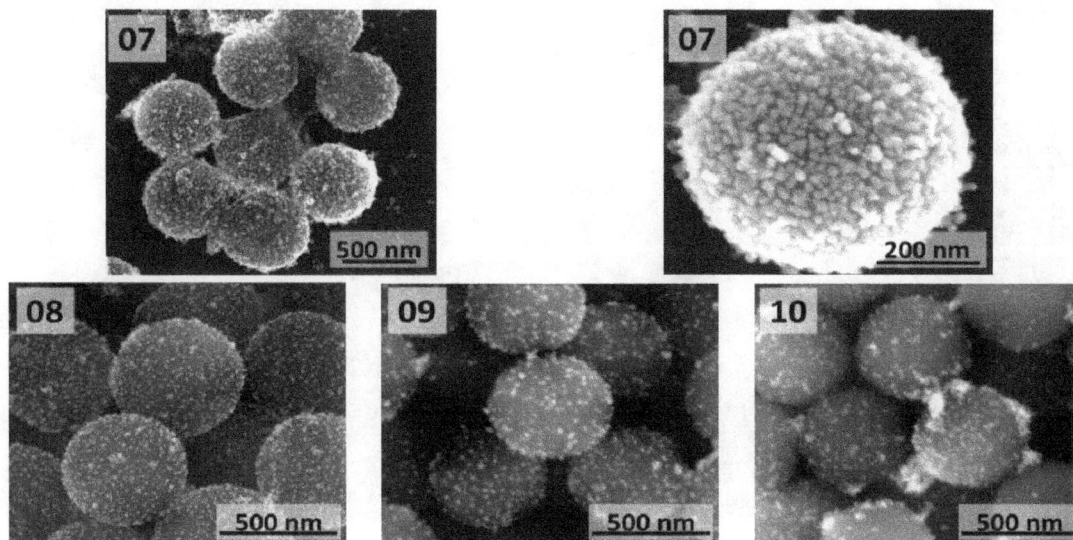

Figure 6. SEM images of SiO_2@Au nanostructures obtained by direct reduction of gold salts in reaction mixture containing different amounts of functionalized SiO_2 particles: (**07**) 2.5 µg/mL; (**08**) 13 µg/mL; (**09**) 50 µg/mL; and (**10**) 125 µg/mL.

The comparison of the results obtained in these studies, syntheses (1)–(10), to previously reported studies [31–33] on SiO_2@Au fabrication using trisodium citrate as reducing agent of gold salts yields some interesting observations. The method described by Zhang *et al.* [31], which closely resembles method reported in this article, yields SiO_2@Au nanostructure with rather low surface coverage with gold, even in case of deposition of AuNPs on functionalized cores. In the studies reported here, surface coverage seems to be higher. These differences could be associated with the methods used for the functionalization of silica core. In our studies, functionalization of core surfaces in toluene allowed obtaining higher coverage with gold nanoparticles than in the case of functionalization in ethanol used in Zhang studies. It is quite difficult to relate the here reported studies to earlier studies [31–33] where trisodium citrate was used as a reducing agent. For all three cases, cores of different sizes, different silica surface functionalization methods, and reaction conditions were used.

2.3. Direct Reduction of Gold Salts on NH2-Functionalized Titanium Dioxide Cores

The TiO2@Au nanostructures obtained by direct reduction of gold salts on functionalized TiO2 particles using trisodium citrate at temperatures in the 50–100 °C range are presented in Figure 7. To the best of our knowledge, it is the first time this method is used for synthesis of TiO2@Au nanostructures. In our studies on the synthesis of TiO2@Au core-shell structures, we have applied many methods that were used for SiO2@Au [16]. However, in many cases methods, which worked fine for SiO2 failed for TiO2. It is related to the fact that opposite to chemically inert SiO2, TiO2 exhibits photocatalytic properties even in amorphous form. Because of that, formaldehyde, most commonly used reducing agent in case of SiO2, did not work in our experiments for TiO2. From the other hand, as shown in studies reported here, use of trisodium citrate yields TiO2@Au core-shell structures with relatively high gold coverage.

Figure 7. SEM images of TiO2@Au nanostructures obtained by direct reduction of gold salts on functionalized TiO2 particles at various temperatures: (**11**) 100 °C; (**12**) 90 °C; (**13**) 80 °C; (**14**) 70 °C; (**15**) 60 °C; and (**16**) 50 °C.

Similarly to studies for silica cores, the aim of lowering temperature was to slow down gold salts reduction process and therefore provide more time for gold seeds to bind to NH2-functionalized titanium dioxide cores. As a result, higher coverage of cores with gold was expected. However, it was observed that by lowering reaction temperature, less gold nanostructures were formed on the surface of TiO2 cores and, in addition, the size of AuNPs increased as the temperature was lowered. Lower number of gold nanostructures attached to cores observed for lower temperature reactions may be related to unbinding of large gold nanostructures from the core surface during the work up of a reaction. It seems that denser surface coverage with gold was observed when AuNPs were deposited on the NH_2-functionalized titanium dioxide cores (Figure 2d).

The UV-Vis spectra of TiO2@Au nanostructures obtained by direct reduction (10–15) (Figure 8) correspond well to spectrum of AuNPs and differ significantly from spectra of TiO2 cores decorated with AuNPs (Figure 3). In all cases, small red shifts of absorption maximum (λ_{max}) as compared to spectrum of AuNPs are observed, which is likely caused by larger size of gold nanostructures on the TiO2 surface compared to Turkevich AuNPs and TiO2 refractive index being higher than that of water.

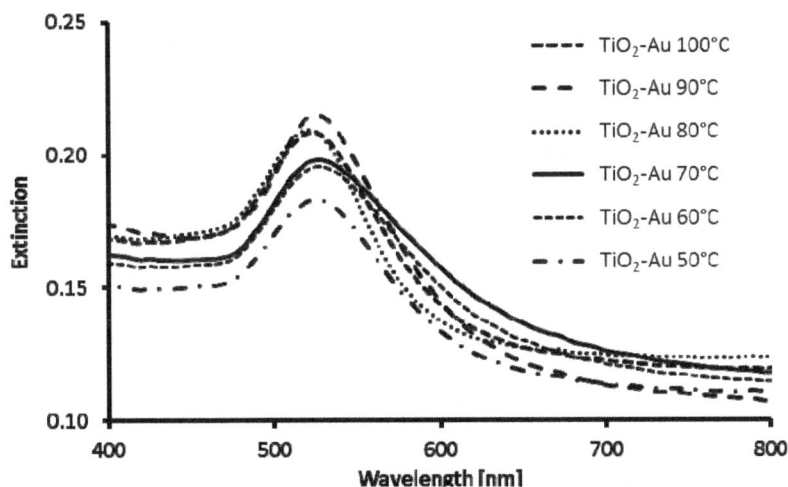

Figure 8. UV-Vis spectra of TiO2@Au nanostructures obtained by direct reduction of gold salts on functionalized TiO2 particles at various temperatures: (**11**) 100 °C; (**12**) 90 °C; (**13**) 80 °C; (**14**) 70 °C; (**15**) 60 °C; and (**16**) 50 °C.

3. Experimental Section

3.1. Reagents

Silica particles (spherical particles with average diameter of 670 nm), titania particles (rutile particles of irregular shape and size in range of 100–200 nm) and trisodium citrate were purchased from POCH S.A. (3-Aminopropyl)-trimethoxysilane (APTMS) was purchased from Sigma Aldrich. Hydrogen tetrachloroaurate(III) trihydrate (HAuCl4) was purchased from Alfa Aesar and was used to prepare 5% dark-aged stock solution of chloroauric acid, later used in the experiments. Other reagents were of analytical grade. All the chemicals were used without further purification. Deionized (DI) water (resistance > 18.2 MΩ) used in all synthesis and washing was prepared by an ultrapure water system HLP 5UV (Hydrolab).

3.2. Functionalization of Silica and Titania Submicroparticles

Silica and titania submicroparticles were functionalized with amine groups using modified method described by Jaroniec *et al.* [17]. The silica or titania particles (0.5 g) were added to 50 mL of anhydrous toluene in Falcon tube, which was then placed in ultrasonic bath and sonicated for 1 h. The suspension was transferred to round bottom flask and allowed to stir vigorously for 1 h. Next, the 2.5 mL of APTMS was added to mixing suspension. Such reaction mixture was boiled under reflux for 24 h.

After completion of the reaction, modified particles were washed several times on a membrane filter with small portions of toluene and ethanol to remove an excessive amount of modifier. The functionalized particles were dried overnight in an oven at ~100 °C.

3.3. Deposition of Turkevich AuNPs on Silica and Titania Particles

Gold nanoparticles were synthesized according to Turkevich method [34]. Briefly, 1 mL of solution of 5 mM $HAuCl_4$ in DI water was added to 18 mL of DI water and mixture was heated until it began to boil. One milliliter of 0.5% trisodium citrate solution was added as soon as boiling commenced. The reaction mixture was heated until evident color change had occurred and then it was removed from heating plate and stirred until it cooled down to room temperature.

Five milligrams of amine-functionalized silica or titania particles were dispersed in 10 mL of DI water under ultrasound for 1 h. Next, 4.5 mL of the silica or titania functionalized particles suspension was added to 30 mL of Turkevich AuNPs solution and mixed for a few hours. In order to remove AuNPs unattached to silica or titania particles surface, the reaction mixture was subjected to a few washing steps by centrifugation and redispersion in 0.5% solution of trisodium citrate.

3.4. Direct Deposition of Au Nanostructures on Silica and Titania Particles Using Turkevich Method

In typical synthesis, 1 mL of 5 mM $HAuCl_4$ aqueous solution was added to 17 mL of DI water and mixed for 5 min before addition of 1 mL of water suspension containing 0.5 mg of amine-functionalized silica or titania particles. Resulting reaction mixture was heated until it reached the selected reaction temperature in the 50–100 °C range and then the preheated 1 mL of 0.5% trisodium citrate solution was added. The reaction mixture was heated until evident color change had occurred and then it was removed from heating plate and continued to stir until it cooled down to room temperature. In order to remove AuNPs unattached to silica or titania particles surface the reaction mixture was subjected to a few washing steps by centrifugation and redispersion in 0.5% solution of trisodium citrate. Conditions of (01)–(16) syntheses described in article are given in Tables 1 and 2.

Table 1. Reaction temperatures and amount of SiO_2-NH_2 particles used in the syntheses 01–10.

Sample Reaction Conditions	01	02	03	04	05	06	07	08	09	10
SiO_2-NH_2 (µg/mL)	25	25	25	25	25	25	2.5	13	50	125
T (°C)	100	90	80	70	60	50	100	90	90	90

Table 2. Reaction temperatures and amount of TiO_2-NH_2 particles used in the syntheses 11–16.

Sample Reaction Conditions	11	12	13	14	15	16
TiO_2-NH_2 (µg/mL)	25	25	25	25	25	25
T (°C)	100	90	80	70	60	50

3.5. Characterization

The UV–Vis extinction spectra were measured at room temperature using Lambda 900 UV-Vis-NIR spectrophotometer (Perkin Elmer, Waltham, MA, USA) in the 400–800 nm spectral range. Suspensions of synthesized nanostructures were measured in 1 cm optical path quartz cuvette.

The morphology of SiO$_2$-Au and TiO$_2$-Au nanostructures was characterized based on the images obtained using a Quanta 3D FEG Dual Beam scanning electron microscope (FEI, Hillsboro, OR, USA). Samples for SEM were prepared by drop-casting of suspensions of core-shell nanostructures on silicon wafer and drying in air.

4. Conclusions

In this paper, we applied Turkevich method for AuNPs synthesis to fabrication of SiO$_2$@Au and TiO$_2$@Au nanostructures. The nanostructures synthesized using this method were compared to SiO$_2$@Au and TiO$_2$@Au nanostructures obtained by deposition of Turkevich AuNPs on amino-functionalized SiO$_2$ and TiO$_2$ cores. The nanostructures fabricated using both methods have different morphologies of gold shell and exhibit different optical properties. In the case of SiO$_2$ cores, both methods may yield core-shell structures with similar optical properties. However, by control of reaction conditions, it is possible to obtain core-shells with semicontinuous gold shell, which are expected to have high extinction in broad spectral range. Interestingly, differences in optical properties were observed for TiO$_2$@Au structures. The significant red shift and broadening of the spectrum in the case of AuNPs decorated TiO$_2$ cores are likely related to high refractive index of TiO$_2$ and dense packing of AuNPs on TiO$_2$ particles surface, resulting in strong AuNPs plasmon-plasmon coupling. Modified Turkevich method yields core-shell nanostructures with dense surface and high reproducibility at temperatures in the 80–100 °C range. To the best of our knowledge, this is the first report describing the use of Turkevich method to direct fabrication of TiO$_2$@Au core-shell nanostructures.

Acknowledgments

Research was supported by project 2011/03/D/ST5/06038 funded by the Polish National Science Centre, project ON507282540 funded by the Polish National Centre for Research and Development (NCBiR) and project OR00005408 funded by the Polish Ministry of Science and Higher Education. We thank Dariusz Zasada for SEM characterization.

Author Contributions

Bartłomiej J. Jankiewicz organized the research and wrote the manuscript; Paulina Dobrowolska, Aleksandra Krajewska and Magdalena Gajda-Rączka performed the experiments; Bartosz Bartosewicz and Piotr Nyga discussed the experiments and participated in writing manuscript. The manuscript was reviewed by all authors.

Conflicts of Interest

The authors declare no conflict of interest.

References

1. Rodriguez-Fernandez, J.; Perez-Juste, J.; Garcia de Abajo, F.J.; Liz-Marzan, L.M. Seeded growth of submicron Au colloids with quadrupole plasmon resonance rodes. *Langmuir* **2006**, *22*, 7007–7010.

2. Kim, F.; Sohn, K.; Wu, J.; Huang, J. Chemical synthesis of gold nanowires in acidic solutions. *J. Am. Chem. Soc.* **2008**, *130*, 14442–14443.

3. Jana, N.R.; Gearheart, L.; Murphy, C.J. Wet Chemical synthesis of silver nanorods and nanowires of controllable aspect ratio. *Chem. Commun.* **2001**, 617–618.

4. Jana, N.R.; Gearheart, L.; Murphy, C.J. Wet chemical synthesis of high aspect ratio gold nanorods. *J. Phys. Chem. B* **2001**, *105*, 4065–4067.

5. Vigderman, L.; Khanal, B.P.; Zubarev, E.R. Functional gold nanorods: Synthesis, self-assembly, and sensing applications. *Adv. Mater.* **2012**, *24*, 4811–4841.

6. Scarabelli, L.; Coronado-Puchau, M.; Giner-Casares, J.J.; Langer, J.; Liz-Marzán, L.M. Monodisperse gold nanotriangles: Size control, large-scale self-assembly, and performance in surface-enhanced raman scattering. *ACS Nano* **2014**, *8*, 5833–5842.

7. Kumar, P.S.; Pastoriza-Santos, I.; Rodriguez-Gonzalez, B.; Garcia de Abajo, F.J.; Liz-Marzan, L.M. High-yield synthesis and optical response of gold nanostars. *Nanotechnol.* **2008**, *19*, doi:10.1088/0957-4484/19/01/015606.

8. Jain, P.K.; Huang, X.; El-Sayed, I.H.; El-Sayed, M.A. Noble metals on the nanoscale: Optical and photothermal properties and some applications in imaging, sensing, biology, and medicine. *Acc. Chem. Res.* **2008**, *41*, 1578–1586.

9. Jain, P.K.; Lee, K.S.; El-Sayed, I.H.; El-Sayed, M.A. Calculated absorption and scattering properties of gold nanoparticles of different size, shape, and composition: Applications in biological imaging and biomedicine. *J. Phys. Chem. B* **2006**, *110*, 7238–7248.

10. Kelly, K.L.; Coronado, E.; Zhao, L.L.; Schatz, G.C. The optical properties of metal nanoparticles: The influence of size, shape, and dielectric environment. *J. Phys. Chem. B* **2003**, *107*, 668–677.

11. Oldenburg, S.J.; Averitt, R.D.; Westcott, S.L.; Halas, N.J. Nanoengineering of optical resonances. *Chem. Phys. Lett.* **1998**, *288*, 243–247.

12. Bardhan, R.; Lal, S.; Joshi, A.; Halas, N.J. Theranostic nanoshells: From probe design to imaging and treatment of cancer. *Acc. Chem. Res.* **2011**, *44*, 936–946.

13. Le, F.; Brandl, D.W.; Urzhumov, Y.A.; Wang, H.; Kundu, J.; Halas, N.J.; Aizpurua, J.; Nordlander, P. Metallic Nanoparticle arrays: A common substrate for both surface-enhanced raman scattering and surface-enhanced infrared absorption. *ACS Nano* **2008**, *2*, 707–718.

14. Bardhan, R.; Grady, N.K.; Cole, J.R.; Joshi, A.; Halas, N.J. Fluorescence enhancement by au nanostructures: Nanoshells and nanorods. *ACS Nano* **2009**, *3*, 744–752.

15. Neumann, O.; Ferontic, C.; Neumann, A.D.; Dong, A.; Schell, K.; Lue, B.; Kime, E.; Quinne, M.; Thompson, S.; Grady, N.; *et al.* Compact solar autoclave based on steam generation using broadband light-harvesting nanoparticles. *Proc. Natl. Acad. Sci. USA* **2013**, *110*, 11677–11681.

16. Jankiewicz, B.J.; Choma, J.; Jamioła, D.; Jaroniec, M. Silica-metal core-shell nanostructures. *Adv. Colloid Interface Sci.* **2012**, *170*, 28–47.

17. Antoschshuk, V.; Jaroniec, M. Adsorption, thermogravimetric, and nmr studies of fsm-16 material functionalized with alkylmonochlorosilanes. *J. Phys. Chem. B* **1999**, *103*, 6252–6261.

18. Xue, J.; Wang, C.; Ma, Z. A facile method to prepare a series of SiO2@Au core/shell structured nanoparticles. *Mater. Chem. Phys.* **2007**, *105*, 419–425.

19. Ashayer, R.; Mannan, S.H.; Sajjadi, S. Synthesis and characterization of gold nanoshells using poly(diallyldimethyl ammonium chloride). *Colloids Surf. A* **2008**, *329*, 134–141.

20. Pham, T.; Jackson, J.B.; Halas, N.J.; Lee, T.R. Preparation and characterization of gold nanoshells coated with self-assembled monolayers. *Langmuir* **2002**, *18*, 4915–4920.

21. Preston, T.C.; Signorell, R. Growth and optical properties of gold nanoshells prior to the formation of a continuous metallic layer. *ACS Nano* **2009**, *3*, 3696–3706.

22. Kim, J.H.; Bryan, W.W.; Lee, T.R. Preparation, Characterization, and optical properties of gold, silver, and gold-silver alloy nanoshells having silica cores. *Langmuir* **2008**, *24*, 11147–11152.

23. Tharion, J.; Satija, J.; Mukherji, S. Glucose mediated synthesis of gold nanoshells: A facile and eco-friendly approach conferring high colloidal stability. *RSC Adv.* **2014**, *4*, 3984–3991.

24. Kim, J.H.; Bryan, W.W.; Chung, H.W.; Park, C.Y.; Jacobson, A.J.; Lee, T.R. Gold, palladium, and gold-palladium alloy nanoshells on silica nanoparticle cores. *ACS Appl. Mater. Interfaces* **2009**, *1*, 1063–1069.

25. Graf, C.; van Blaaderen, A. Metallodielectric colloidal core-shell particles for photonic applications. *Langmuir* **2002**, *18*, 524–534.

26. Brinson, B.E.; Lassiter, J.B.; Levin, C.S.; Bardhan, R.; Mirin, N.; Halas, N.J. Nanoshells made easy: Improving Au layer growth on nanoparticle surfaces. *Langmuir* **2008**, *24*, 14166–14171.

27. English, M.D.; Waclawik, E.R. A novel method for the synthesis of monodisperse gold-coated silica nanoparticles. *J. Nanopart. Res.* **2012**, *14*, 650–660.

28. Lim, Y.T.; Park, O.O.; Jung, H.T. Gold nanolayer-encapsulated silica particles synthesized by surface seeding and shell growing method: Near infrared responsive materials. *J. Colloid Interface Sci.* **2003**, *263*, 449–453.

29. Pol, V.G.; Gedanken, A.; Calderon-Moreno, J. Deposition of gold nanoparticles on silica spheres: A Sonochemical approach. *Chem. Mater.* **2003**, *15*, 1111–1118.

30. Choma, J.; Dziura, A.; Jamioła, D.; Nyga, P.; Jaroniec, M. Preparation and properties of silica-gold core-shell particles. *Colloids Surf. A* **2011**, *373*, 167–171.

31. Zhang, L.; Feng, Y.G.; Wang, L.Y.; Zhang, J.Y.; Chen, M.; Qian, D.J. Comparative studies between synthetic routes of SiO2@Au composite nanoparticles. *Mater. Res. Bull.* **2007**, *42*, 1457–1467.

32. Storti, B.; Elisei, F.; Abbruzzetti, S.; Viappiani, C.; Latterini, L. One-pot synthesis of gold nanoshells with high photon-to-heat conversion efficiency. *J. Phys. Chem. C* **2009**, *113*, 7516–7521.

33. Zhang, P.; Guo, Y. Surface-enhanced raman scattering inside metal nanoshells. *J. Am. Chem. Soc.* **2009**, *131*, 3808–3809.

34. Turkevich, J.; Stevenson, P.L.; Hillier, J. A study of the nucleation and growth process in the synthesis of colloidal gold. *Discuss. Faraday Soc.* **1951**, *11*, 55–75.

35. Frens, G. Controlled nucleation for the regulation of the particle size in monodisperse gold suspensions. *Nature Phys. Sci.* **1973**, *241*, 20–22.

36. Westcott, S.L.; Oldenburg, S.J.; Lee, T.R.; Halas, N.J. Formation and adsorption of clusters of gold nanoparticles onto functionalized silica nanoparticle surfaces. *Langmuir* **1998**, *14*, 5396–5401.

37. Mueller, R.; Kammler, H.K.; Wegner, K.; Pratsinis, S.E. OH surface density of SiO2 and TiO2 by thermogravimetric analysis. *Langmuir* **2003**, *19*, 160–165.

38. Barabanova, A.I.; Pryakhina, T.A.; Afanas'ev, E.S.; Zavin, B.G.; Vygodskii, Y.S.; Askadskii, A.A.; Philippova, O.E.; Khokhlov, A.R. Anhydride modified silica nanoparticles: Preparation and characterization. *Appl. Surface Sci.* **2012**, *258*, 3168–3172.

39. Takahashi, J.; Itoh, H.; Motai, S.; Shimada, S. Dye adsorption behavior of anatase- and rutile-type TiO$_2$ nanoparticles modified by various heat-treatments. *J. Mater. Sci.* **2003**, *38*, 1695–1702.

40. De Silva, V.C.; Nyga, P.; Drachev, V.P. Scattering suppression of silica microspheres with semicontinuous plasmonic shell. *ArXiv E-Prints* **2015**, arXiv:1501.00233v1.

41. Rohde, C.A.; Hasegawa, K.; Deutsch, M. Coherent light scattering from semicontinuous silver nanoshells near the percolation threshold. *Phys. Rev. Lett.* **2006**, *96*, doi:10.1103/PhysRevLett.96.045503.

13

Bulk Heterojunction Solar Cell with Nitrogen-Doped Carbon Nanotubes in the Active Layer: Effect of Nanocomposite Synthesis Technique on Photovoltaic Properties

Godfrey Keru [1], Patrick G. Ndungu [1,2], Genene T. Mola [1] and Vincent O. Nyamori [1,*]

[1] School of Chemistry and Physics, University of KwaZulu-Natal, Private Bag X54001, Durban 4000, South Africa; E-Mails: 212561611@stu.ukzn.ac.za (G.K.); Mola@ukzn.ac.za (G.T.M.)

[2] Department of Applied Chemistry, University of Johannesburg, P.O. Box 17011, Doornfontein, Johannesburg 2028, South Africa; E-Mail: pndungu@uj.ac.za

* Author to whom correspondence should be addressed; E-Mail: nyamori@ukzn.ac.za;

Academic Editor: Klara Hernadi

Abstract: Nanocomposites of poly(3-hexylthiophene) (P3HT) and nitrogen-doped carbon nanotubes (N-CNTs) have been synthesized by two methods; specifically, direct solution mixing and *in situ* polymerization. The nanocomposites were characterized by means of transmission electron microscopy (TEM), scanning electron microscopy (SEM), X-ray dispersive spectroscopy, UV-Vis spectrophotometry, photoluminescence spectrophotometry (PL), Fourier transform infrared spectroscopy (FTIR), Raman spectroscopy, thermogravimetric analysis, and dispersive surface energy analysis. The nanocomposites were used in the active layer of a bulk heterojunction organic solar cell with the composition ITO/PEDOT:PSS/P3HT:N-CNTS:PCBM/LiF/Al. TEM and SEM analysis showed that the polymer successfully wrapped the N-CNTs. FTIR results indicated good π-π interaction within the nanocomposite synthesized by *in situ* polymerization as opposed to samples made by direct solution mixing. Dispersive surface energies of the N-CNTs and nanocomposites supported the fact that polymer covered the N-CNTs well. J-V analysis show that good devices were formed from the two nanocomposites, however, the *in situ* polymerization nanocomposite showed better photovoltaic characteristics.

Keywords: polythiophene; nitrogen-doped carbon nanotubes; nanocomposites; photovoltaic properties

1. Introduction

The electrical conductivity of a linear chain organic polymer (polyacetylene) was first discovered by Shirikawa *et al.* in 1977 [1]. Later, a number of other organic molecules particularly polythiophene groups emerged as alternative conducting polymers (CPs) with potential applications in the area of opto-electronic devices. The advantages of CPs over inorganic semi-conductors in opto-electronic applications include the ease of processability, tuneable properties of the molecules, flexibility, and being light weight [2]. CPs consist of alternating single and double bonds; the π-electrons in their double bonds are mobile due to overlap of π-orbitals [3]. Electronic properties of CPs can be tuned during synthesis and they also have good magnetic and optical properties [4]. Of late, CPs have been considered as one of the best alternatives for the production of solar cells. This realization stems from the fact that solar cells fabricated by use of inorganic materials can be expensive and also the processes that are utilized during their manufacture can be extensively energy intensive [5]. For CPs to function well in solar cells they should possess the following characteristics: soluble in common organic solvents, can form a thin film on substrates, low band-gap to enhance absorption, partially miscible with electron acceptors, a good hole conductor, and chemically stable in ambient conditions [3]. Polythiophenes not only have several of the characteristics named above, but also have efficient electronic conjugation, as well as synthetic versatility [3]. Polythiophenes are made by linking thiophene rings, which are insoluble, and then improving their solubility by introducing alkyl chains, *i.e.*, hexyl and octyl groups. Polythiophenes have been studied for various applications, which include organic field effect transistors [6], solar cells [7], sensors, and light emitting diodes [8].

Although CPs have unique properties their applications in various fields are hampered by their poor environmental stability and mechanical strength. This problem can be overcome by introducing different fillers within the CPs to form nanocomposites [9,10]. Among the different fillers available, carbon nanotubes (CNT) have attracted a great deal of interest due to their unique structural, electrical, and mechanical properties [11]. Further enhancement of these unique properties can be achieved by doping CNTs with boron or nitrogen to form boron- and nitrogen-doped CNTs (B-CNTs or N-CNTs). Nitrogen-doping creates defects on the walls of CNTs that improves the ability of the surfaces to undergo various covalent chemistries, provide good anchoring sites for nanoparticles, introduces a variety of functional groups, and more importantly, it also improves the electrical conductivity of CNTs [12]. For example, Panchakarla *et al.* [13] reported higher electrical conductivity for N-CNTs than for pristine CNTs. CNTs/N-CNTs in polymer nanocomposites enhance Young's modulus, the tensile strength, and electrical conductivity [14]. Additionally, the incorporation of CNTs can be an effective route to synthesize low density, high performance thermoelectric materials [15]. Nanocomposites of CNTs coated with CPs have found use in organic field emission devices [16], light emitting diodes [8,17], electronic devices and sensors [18], and organic solar cells (OSC) [19]. However, effective utilization of CNTs in the polymer nanocomposites strongly depends on the dispersion. For example, poorly dispersed CNTs in the active layer of OSC can act as recombination sites and can also lead to short-circuiting [20].

OSCs have gained a great deal of attention in recent years due to the high expectation of producing relatively cheap devices for converting solar energy directly to electricity [10]. Some of the advantages of OSCs over inorganic solar cells include low-cost manufacturing, high-throughput production, and high flexibility, and, therefore, OSCs can be cast on flexible substrates or on curved surfaces [19]. Although CNTs have many advantages when well dispersed in the polymer matrix, the performance of OSCs with CNTs in the active layer have continued to perform poorly when compared to polymer/fullerene systems, e.g., poly(3-hexylthiophine) (P3HT) and [6,6]-phenyl-C_{61}-butyric acid methyl ester (PCBM). This has been attributed to short-circuiting as a result of a mixture of semi-conducting and metallic CNTs [21], and filamentary short-circuiting due to CNTs extending outside the active layer [22]. Poor performance could also be due to unbalanced charge mobility for a device with CNTs, as one charge carrier will be transferred very fast while the other is subjected to hopping in disordered organic materials [23].

Jun *et al.* [24] fabricated a solar cell with CNTs functionalized with alkyl-amides in an active layer of P3HT:PCBM. The efficiency of this device increased by 30% from 3.2% to 4.4% compared with a device without CNTs. They attributed this increase to wide band absorption, high charge carrier mobility and improved dispersion in the polymer matrix. Kalita *et al.* [25] used plasma oxygen-functionalized CNTs in the active layer of P3HT:PCBM and reported an 81.8% efficiency increase from 1.21% to 2.2% compared with a device without CNTs. They attributed this increase to improved hole mobility and increased surface area for excitons dissociation.

In this paper we compare the effect of synthesis technique for nanocomposites on their photovoltaic properties. The two techniques compared are: oxidative *in situ* polymerization, and direct solution mixing of P3HT and N-CNTs. We also report on a unique characterization technique whereby dispersive surface energy was used to determine how effective the polymer wrapped/covered the walls of N-CNTs. Finally, the results on the use of nanocomposites in the active layer of organic solar cells (OSC) is presented and discussed.

2. Experimental Section

2.1. Materials

Chemicals used in this study were of analytical grade and were used as received unless stated otherwise. Anhydrous ferric chloride (99%), [6,6]-phenyl-C_{61}-butyric acid methyl ester (PCBM) (98%), regioregular poly (3-hexylthiophene-2,5-diely (99%), and 3-hexylthiophene (99%) were purchased from Sigma Aldrich (St. Louis, MO, USA) while, chloroform (99%) was sourced from an alternative supplier (Merck Chemicals, S.A). Indium tin oxide (ITO) coated glasses slide was purchased from Merck, Germany. Chloroform was dried before being used.

2.2. Synthesis of Nitrogen-Doped CNTs and Nanocomposites

N-CNTs were synthesized in our laboratory by a chemical vapor deposition floating catalyst method as reported elsewhere [26]. Briefly, 2.5 wt% of 4-pyridinyl-4-aminomethylidinephenylferrocene catalyst was dissolved in acetonitrile solvent to make 100 wt% solution. This was followed by pyrolysis at 850 °C. The crude N-CNTs obtained were purified by, firstly, calcining at 400 °C in air to remove amorphous carbon, and then refluxing with 6 M HNO_3 for 24 h at 80 °C to remove any iron residue used as catalyst during the N-CNTs synthesis.

The nanocomposites were synthesized by the means of two techniques, namely, oxidative *in situ* polymerization and direct solution mixing. For oxidative *in situ* polymerization of 3-hexylthiophene monomers on the walls of N-CNTs, this was achieved by use of a similar method as reported by Karim [27]. In brief, 1 wt% (of the weight of 3HT monomers) of N-CNTs (6.8 mg) was weighed, 50 mL of dry chloroform was added and the mixture was placed in a two-necked round-bottomed flask with a stirrer. The mixture was sonicated for 1 h to disperse the N-CNTs. Thereafter, 0.648 g (4 mmol) of anhydrous ferric chloride in 50 mL of dry chloroform was added to the above dispersion and further sonicated for 30 min. Then 673.2 mg (2 mmol) of 3-hexylthiophene monomers in 25 mL of dry chloroform solution was placed in a pressure-equalized funnel and added dropwise to the above mixture with constant stirring. Stirring continued under the same conditions for the next 24 h. The nanocomposite was precipitated with methanol; vacuum filtered, washed with methanol, 0.1 M HCl, deionized water, acetone, and then eventually vacuum dried for 24 h at room temperature.

The direct solution mixing method to synthesize the nanocomposite was adapted from a method reported by Lee *et al.* [28]. In brief, P3HT was dissolved in dry chloroform to make 20 $mg \cdot mL^{-1}$ solution, and 1 wt% of N-CNT (of the weight of P3HT) were added to the solution of P3HT in chloroform. The mixture was sonicated for 1 h then stirred for 12 h in the dark to protect it from light. This solution was spin coated directly on ITO glass substrate.

The active layer for the solar cell device was prepared by mixing the N-CNTs/P3HT nanocomposite with [6,6]-phenyl-C_{61}-butyric acid methyl ester (PCBM) at a ratio of 1:0.8 by mass to make 20 $mg \cdot mL^{-1}$ solution in chloroform. The mixture was sonicated for two hours before spin-coating onto ITO coated glass substrates.

2.3. Characterization of the Nanocomposites

The morphology and structure of the nanocomposites was characterized by; transmission electron microscopy (TEM), (JEOL JEM 1010, JEOL Ltd., Tokyo, Japan) at 200 kV. The nanocomposite samples were dispersed in ethanol by sonication before being deposited on carbon-coated copper grids. A scanning electron microscope (Carl Zeiss ultra plus field emission electron microscopy (FEGSEM), Carl Zeiss, Cambridge, UK) was used at 5 kV accelerating voltage. Samples were placed on aluminum stubs by using carbon tape.

Raman spectra of N-CNTs and the nanocomposites were recorded with a DeltaNu Advantage 532TM Raman spectrometer (DeltaNu, Vancouver, BC, Canada). The excitation source was a Nd:YAG solid state crystal class 3b diode laser at 532 nm excitation wavelength. Nuspec TM software was used to capture generated spectra. Thermogravimetric analysis (TGA) was performed with a TA Instruments Q series™ thermal analyzer DSC/TGA (Q600). The P3HT, nanocomposites and N-CNTs were heated at a rate of 10 $°C \cdot min^{-1}$ under an air flow rate of 50 $mL \cdot min^{-1}$ and the data was captured and analyzed using the TA Instrument Universal Analysis 2000 software (TA Instrument, New Castle, DE, USA).

An inverse gas chromatography (IGC) surface energy analyzer (SEA) (Cirrus, Dublin, Ireland) was used to determine the surface energy properties of N-CNTs and the nanocomposites. About 30 mg of the sample was packed in an IGC salinized glass column of 300 mm length and 4 mm internal diameter. The column was filled with salinized glass wool on both ends of the sample until it was well packed. Cirrus control software was used to control the analysis and Cirrus plus software was used for data analysis. FTIR spectra were recorded with KBr pellets on a Perkin-Elmer Spectrum 1 FTIR spectrometer (PerkinElmer, Waltham, MA, USA) equipped with spectrum Rx software. Photoluminescence spectra were obtained with a Perkin Elmer Spectro Fluorimeter equipped with FL Winlab software at an excitation wavelength of 298 nm in chloroform solution. UV-Vis spectra were recorded in a chloroform solution with a Perkin Elmer Lamba 35 dual-beam UV-Vis spectrophotometer and data analyzed with FL Winlab software. Samples for the Photoluminescence and UV-Vis were prepared using a modified method as reported by Goutam *et al.* [14], briefly, 10 mg of P3HT and nanocomposites were dissolved in dry chloroform to make 100 mL solutions. Necessary equivalent dilutions were made using micropipette to record spectra (Cleveland, QC, Canada).

2.4. Device Preparation

Devices were prepared in ambient conditions on ITO-coated glass substrates with a shunt resistance of 15 Ω. Half of the ITO coat was etched with a mixture of water, HCl and HNO_3 in the ratio of 12:12:1 by volume, respectively, and then placed in deionized water. The etched substrate was thereafter cleaned by sonication for 10 min each with separate solution of detergent, distilled water, acetone, and finally with isopropanol. Thereafter, the substrate was dried at 50 °C for 10 min. The hole transport layer poly(3,4-ethylenedioxythiophene): poly(styrenesulfonate) (PEDOT:PSS) was spin-coated on the clean substrate at 3000 rpm for 50 s, and then annealed for 10 min at 120 °C. The active layer, a mixture of P3HT/N-CNTs:PCBM was spin-coated at 1500 rpm for 30 s, and then annealed for 20 min at 120 °C. Before vacuum evaporation of the counter electrode, 0.6 nm of lithium fluoride (LiF) was evaporated on top to serve as a hole-blocking layer. A 60 nm counter electrode consisting of Al metal was thermally evaporated at 2.22×10^{-7} mbar in an HHV Auto 306 vacuum evaporator equipped with INFICON SQM-160 thin film deposition thickness and rate monitor. Current-voltage characterization was determined by using a standard solar simulator model #SS50AAA (Pet Photoemission Tech. Inc., Camarillo, CA, USA), with a Keithley 2420 source meter.

3. Results and Discussion

Scheme 1 illustrates how the synthesis of the nanocomposites was achieved. In the *in situ* polymerization technique, monomers were polymerized directly on the surface of the N-CNTs. However, in the direct solution mixing, a solution of N-CNTs was mixed with a solution of the polymer.

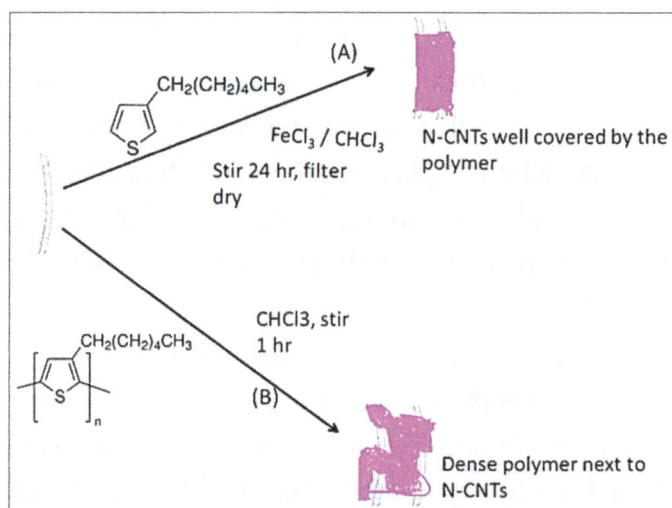

Scheme 1. Synthesis of the nanocomposites (**A**) *in situ* polymerization; and (**B**) direct mixing.

3.1. Morphology and Structure of the Nanocomposite

Figure 1 shows the structure of the N-CNTs before and after formation of the nanocomposites with poly(3-hexylthiophene). From the TEM images it was observed that the polymer coated the surface of the N-CNTs. From the bamboo structures observed, it can be deduced that the tubular inner part consist mainly of N-CNTs and the coated surface is conducting P3HT. The smooth surfaces of the N-CNTs (Figure 1A) became rough after they were covered by the polymer (Figure 1B).

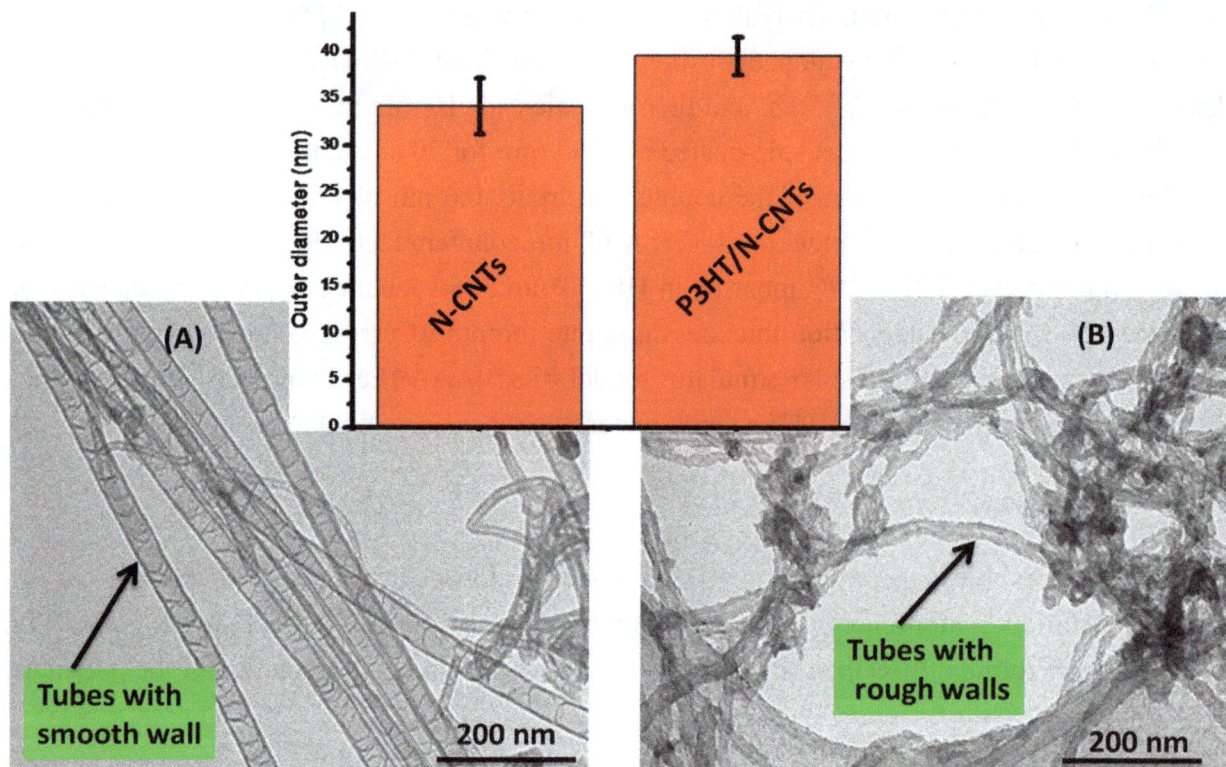

Figure 1. TEM images of (**A**) purified N-CNTs; and (**B**) N-CNT/P3HT nanocomposite synthesized by *in situ* polymerization (inset comparison of outer diameters).

Figure 2 shows the morphology of the N-CNTs and nanocomposite synthesized by *in situ* polymerization as observed in SEM.

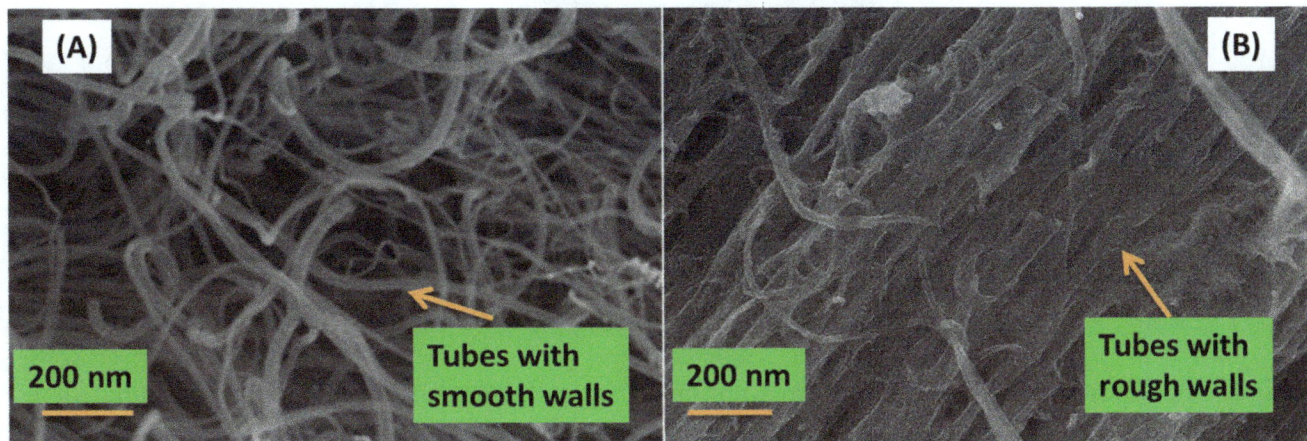

Figure 2. Morphology of (**A**) N-CNTs; and (**B**) nanocomposite of N-CNTs/P3HT synthesized by *in situ* polymerization.

From Figure 2A, the N-CNTs appear as an entangled mat of tubular structures with smooth surfaces which is a characteristic of carbon-based nanotubes. However, in Figure 2B a few thick tubular structures with rough surfaces and agglomerated mat-like structures were observed, and this is indicative of the polymer wrapping onto the nanotubes to form a nanocomposite. EDX, which was coupled with FEGSEM, provided further evidence of the nanocomposite elemental composition, which consisted of carbon, sulfur and oxygen. Oxygen observed in the nanocomposite could be due to the introduction of oxygenated groups during acid purification and functionalization of N-CNTs. The at% of carbon increased with the formation of the nanocomposite as compared to that in polymer. Karim [27] reported similar results when they synthesized P3HT/MWCNTs nanocomposites by *in situ* polymerization.

Further evidence of the polymer wrapping the N-CNTs was obtained by measuring the outer diameters of the N-CNTs and nanocomposite from their TEM images, (inset Figure 1). This was determined from not less than 50 TEM images and over 200 tubes per sample. The diameters were observed to increase with formation of the nanocomposites. An increase of ≈15.9% was observed; a good indication that the N-CNTs were wrapped by the polymer.

Figure 3 further confirms formation of a nanocomposite whereby, the brown color of pristine P3HT changed to dark brown. Our observations concur with what is reported elsewhere in the literature [29].

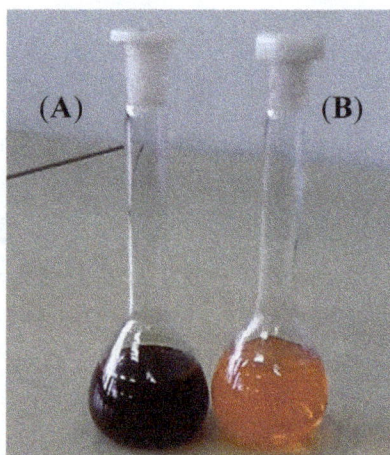

Figure 3. Color of P3HT in chloroform solution (**A**) after formation of nanocomposite and (**B**) pristine P3HT under white light.

3.2. Vibrational and Spectral Characteristics of P3HT and the Nanocomposite

Interaction between P3HT and N-CNTs in the two nanocomposites was assessed by means of Raman spectroscopy. Figure 4 shows Raman vibration peaks of the nanocomposites.

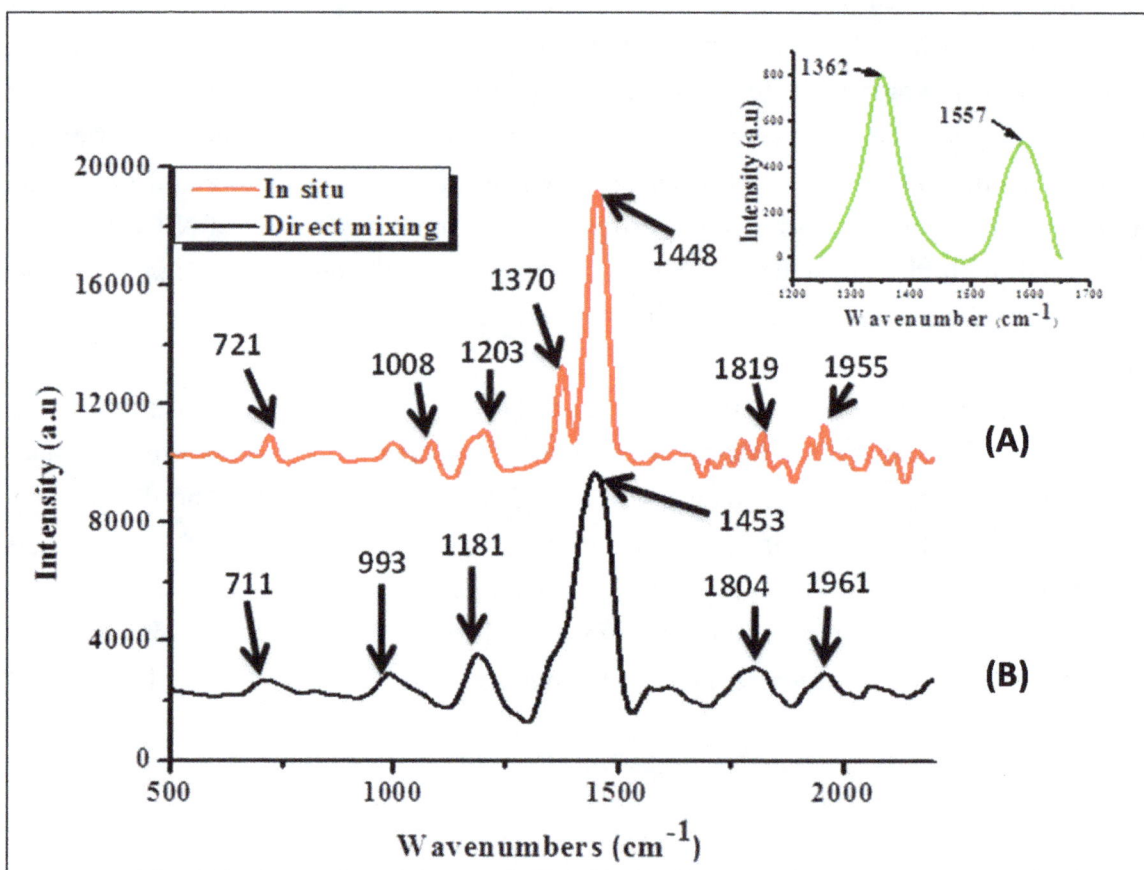

Figure 4. Raman spectroscopy results of nanocomposites, (A) direct mixing and (B) *in situ* polymerization (inset position of D-band and G-band for N-CNTs).

For the N-CNTs (Figure 4 inset) the peak at 1593 nm represents the G-band, which originates from the Raman E_{2g} mode while the one at 1356 cm^{-1} is the disorder-induced band. The I_D/I_G ratio for N-CNTs was 1.55, which was an indication of high disorder due to nitrogen-doping. Both peaks were consequently absent in the nanocomposites. The observed peaks for both nanocomposites were almost in similar position and can be assigned as follows; peak at 704–719 cm^{-1} is the C–S–C ring deformation for thiophene rings while that at around 1198 cm^{-1} is the C–C symmetric stretching and C–H bending vibrations. The peak in the range of 1373–1377 cm^{-1} is the C–C stretch deformation in organic thiophene rings while that around 1438–1454 cm^{-1} is the symmetric C–C stretch deformations in alkyl chain and 1800–1850 cm^{-1} is the asymmetric C–C stretching deformation for of thiophene ring [30].

FTIR spectroscopy results for P3HT and P3HT/N-CNTs nanocomposites synthesized by both techniques are presented in Figure 5. P3HT shows a peaks at 2925 and 2844 cm^{-1} assigned to C–H stretching vibrations, the peak at 1645 cm^{-1} is assigned to the aryl substituted C=C of the thiophene ring, the peak at 1440 cm^{-1} is attributed to the vibrational stretch of the thiophene ring and peaks between 900 and 670 cm^{-1} are due to the C–H out-of-plane deformation of thiophene [30].

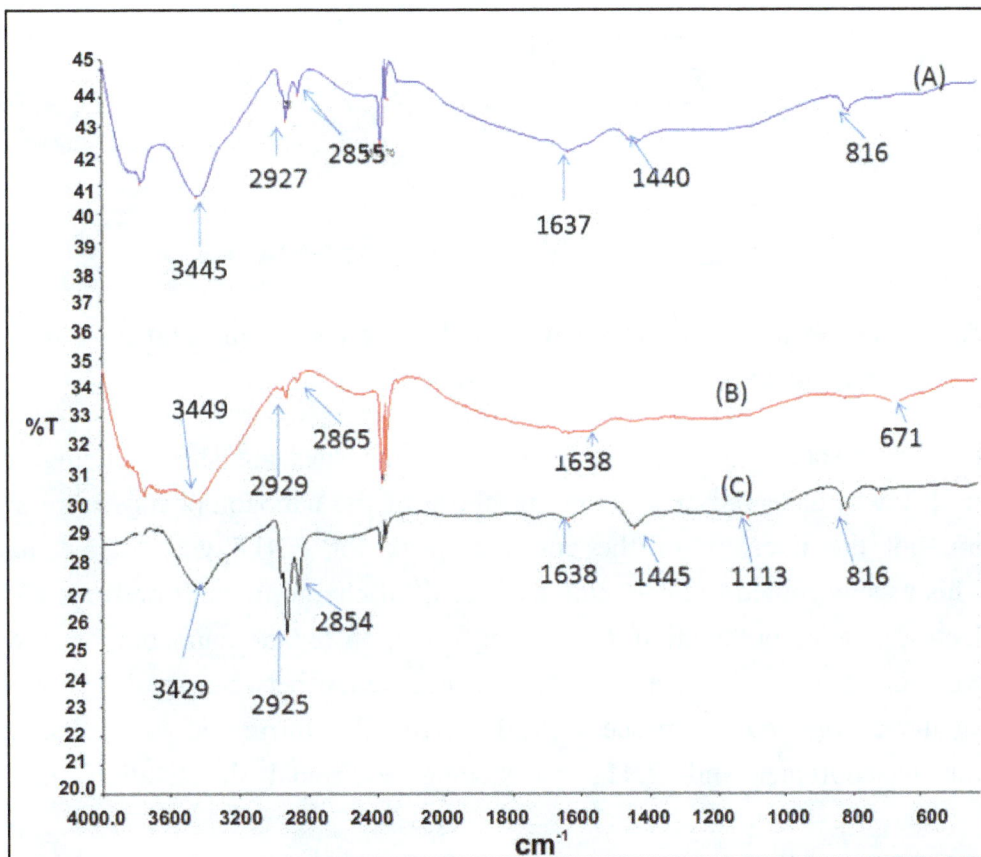

Figure 5. FTIR absorption frequencies for P3HT and nanocomposites (A) P3HT; (B) *in situ*; (C) direct mixing.

The nanocomposites synthesized by direct solution mixing gave almost all the peaks as for P3HT which was an indication of poor interaction between the polymer and N-CNTs. However, for the nanocomposite synthesized by *in situ* polymerization the C–H stretch vibration peak shifted slightly to higher wavenumbers from 2925 to 2929 cm^{-1}. A slight shift to longer wavenumbers can be attributed to CH-π interaction between N-CNTs and P3HT [31]. Additionally, the peak at 1440 cm^{-1} was not observed for this nanocomposite, which was an indication that stretching vibration of thiophene ring was interfered with and also, evidence of π-π interaction between N-CNTs and the thiophene rings of P3HT [31].

UV-visible absorption spectra of pristine P3HT and the nanocomposites in chloroform solution are presented in Figure 6A. The absorption maximum (λ_{max}) for P3HT was observed at 442 nm, an indication of extensive π-conjugation [14]. The absorption peak for P3HT that we observed compared well with values reported in literature [27]. From the figure it was noted that N-CNTs did not make significant contribution to the spectra but a small red shift was noted to λ_{max} of 445 nm for the nanocomposites synthesized with both techniques. Slight red shift of λ_{max} can be attributed to an increased conjugation length of the polymer due to strong π–π interaction with N-CNTs as a result of increased organization of the polymer chains on the nanotube surface [32].

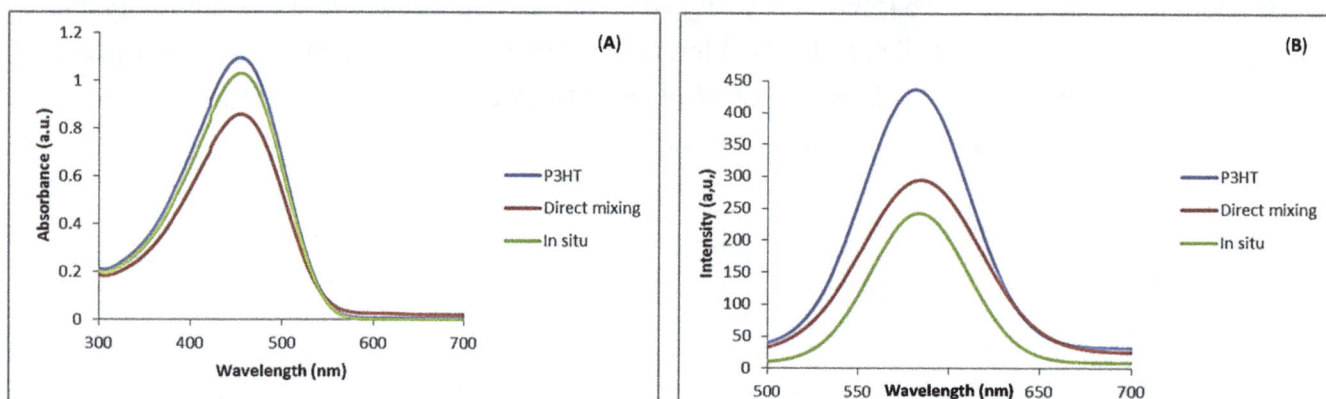

Figure 6. (A) Uv-Vis absorption spectra; **(B)** photoluminescence emission for P3HT and P3HT/N-CNTs nanocomposites in chloroform solution.

Photoluminescence (PL) spectra of P3HT and the nanocomposites are shown in Figure 6B. The emission peak of P3HT was observed at 581 nm and those of the nanocomposites were slightly red shifted to 585 nm, but the intensity of the emission peak for P3HT was higher than that of nanocomposites. This was attributed to quenching as a result of charge transfer between N-CNTs and P3HT reducing electron-hole recombination. Quenching was high for the nanocomposite synthesized by *in situ* polymerization. This can be due to better π–π interaction between the P3HT and N-CNTs surfaces enhancing the charge transfer process. Kuila *et al.* [31] attributed PL quenching to π–π interaction between the polymer and CNTs introducing additional deactivation paths for the excited electrons.

3.3. Thermal Stability of the Nanocomposites

The TGA thermograms for N-CNTs and the nanocomposites are presented in Figure 7. Both nanocomposites exhibit a two-step weight loss process and this could suggest that they are mixtures. The initial decomposition temperature for the nanocomposites is lower than for N-CNTs and indicates that they are less thermally stable.

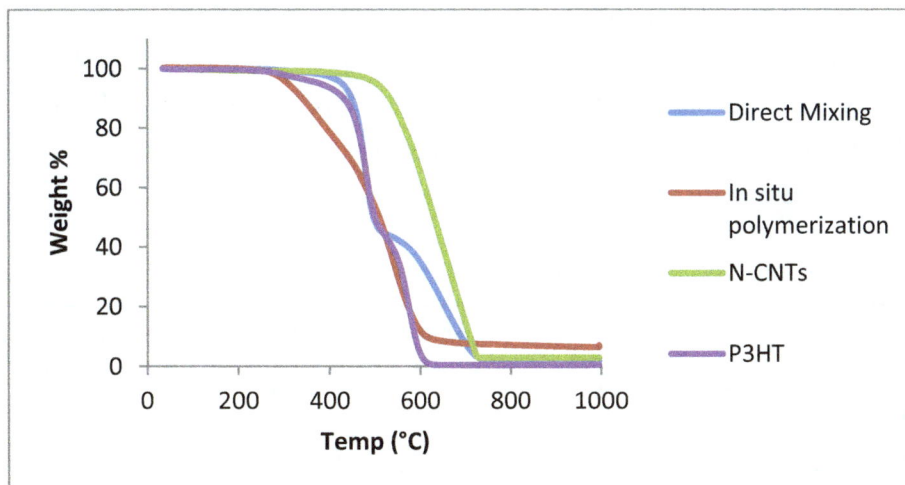

Figure 7. Thermogravimetric analysis of N-CNTs, P3HT and P3HT/N-CNTs nanocomposites.

The nanocomposite formed by *in situ* polymerization is the least thermally stable. The possible reason for this could be due to high reactivity of monomers polymerizing and thus, forming more polymer on the surface of N-CNTs. The high amount of residue for *in situ* polymerization nanocomposite could be due to some remnant ferric oxide initiator remaining entrapped as the nanocomposite even after washing.

3.4. Surface Energy Analysis of Nanocomposites

The effectiveness of the wrapping/covering of N-CNTs with the polymer was determined by comparing dispersive components of surface energy of the N-CNTs and the nanocomposites by using inverse gas chromatography equipped with a flame ionization detector. The dispersive component of surface energy (γ_s^d) can be a useful tool to examine surfaces of a solid whereby changes in surface properties can easily be detected. The γ_s^d was obtained from the retention time (t_R) of a given volume of a series of normal alkanes (C5–C9) at a flow rate of 10 mL·min^{-1} and 0.05 surface coverage. The γ_s^d was calculated from the slope of a straight line drawn from $RT\ln V_N$ against the number of carbon atoms in the n-alkanes by using the Doris and Gray method at peak maximum time [33]. R is the gas constant, T is the absolute column temperature and V_N is the net retention volume of non-polar probes as well as polar probes and can be calculated from Equation (1):

$$V_N = Fj(t_R - t_M)\left(\frac{P_0 - P_w}{P_0}\right)\left(\frac{T_c}{T_{meter}}\right) \qquad (1)$$

where F is the flow rate; t_R and t_M are the retention and dead times measured with a specific probe and a non-adsorbing probe (such as methane) respectively; P_0 is the pressure at the flow meter; P_w is the vapor pressure of pure water at the temperature of the flow meter (T_{meter}); and T_c is the column temperature [34]. For j it was James-Martin correction factor of gas compressibility when the column inlet (P_i) and outlet (P_0) pressures are different given by Equation (2):

$$j = {^3/_2}\frac{\left[\left(P_i/P_0\right)^2 - 1\right]}{\left[\left(P_i/P_0\right)^3 - 1\right]} \qquad (2)$$

The polar probes acetone, acetonitrile, ethyl acetate and dichloromethane were used to determine the acid/base properties of N-CNTs and the nanocomposites surfaces. The γ_s^d was determined at 100 °C for the N-CNTs and the nanocomposites by using five alkanes namely, pentane, hexane, heptane, octane and nonane. Table 1 presents the data on the γ_s^d, the acid and base constants determined from the interactions with the polar probes, and the acid base ratio.

Table 1. γ_s^d, acid-base constants determined from interaction with polar probes.

Sample	Surface energy (mJ·m^{-2})	Acid Constant-Ka	Base Constant-Kb	Acid-Bascity ratio	Specific (Acid-Base) Free Energy (kJ·Mol^{-1})			
					Acetone	Aceto-Nitrile	Ethyl Acetate	Dichloro-Methane
N-CNTs	49.02	0.0275	0.4975	0.0552	6.75	11.8	5.20	8.14
In Situ Nano-composite	56.53	0.0622	0.5738	0.1084	11.1	15.7	7.74	8.49
Direct mixing Nano-composite	46.68	0.3031	0.4446	0.6819	26.3	26.4	24.5	7.35

The γ_s^d of N-CNTs was higher than that of the nanocomposite made by the direct mixing method and lower than those samples synthesized by *in situ* polymerization. The differences can be attributed to the slight difference in the final morphology of the nanocomposites. When comparing the two nanocomposites only, the direct mixing method reduces the dispersive component of the surface energy by either effectively wrapping CNT bundles, leaving fewer exposed CNT tips, or through a combination of both factors. The *in situ* method has a slightly larger dispersive component than the N-CNTs due to the polymerization process and the combination of ultra sound effectively de-bundling the N-CNTs. This allows for a greater amount of individual N-CNTs to be wrapped by the polymer and also, allow for greater exposure of surface groups at the tips of the N-CNTs. The specific free energy ΔG^{AB} of adsorption for acid-base specific interaction was high for bi-functional acetonitrile for both N-CNTs and the nanocomposites showing the surfaces are covered by donor groups. N-CNTs have extra electrons due to the lone pair of electrons on nitrogen and P3HT contain conjugated π–electrons, which make them donor groups. The Gutmann acid (K_a) and base (K_b) constants were used to determine the surface chemistry of the samples. K_b values were higher for both N-CNTs and nanocomposites than K_a showing that the surfaces were covered by donor groups.

3.5. Photovoltaic Properties

Several organic solar cells were fabricated by using a bulk heterojunction design in which the photoactive layer was composed of a blend of donor and acceptor molecules. Figure 8 shows a schematic diagram of the OSC device structure employed in this investigation. The electrical properties of the devices were studied by measuring the current-voltage (J-V) characteristics from each diode in the sample.

Figure 8. Diagram of the OSC showing the arrangements of the thin layers.

The important parameters of the cell are derived from the diode equation, which often describes the J-V characteristics of a diode. The fill factor (FF), which determines the quality of the device, and the power conversion efficiency (PCE), which provide the device output are defined as:

$$FF = \frac{J_{Max} \times V_{Max}}{J_{sc} \times V_{oc}} \tag{3}$$

$$PCE = FF \frac{J_{SC} \times V_{OC}}{P_{In}} \tag{4}$$

where J_{MAX} and V_{MAX} are current density and voltage at maximum power point; J_{SC} is short circuit current density; V_{OC} is open circuit voltage; and P_{in} is incident light power [35].

Characterization of the cell under light illumination was performed by using a solar simulator operating at AM 1.5, 100 mW·cm^{-2}. The photoactive layers of the devices were fabricated from the two different nanocomposites obtained by *in situ* polymerization and direct solution mixing. Figure 9 shows the measured J-V curves of the devices produced under the two types of photoactive layers. The parameters of the solar cells derived from the data indicate that the devices prepared by *in situ* polymerization generally out performed those fabricated by direct solution mixing. According to the summary given in Table 2 the J_{SC}, FF and PCE of the devices based on the nanocomposite synthesized by *in situ* polymerization are higher than those by direct mixing method.

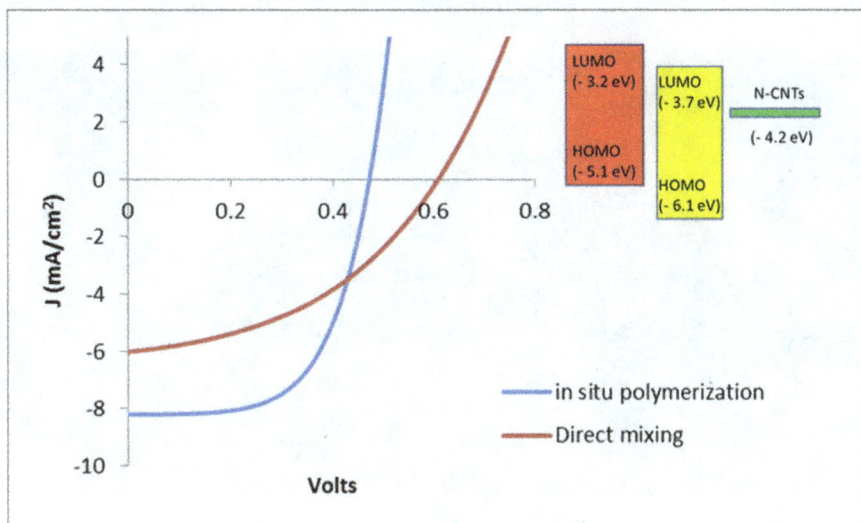

Figure 9. J-V curves of the devices from the two nanocomposites.

The measured parameters of the cells are summarized in Table 2.

Table 2. Measured cell parameters.

Method of Synthesis	V_{OC} (volts)	Jsc (mA·cm^{-2})	FF	Efficiency (%)
Direct mixing	0.61	5.084	29.26	0.51
In situ polymerization	0.48	7.731	41.63	1.66
Reference device (P3HT:PCBM)	0.55	6.97	54.4	2.09

The higher Jsc suggests that better charge transport properties are exhibited in the photoactive layer prepared by *in situ* polymerization. In other words, the medium has better carrier mobility of charge carriers, which increases device performance. In fact, the SEM images of the two active layers given below partially explain difference in the devices performances. However, the open circuit voltage of the OSC prepared by direct solution mixing is higher than the former by nearly 135 mV, but, it is very close to the value of V_{OC} of the reference cell. This is an indication for the existence of high non-radiative recombination of the free charge carriers in the medium formed by *in situ* polymerization. Despite the higher solar cell performance of the *in situ* polymerization it suffers from high charge recombination process, which limits the potential ability of the medium for higher power conversion. The scanning electron microscopy (SEM) images were taken from various samples to investigate the morphologies of the active layers.

Figure 10 shows the surface morphologies captured from samples coated with the solutions of the active layer prepared both by direct solution mixing (Figure 10A) and *in situ* polymerization (Figure 10B). The SEM images clearly showed that the dispersion of the N-CNTs in the polymer matrix was not good for the nanocomposite formed by direct solution mixing. This composite favors CNT agglomeration and entanglement that form various CNT clusters Figure 10A. On the other hand, good dispersion of N-CNTs was observed for the nanocomposite prepared by *in situ* polymerization (Figure 10B) which can be ascribed to the fact that the monomers were polymerized on the surface of the N-CNTs and this inhibits π–π interaction of the tubes and thereby decrease agglomeration. Furthermore, good dispersion meant there was a continuous percolation path for free charge carriers and eventual collection at the electrodes thereby improving cell performance.

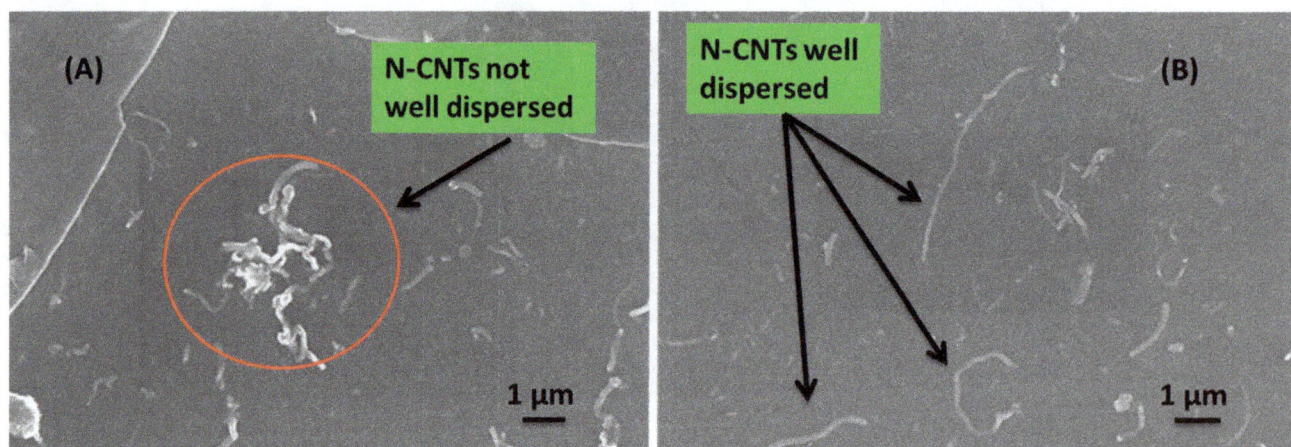

Figure 10. SEM images of the cell morphology showing N-CNTs (**A**) not well dispersed in the direct mixing nanocomposite; and (**B**) well dispersed for the *in situ* polymerization nanocomposite.

The high PCE and FF from devices prepared by *in situ* polymerization can be attributed to the small size of monomer molecules making the nanocomposite adduct more homogeneous compared with the one prepared by mixing solutions of polymer and N-CNTs [5]. Lee *et al.* [28] reported an efficiency of 3.8% for a nanocomposites of N-CNTs and P3HT prepared by direct mixing. The high efficiency of their device compared with ours could be due to preparation conditions. The nanocomposite in this work was prepared under ambient conditions whereas theirs was prepared in an inert gas atmosphere. Javier and Werner [32] used direct mixing to prepare a nanocomposite of pristine multi-wall CNTs (MWCNTs) and P3HT in ambient conditions. Their device recorded lower J_{SC} and FF than what we observed in our device from nanocomposites by direct solution mixing. We attribute the high J_{SC} and FF of our device to improved charge transfer by N-CNTs. Wu *et al.* [36] mixed pristine MWCNTs with P3HT and reported a high efficiency of 3.47% for devices prepared in an inert gas atmosphere.

From the above examples it is to be noted that most of the nanocomposites used in solar cells are more often prepared by direct solution mixing. However, according to the results found from the current synthesis and characterization, using FTIR, PL and J-V, it appeared to us that *in situ* polymerization would be the best technique for preparation nanocomposite. The poor performance of the devices prepared from direct solution mixing could be due to energetic agitation brought about by shear intensive mechanical stirring and ultrasonication used during nanocomposite preparation which initiates polymer chain breakage and degradation of opto-electrical properties [24].

4. Conclusions

In situ polymerization and direct solution mixing techniques have been used successfully to synthesize conducting nanonanocomposites of P3HT and N-CNTs. N-CNTs formed extra exciton dissociation sites which were observed by PL quenching. The diodes formed from the nanonanocomposites had positive rectification confirming their conductive nature but, the efficiency observed was very low. The *in situ* polymerization technique was observed to be better method for synthesising nanonanocomposites for organic solar cells. More investigation is required to determine why nitrogen-doping of the N-CNTs that is expected to improve the conductivity of the nanonanocomposites, did not improve the cell efficiency as anticipated.

Acknowledgments

The authors wish to thank the University of KwaZulu-Natal (UKZN), the National Research Foundation (NRF) and India, Brazil and South Africa (IBSA) energy project for financial assistance. Godfrey Keru also thanks the UKZN College of Agriculture, Engineering and Science for the award of a postgraduate bursary.

Author Contributions

Synthesis and characterization of the CNTs and nanocomposites were conducted by Godfrey Keru. Genene Mola and Godfrey Keru were responsible for device assembly and characterization. All authors contributed equally to the analysis and interpretation of the data, drafting of the manuscript, and approved the final version submitted.

Conflicts of Interest

The authors declare no conflict of interest.

References

1. Shirakawa, H.; Louis, E.J.; Macdiarmid, A.G.; Chiang, C.K.; Heeger, A.J. Synthesis of electrically conducting organic polymers-halogen derivatives of polyacetylene, (CH)X. *J. Chem. Soc. Chem. Commun.* **1977**, 578–580.

2. Giulianini, M.; Waclawik, E.R.; Bell, J.M.; Scarselli, M.; Castrucci, P.; de Crescenzi, M.; Motta, N. Microscopic and spectroscopic investigation of poly(3-hexylthiophene) interaction with carbon nanotubes. *Polymers* **2011**, *3*, 1433–1446.

3. Bounioux, C.; Katz, E.A.; Yerushalmi-Rozen, R. Conjugated polymers-carbon nanotubes-based functional materials for organic photovoltaics: A critical review. *Polym. Adv. Technol.* **2012**, *23*, 1129–1140.

4. Karim, M.R.; Lee, C.J.; Lee, M.S. Synthesis and characterization of conducting polythiophene/carbon nanotubes composites. *J. Polym. Sci. Part A Polym. Chem.* **2006**, *44*, 5283–5290.

5. Keru, G.; Ndungu, P.G.; Nyamori, V.O. A review on carbon nanotube/polymer composites for organic solar cells. *Int. J. Energy Res.* **2014**, *38*, 1635–1653.

6. Tsumura, A.; Koezuka, H.; Ando, T. Polythiophene field-effect transistor: Its characteristics and operation mechanism. *Synth. Met.* **1988**, *25*, 11–23.

7. Kim, Y.; Cook, S.; Tuladhar, S.M.; Choulis, S.A.; Nelson, J.; Durrant, J.R.; Bradley, D.D.; Giles, M.; McCulloch, I.; Ha, C.-S. A strong regioregularity effect in self-organizing conjugated polymer films and high-efficiency polythiophene: Fullerene solar cells. *Nat. Mater.* **2006**, *5*, 197–203.

8. Burroughes, J.H.; Bradley, D.D.C.; Brown, A.R.; Marks, R.N.; Mackay, K.; Friend, R.H.; Burns, P.L.; Holmes, A.B. Light-emitting diodes based on conjugated polymers. *Nature* **1990**, *347*, 539–541.

9. Choudhary, V.; Gupta, A. Polymer/carbon nanotube nanocomposites. In *Material Sciences, Polymer*; Yellampalli, S., Ed.; In Tech: Rijeka, Croatia, 2011; pp. 65–90.

10. Ltaief, A.; Bouazizi, A.; Davenas, J. Charge transport in carbon nanotubes-polymer composite photovoltaic cells. *Materials* **2009**, *2*, 710–718.

11. Karim, M.R.; Yeum, J.H.; Lee, M.S.; Lim, K.T. Synthesis of conducting polythiophene composites with multi-walled carbon nanotube by the γ-radiolysis polymerization method. *Mater. Chem. Phys.* **2008**, *112*, 779–782.

12. Ayala, P.; Arenal, R.; Rümmeli, M.; Rubio, A.; Pichler, T. The doping of carbon nanotubes with nitrogen and their potential applications. *Carbon* **2010**, *48*, 575–586.

13. Panchakarla, L.S.; Govindaraj, A.; Rao, C.N.R. Boron- and nitrogen-doped carbon nanotubes and graphene. *Inorg. Chim. Acta* **2010**, *363*, 4163–4174.

14. Goutam, P.J.; Singh, D.K.; Giri, P.K.; Iyer, P.K. Enhancing the photostability of poly(3-hexylthiophene) by preparing composites with multiwalled carbon nanotubes. *J. Phys. Chem. B* **2011**, *115*, 919–924.

15. Du, Y.; Shen, S.Z.; Yang, W.D.; Chen, S.; Qin, Z.; Cai, K.F.; Casey, P.S. Facile preparation and characterization of poly (3-hexylthiophene)/multiwalled carbon nanotube thermoelectric composite films. *J. Electron. Mater.* **2012**, *41*, 1436–1441.

16. Connolly, T.; Smith, R.C.; Hernandez, Y.; Gun'ko, Y.; Coleman, J.N.; Carey, J.D. Carbon-nanotube-polymer nanocomposites for field-emission cathodes. *Small* **2009**, *5*, 826–831.

17. Kim, J.-Y.; Kim, M.; Kim, H.; Joo, J.; Choi, J.-H. Electrical and optical studies of organic light emitting devices using SWCNTs-polymer nanocomposites. *Opt. Mater.* **2003**, *21*, 147–151.

18. Philip, B.; Xie, J.; Chandrasekhar, A.; Abraham, J.; Varadan, V.K. A novel nanocomposite from multiwalled carbon nanotubes functionalized with a conducting polymer. *Smart Mater. Struct.* **2004**, *13*, 295.

19. Lee, I.; Lee, S.; Kim, H.; Lee, H.; Kim, Y. Polymer solar cells with polymer/carbon nanotube composite hole-collecting buffer layers. *Open. Phys. Chem. J.* **2010**, *4*, 1–3.

20. Miller, A.J.; Hatton, R.A.; Silva, S.R.P. Water-soluble multiwall-carbon-nanotube-polythiophene composite for bilayer photovoltaics. *Appl. Phys. Lett.* **2006**, *89*, 123115:1–123115:3.

21. Kanai, Y.; Grossman, J.C. Role of semiconducting and metallic tubes in P3HT/carbon-nanotube photovoltaic heterojunctions: Density functional theory calculations. *Nano Lett.* **2008**, *8*, 908–912.

22. Berson, S.; de Bettignies, R.; Bailly, S.; Guillerez, S.; Jousselme, B. Elaboration of P3HT/CNT/PCBM composites for organic photovoltaic cells. *Adv. Funct. Mater.* **2007**, *17*, 3363–3370.

23. Liming, L.; Stanchina, W.E.; Guangyong, L. Enhanced performance of bulk heterojunction solar cells fabricated by polymer:Fullerene:Carbon-nanotube composites. In Proceedings of the NANO 8th IEEE Conference on Nanotechnology, Arlington, TX, USA, 18–21 August 2008; pp. 233–236.

24. Jun, G.H.; Jin, S.H.; Park, S.H.; Jeon, S.; Hong, S.H. Highly dispersed carbon nanotubes in organic media for polymer: Fullerene photovoltaic devices. *Carbon* **2012**, *50*, 40–46.

25. Kalita, G.; Wakita, K.; Umeno, M. Efficient bulk heterojunction solar cells incorporating carbon nanotubes and with electron selective interlayers. In Proceedings of the 35th IEEE Photovoltaic Specialists Conference, Honolulu, HI, USA, 20–25 June 2010; pp. 90–94.

26. Keru, G.; Ndungu, P.G.; Nyamori, V.O. Nitrogen-doped carbon nanotubes synthesised by pyrolysis of (4-{[(pyridine-4-yl)methylidene]amino}phenyl)ferrocene. *J. Nanomater.* **2013**, *2013*, 1–7.

27. Karim, M.R. Synthesis and characterizations of poly(3-hexylthiophene) and modified carbon nanotube composites. *J. Nanomater.* **2012**, *2012*, 1–8.

28. Lee, J.M.; Park, J.S.; Lee, S.H.; Kim, H.; Yoo, S.; Kim, S.O. Selective electron- or hole-transport enhancement in bulk-heterojunction organic solar cells with N- or B-doped carbon nanotubes. *Adv. Mater.* **2011**, *23*, 629–633.

29. Goutam, P.J.; Singh, D.K.; Iyer, P.K. Photoluminescence quenching of poly(3-hexylthiophene) by carbon nanotubes. *J. Phys. Chem. C* **2012**, *116*, 8196–8201.

30. Louarn, G.; Trznadel, M.; Buisson, J.P.; Laska, J.; Pron, A.; Lapkowski, M.; Lefrant, S. Raman spectroscopic studies of regioregular poly(3-alkylthiophenes). *J. Phys. Chem.* **1996**, *100*, 12532–12539.

31. Kuila, B.K.; Malik, S.; Batabyal, S.K.; Nandi, A.K. *In-situ* synthesis of soluble poly(3-hexylthiophene)/multiwalled carbon nanotube composite: Morphology, structure, and conductivity. *Macromolecules* **2007**, *40*, 278–287.

32. Arranz-Andrés, J.; Blau, W.J. Enhanced device performance using different carbon nanotube types in polymer photovoltaic devices. *Carbon* **2008**, *46*, 2067–2075.

33. Dorris, G.M.; Gray, D.G. Adsorption of n-alkanes at zero surface coverage on cellulose paper and wood fibers. *J. Colloid Interface Sci.* **1980**, *77*, 353–362.

34. Zhang, X.; Yang, D.; Xu, P.; Wang, C.; Du, Q. Characterizing the surface properties of carbon nanotubes by inverse gas chromatography. *J. Mater. Sci.* **2007**, *42*, 7069–7075.

35. Günes, S.; Neugebauer, H.; Sariciftci, N.S. Conjugated polymer-based organic solar cells. *Chem. Rev.* **2007**, *107*, 1324–1338.

36. Wu, M.-C.; Lin, Y.-Y.; Chen, S.; Liao, H.-C.; Wu, Y.-J.; Chen, C.-W.; Chen, Y.-F.; Su, W.-F. Enhancing light absorption and carrier transport of P3HT by doping multi-wall carbon nanotubes. *Chem. Phys. Lett.* **2009**, *468*, 64–68.

14

Revisiting the Hydrogen Storage Behavior of the Na-O-H System

Jianfeng Mao [1], Qinfen Gu [2] and Duncan H. Gregory [1],*

[1] WestCHEM, School of Chemistry, Joseph Black Building, University of Glasgow,
Glasgow G12 8QQ, UK; E-Mail: jeff.mao@hotmail.com

[2] Australian Synchrotron, Clayton, Victoria 3168, Australia; E-Mail: Qinfen.Gu@synchrotron.org.au

* Author to whom correspondence should be addressed; E-Mail: Duncan.Gregory@Glasgow.ac.uk;

Academic Editor: Umit Demirci

Abstract: Solid-state reactions between sodium hydride and sodium hydroxide are unusual among hydride-hydroxide systems since hydrogen can be stored reversibly. In order to understand the relationship between hydrogen uptake/release properties and phase/structure evolution, the dehydrogenation and hydrogenation behavior of the Na-O-H system has been investigated in detail both *ex-* and *in-situ*. Simultaneous thermogravimetric-differential thermal analysis coupled to mass spectrometry (TG-DTA-MS) experiments of NaH-NaOH composites reveal two principal features: Firstly, an H_2 desorption event occurring between 240 and 380 °C and secondly an additional endothermic process at around 170 °C with no associated weight change. *In-situ* high-resolution synchrotron powder X-ray diffraction showed that NaOH appears to form a solid solution with NaH yielding a new cubic complex hydride phase below 200 °C. The Na-H-OH phase persists up to the maximum temperature of the *in-situ* diffraction experiment shortly before dehydrogenation occurs. The present work suggests that not only is the *inter*-phase synergic interaction of protic hydrogen (in NaOH) and hydridic hydrogen (in NaH) important in the dehydrogenation mechanism, but that also an *intra*-phase $H^{\delta+}... H^{\delta-}$ interaction may be a crucial step in the desorption process.

Keywords: hydrogen storage; sodium oxide; sodium hydride; sodium hydroxide; *in-situ* synchrotron powder diffraction

1. Introduction

With the continued depletion of fossil fuel resources and the increasing impact of environmental pollution, renewable and clean energy sources such as wind and solar technologies have become a major priority and been the subject of increased research interest. However, many renewable energy sources are intermittent and so a means by which energy can be stored and transported is vital [1]. One approach is to store energy chemically as a clean fuel and hydrogen is regarded as one of the best options due to its abundance, high gravimetric energy density and capacity to be sustainably generated. However, a means to store hydrogen safely at high capacity and low cost is key to its successful implementation [2,3].

Compared to storage of hydrogen as a compressed gas or a cryogenic liquid, solid-state hydrogen storage is more effective volumetrically and alleviates safety and cost concerns associated with high pressure and/or low temperature. For more than 10 years materials such as light metal hydrides [4], complex hydrides [5–7] and chemical hydrides [8] have been investigated in this capacity and significant advances in understanding and performance have been made. However, no single hydride system yet fulfills all the necessary criteria for mobile applications (principally gravimetric and volumetric capacity, sorption enthalpy and kinetics) [9]. When compared to mobile applications, the requirements for stationary applications are rather different and cycle life/longevity, cost and safety can become the most important parameters [10]. Given the technical demands for portable storage, static medium-large scale hydrogen storage may be at the forefront of initial efforts in energy storage and in shifting electrical energy from peak to off-peak periods to achieve smart grid management. For this purpose, an abundant, low cost and non-toxic hydrogen storage material becomes increasingly attractive to ensure large-scale and long-term applications with the minimum of financial burden.

Among various hydrogen storage materials, sodium based materials are very promising as hydrogen storage media for stationary applications since sodium is one of the most abundant elements on Earth (2.64 wt%), and is relatively cheap [11–15]. Moreover, sodium resources are geographically ubiquitous (e.g. from the sea and from underground deposits). Recently, Xu et $al.$ found that sodium oxide, Na_2O can absorb hydrogen easily at close to ambient temperature (~60 °C) to form NaH and NaOH. Further, the hydrogenated products, NaH and NaOH, can be readily converted back to Na_2O by thermal treatment [16,17]:

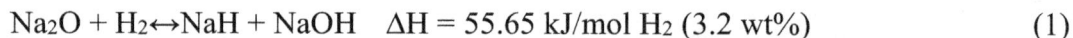

$$Na_2O + H_2 \leftrightarrow NaH + NaOH \quad \Delta H = 55.65 \text{ kJ/mol } H_2 \text{ (3.2 wt\%)} \qquad (1)$$

Dehydrogenation in the NaH-NaOH system occurs at a lower temperature than that for NaH alone and given that NaOH and NaH contain $H^{\delta+}$ and $H^{\delta-}$, respectively, the interaction of these two hydrogen species could be responsible for the relative decrease in dehydrogenation temperature between Na-O-H and Na-H [16,17]. The dehydrogenation process and mechanism are still not clear, however. Interestingly a number of much earlier studies of the Na-O-H system have indicated that the alkali metal hydride and corresponding hydroxide may be miscible [18–22]. More recently, during a study of the effect of NaOH as an additive in the Na-H system, it was observed that $ca.$ 10 mol% of NaOH could apparently be incorporated into the NaH structure and resulted in enhanced hydrogen motion in NaH above 150 °C [23]. These previous results suggest that OH$^-$ can be substituted for H$^-$ within the NaH structure above 150 °C. Given the connectivity between proton conduction and hydrogen uptake/release kinetics [24], an investigation of the structural and phase composition changes in the Na-O-H system during heating should provide considerable insight into the desorption mechanism and ultimately in how the performance of the system might be improved. In this paper, we present a detailed study of hydrogenation and dehydrogenation in the Na-O-H system, particularly exploiting *in-situ* synchrotron X-ray powder diffraction methods to elucidate the subtle changes in composition and structure immediately prior to hydrogen release.

2. Results and Discussion

2.1. Hydrogenation of Na₂O

The time-resolved hydrogenation profile of as-milled Na_2O with temperature (at 18 bar) is shown in Figure 1. On heating at 3 °C min^{-1}, hydrogen uptake is initiated close to room temperature (30–50 °C). Upon heating to 140 °C, the hydrogen uptake rate increases significantly and 2.5 wt% H_2 can be stored under these conditions. Subsequently, hydrogenation is relatively slow and a total of 3.3 wt% hydrogen is absorbed by 400 °C. These uptake results are consistent with previous studies [16,17], although it should be noted that the experimental figure of 3.3 wt% slightly exceeds the theoretical capacity. In principle, this could be possible due to the partial reaction of Na_2O with moisture during hydrogen uptake, but powder X-ray diffraction (PXD) patterns before and after hydrogenation suggest that the reaction of an impurity in the starting material with hydrogen may also contribute as discussed below.

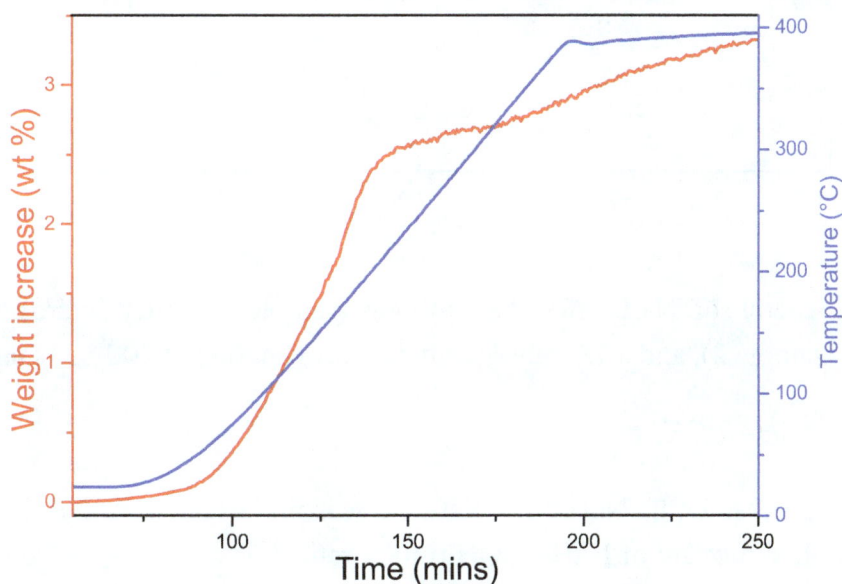

Figure 1. Hydrogen uptake of as-milled Na_2O under 18 bar of hydrogen.

PXD was performed to clarify the chemical reactions that occur on hydrogenation. Figure 2 shows the PXD patterns of the Na_2O starting material after mechanical milling (sample 1), the material after hydrogenation (sample 2) and finally after subsequent dehydrogenation (sample 3). Although Na_2O was the main phase in the ball milled starting material as expected, an impurity phase of Na_2O_2 was also detected, and analysis of the diffraction pattern yielded a phase fraction of *ca.* 10 wt%. After hydrogenation at 400 °C, NaH and NaOH were formed and the diffraction peaks from Na_2O and Na_2O_2 were no longer present. Upon dehydrogenation at 400 °C, Na_2O reformed and was present as the majority phase. Overall, therefore, the material system demonstrated a reversible reaction as described by Equation 1. The presence of NaOH in the dehydrogenated product, sample 3 could suggest partial hydrolysis of Na_2O, but the more likely origin of the hydroxide is from hydrogenation of the original Na_2O_2 impurity in the starting material (since $Na_2O_2 + H_2 \rightarrow 2NaOH$). This leads to an excess of NaOH in the hydrogenated product, sample 2 (*i.e.*, NaH:NaOH is not in the expected 1:1 molar ratio following hydrogen uptake) which persists during the dehydrogenation process at 400 °C However, Rietveld refinement for hydrogenated Na_2O shows that the molar ratio of NaOH and NaH in the sample is 1.66:1. The phase fraction of NaOH exceeds the amount expected solely from the hydrogenation of 10 wt% Na_2O_2. Thus, both the hydrolysis of Na_2O (and possibly of NaH) and the hydrogenation of the original Na_2O_2 impurity lead to the observed excess of NaOH.

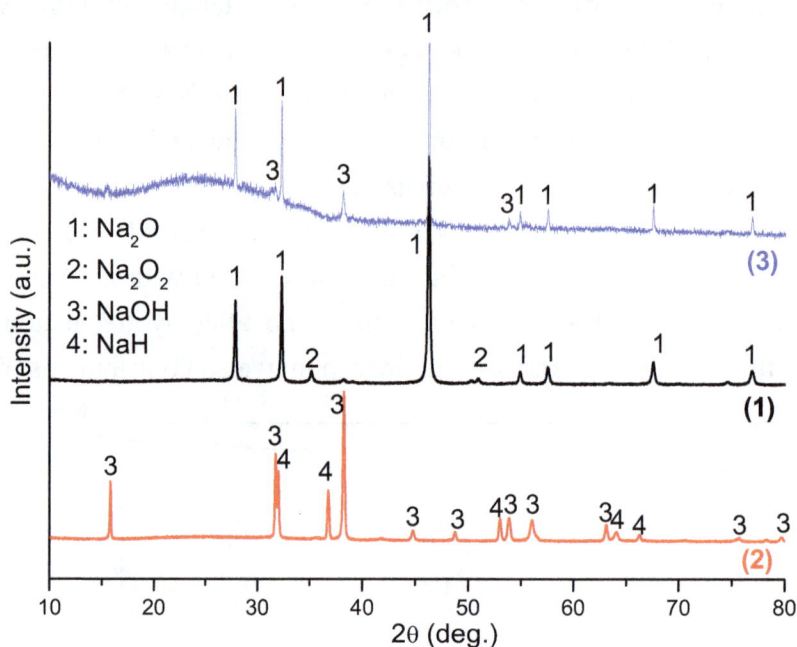

Figure 2. XRD patterns of Na_2O after ball milling (sample 1), after hydrogenation at 400 °C/18 bar H_2 (sample 2), and after subsequent dehydrogenation at 400 °C (sample 3).

2.2. Dehydrogenation of NaH-NaOH

The dehydrogenation process in the NaH-NaOH system was investigated further. Figure 3 shows a PXD pattern of the as-milled mixture of NaH-NaOH (molar ratio 1:1; sample 4) as compared with the individual commercial starting materials NaH and α-NaOH. It can be seen that the as-received NaH contains a small amount of NaOH impurity, while the as-received NaOH contains a minor phase of the hydrated hydroxide, $NaOH \cdot H_2O$. By comparison, only NaH and NaOH were detected in sample 4. The absence of the $NaOH \cdot H_2O$ impurity in sample 4 is attributed to dehydration during ball milling.

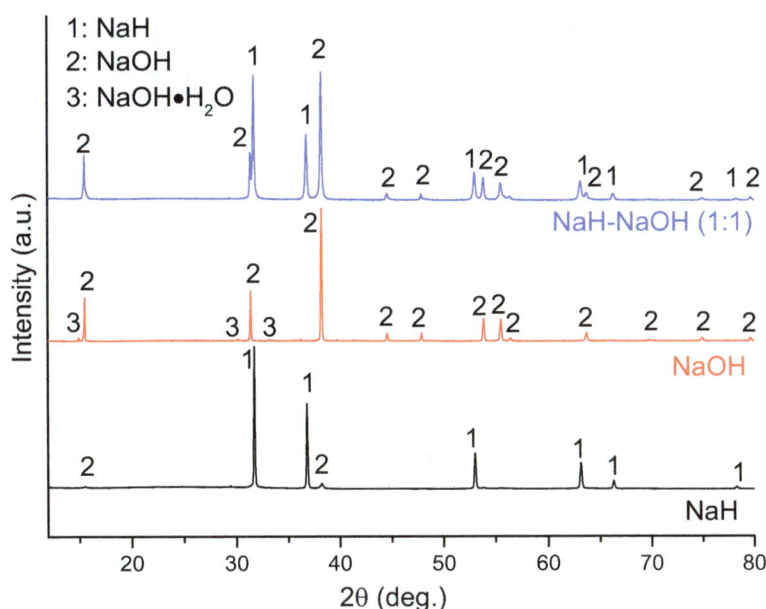

Figure 3. PXD patterns of as-received NaH and NaOH, and as-milled NaH-NaOH (1:1), sample 4.

The thermal decomposition behavior of sample 4 compared to NaH and NaOH was investigated by DTA, as shown in Figure 4. The DTA profile for NaH shows one endothermic peak at *ca.* 360 °C, which can be assigned to the decomposition of NaH to Na metal and hydrogen. For NaOH, the DTA profile shows two endothermic peaks at 299 and 319 °C, which can be assigned to the α-β (orthorhombic-monoclinic) phase transition and the melting point, respectively [25,26]. By contrast, sample 4 shows different features, displaying multiple endothermic peaks. The first endothermic event occurs at 171 °C and is relatively well-defined while the second is more complex (consisting of perhaps three or more individual processes) and reaches a maximum in the DTA profile at 333 °C. The second endothermic event can be attributed to the reaction of NaH and NaOH culminating in the formation of Na_2O and hydrogen. The results further confirmed that the dehydrogenation pathway of the NaH-NaOH mixture is entirely different to that observed for NaH. The first endothermic event is not observed in either NaH or NaOH. It would appear therefore that the synergic interaction of hydrogen species in NaH and NaOH not only contributes to the lower dehydrogenation temperature *vs.* NaH itself, but also leads to likely reaction in the solid state and structural changes during the heating period prior to dehydrogenation. To clarify the structural changes that occur before dehydrogenation, variable temperature, *in-situ* synchrotron PXD experiments were conducted on the NaH-NaOH mixture and are discussed below.

Ex-situ XRD characterization was performed on several dehydrogenated hydride-hydroxide mixtures of varying molar ratio after heating each to 400 °C (Figure 5). Clearly, Na_2O was the main phase in the dehydrogenated NaH-NaOH (1:1) sample, but some residual NaOH remains also detected. In this case, the presence of residual NaOH can be attributed to the fact that the NaH and NaOH starting materials contain low levels of NaOH and $NaOH \cdot H_2O$ impurities, respectively. Hence the NaH:NaOH ratio departs from the ideal stoichiometric 1:1 value in the mixture. In order to obtain single phase Na_2O as a dehydrogenation product, we optimized the quantity of NaH and NaOH in the starting mixture. As shown in Figure 5, NaOH impurity reflections diminished as the NaH:NaOH ratio increased, reaching a minimum when a 23 wt% excess of NaH was used. On adding a 25 wt% excess of NaH, no NaOH was observed in diffraction patterns but reflections originating from Na metal appeared. Therefore, use of 23 wt% excess of NaH was found to be the optimum reactant composition required to obtain Na₂O with minimal impurities (sample 5).

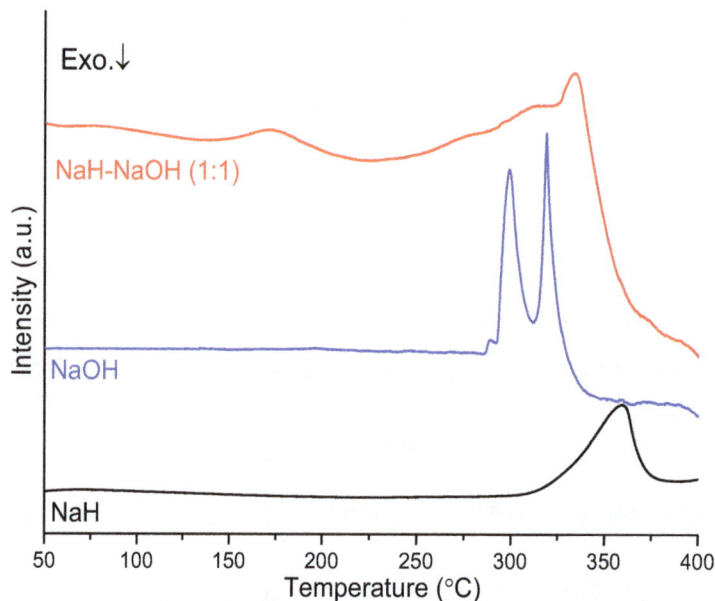

Figure 4. DTA profiles of NaH, NaOH, and as-milled NaH-NaOH (1:1) mixture.

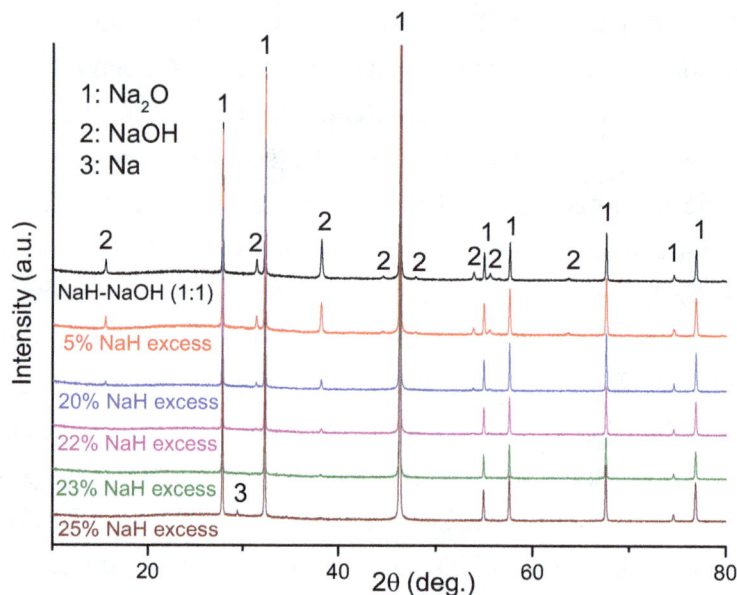

Figure 5. PXD patterns of NaH-NaOH with varying hydride:hydroxide molar ratio after dehydrogenation at 400 °C.

The thermal behavior of sample 5 was investigated by TG-DTA-MS (Figure 6). TG results showed that the mixture released gas from 200–378 °C with a corresponding weight loss of 3.1 wt%. Moreover, it was evident from mass spectra collected simultaneously while heating that hydrogen was the only evolved gaseous product, consistent with the reversible reaction described by equation 1. The DTA profile for sample 5 reveals a similar feature with the sample NaH-NaOH (1:1). The first endothermic peak was observed at 170 °C, while the second cluster of overlapping endothermic peaks with a maximum at 342 °C is attributed to the dehydrogenation reaction of NaH with NaOH and might be expected to incorporate first the α-β phase transition and second the melting of NaOH.

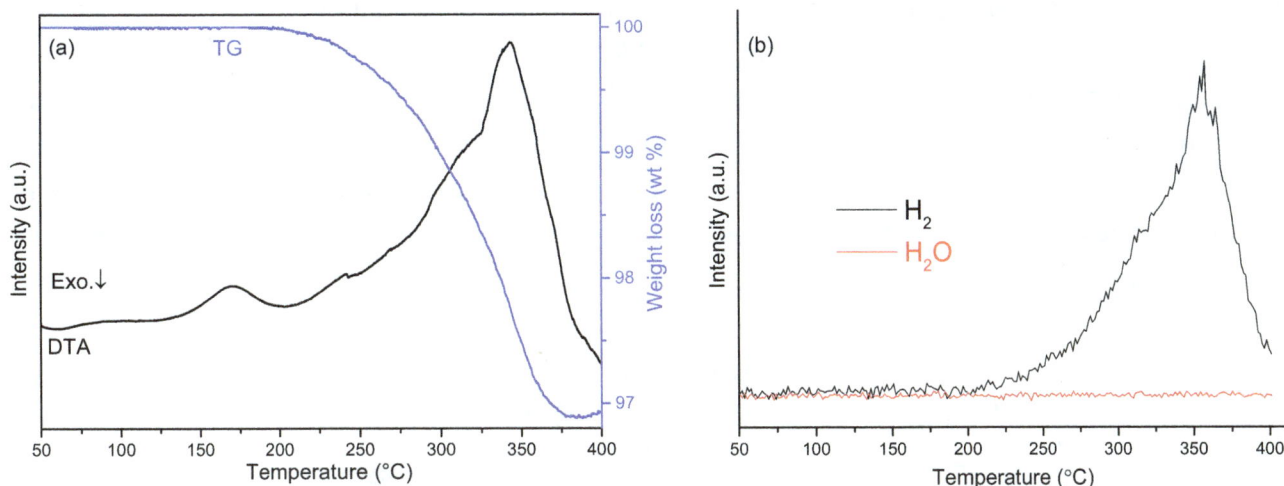

Figure 6. Simultaneous TG-DTA-MS profiles of sample 5: (**a**) TG-DTA traces; (**b**) mass spectra for hydrogen and water.

2.3. In-Situ Synchrotron Powder Diffraction

In-situ synchrotron PXD methods were used to clarify the reaction mechanism underpinning the first endothermic event at *ca.* 170 °C by heating sample 5 from room temperature to 260 °C. Figure 7 shows selected regions of the *in-situ* diffraction patterns for sample 5, the relevant peak indices and a plot of the resulting cell volume changes of the NaH and NaOH phases, respectively, as a function of temperature. Structure refinements performed against the synchrotron PXD data were conducted for each data set (see supplementary information). A fundamental parameter (FP) approach was employed in TOPAS to perform whole-pattern profile fitting of the diffraction data collected in transmission mode. The diffraction background was fitted with Chebychev functions. Structural data from the ICDD PDF4 (2014) database for NaH (02-0809) and α-NaOH (078-0188) were used as starting models in TOPAS. At room temperature NaH crystallizes in cubic space group $Fm\overline{3}m$ (No.225; $a = 4.8826(1)$ Å), while α-NaOH is orthorhombic (*Cmcm*, No. 63; $a = 3.4039(1)$ Å, $b = 3.4011(1)$ Å, $c = 11.3901(1)$ Å).

Before heating, room temperature PXD data show that sample 5 consists only of NaH (44 wt%) and NaOH (56 wt%). There are no impurity diffraction peaks observed in the PXD pattern. As shown in Figure 7, below 110 °C, reflections for both NaH and NaOH shift slightly to lower angle with increasing temperature. This shift corresponds to the expected thermal expansion of the NaH and NaOH lattices. The cell volume for NaH increases from *ca.* 116.4 Å^3 to *ca.* 118.5 Å^3 over this temperature range, corresponding to a volume expansion of approximately 1.8%. Over the same temperature range, the cell volume for NaOH increases by approximately 0.99% and the phase fractions of NaH and NaOH remain effectively constant, indicating that no reaction occurs between the hydride and hydroxide.

From 110 °C, all the NaH peaks become increasingly asymmetric with some evidence of reflections emerging at slightly higher 2θ values to the main peaks as manifested by a broadening tail to all NaH reflections. Meanwhile, the NaH peaks shift continuously to lower 2θ angles and the hydride cell volume increases significantly such that at 180 °C the value is approximately 7.2% larger compared to that at room temperature (Figure 7e). Most significantly, however, starting from 190 °C, the NaH peaks become asymmetric at lower diffraction angle while simultaneously the NaOH diffraction peaks become weaker. As can be seen clearly for the NaH 200 reflection at $2\theta \sim 22.4°$ (Figure 7b), as the temperature increases a new peak starts to appear at a slightly lower 2θ angle. Given that every NaH peak splits in the same sense and that all the original reflections are also retained, the phenomenon indicates the formation of a slightly larger unit cell with a structure type common to the original NaH phase. This indicates that a new NaH-like phase with slightly larger lattice parameters is formed as NaOH becomes depleted.

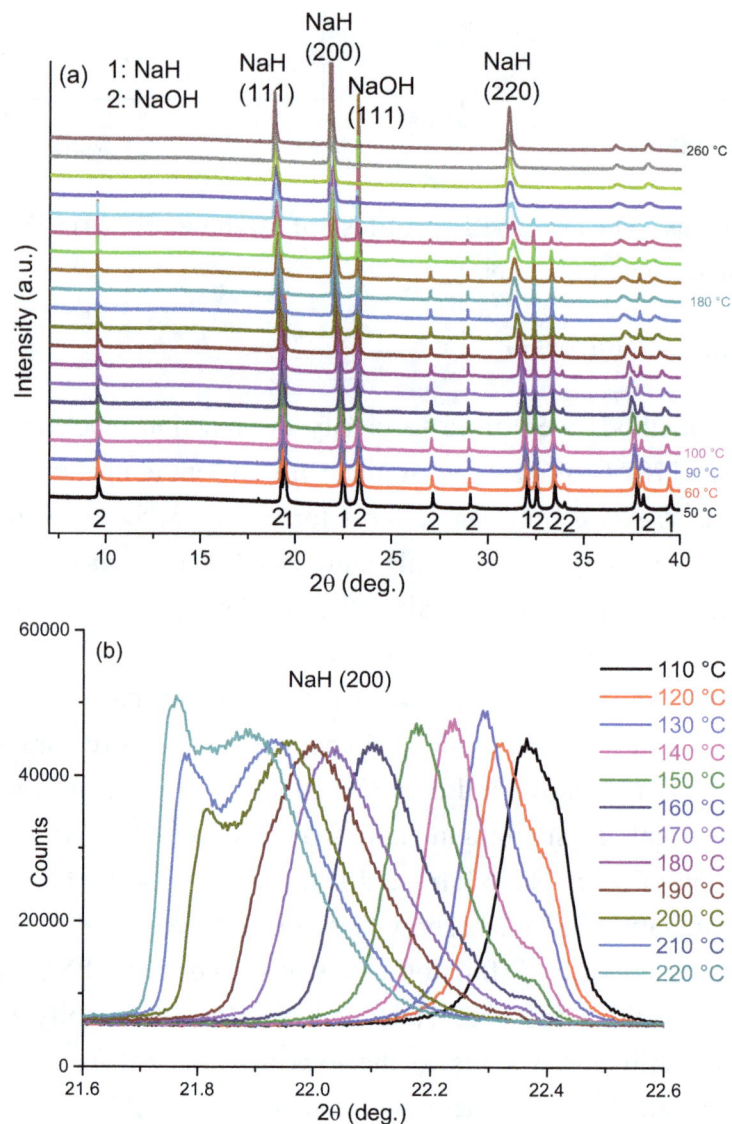

Figure 7. *Cont.*

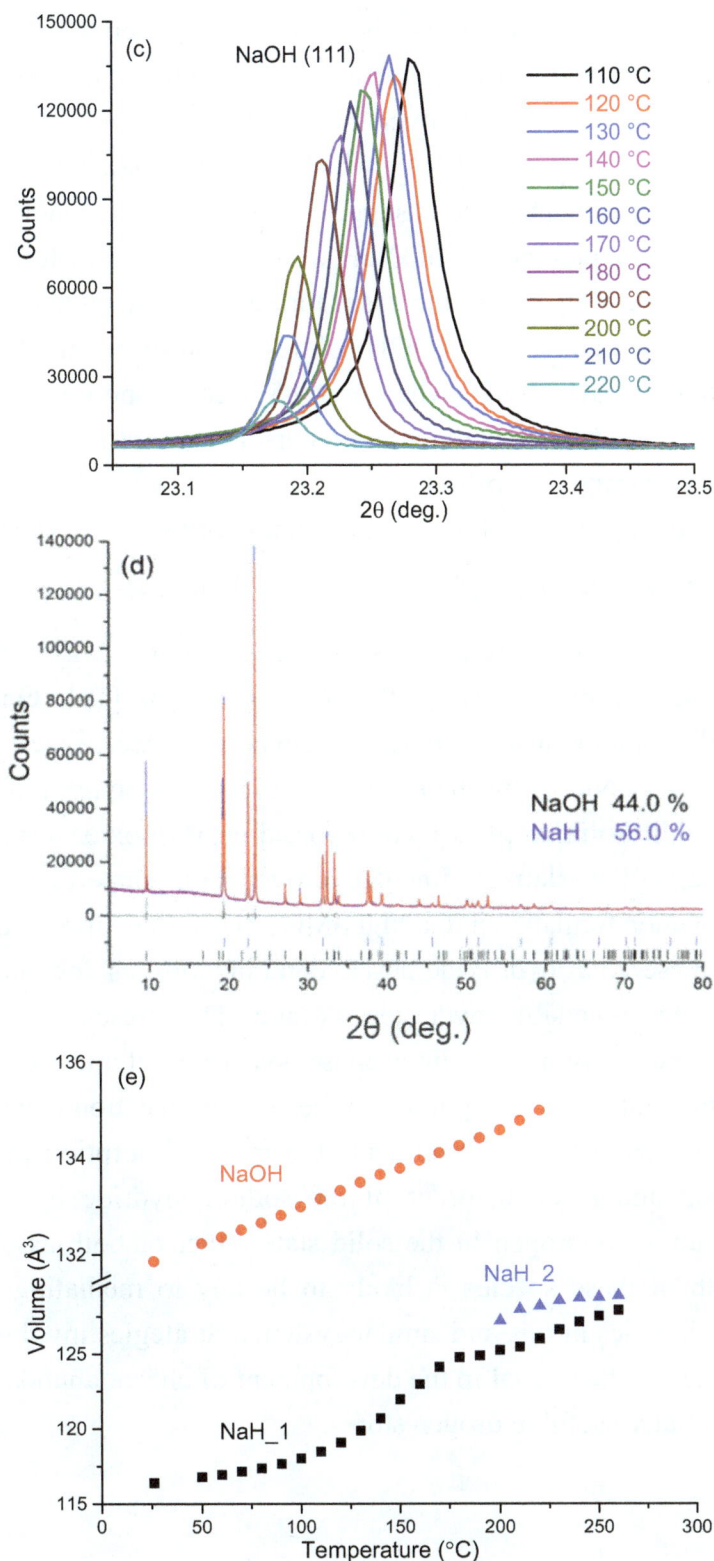

Figure 7. *In-situ* synchrotron PXD patterns for sample 5 heated from room temperature to 260 °C with a constant heating rate of 10 °C min^{-1} in a closed quartz capillary under Ar atmosphere ($\lambda = 0.9533$ Å) showing: (**a**) the region from $7° \leq 2\theta \leq 40°$; (**b**) the change in the NaH (200) reflections from 110 to 220 °C; (**c**) the change in the NaOH (111) reflections from 110 to 220 °C; (**d**) the refinement profile at 100 °C; and (**e**) cell volume against temperature for NaH ("NaH_1"), "Na-O-H" ("NaH_2") and NaOH, respectively, from room temperature to 260 °C.

By 240 °C there is no evidence of NaOH in the diffraction patterns and the NaH-like phase reflections become more symmetric and sharper, indicative of a single phase. Also noteworthy is that there is no evidence for the formation of Na_2O over the entire temperature range of the experiment. The cell volume of the hydride at 240 °C is 127.1 $Å^3$, which is approximately 9.2% larger as compared to the room temperature NaH structure. The results suggest that a structural change begins from 110 °C and that by 190 °C NaOH reacts appreciably with NaH in the solid state to form a complex hydride with an NaH-type structure. As the temperature increases further, a single composition of the NaH-like phase is formed. Thus we propose that an NaH-NaOH solid solution forms in which up to 50% of the hydride is replaced by hydroxide; $NaH_{1-x}(OH)_x$ where $x \leq 0.5$. Hence, one might reasonably speculate that the formation of the hydride-hydroxide is the vital precursor to an *intraphase* $H^{\delta}\ldots H^{\delta-}$ interaction and the hydrogen evolution step during desorption.

The process could thus be represented by the modified version of the reaction equation below:

$$NaH + NaOH \rightarrow 2NaH_{0.5}(OH)_{0.5} \rightarrow Na_2O + H_2 \qquad (2)$$

When compared with other $AH\text{-}AOH(AOH)_2$ systems (A = Li, K, Mg), similar solid solution behavior has only been observed previously in the K-O-H system [19]. On heating, however, KOH-KH does not apparently combine to form K_2O, but rather KH decomposes independently while KOH remains to 500 °C [17]. By contrast, in the LiOH-LiH system, although the final decomposition reaction product is Li_2O, no solid solution phases are reported prior to oxide formation upon heating, which might be a consequence of the relatively low dehydrogenation temperature [17,27,28]. There is no strong evidence for solid solution formation in the $MgH\text{-}Mg(OH)_2$ system prior to dehydrogenation [29]. Na-O-H is unique among these four hydroxide-based combinations as the only system with the appropriate thermodynamics for reversible hydrogen storage. The present work suggests that for hydroxide-hydride materials not only is the inter-phase synergic interaction of protic hydrogen (in NaOH) and hydridic hydrogen (in NaH) important in the dehydrogenation mechanism, but that also an intra-phase $H^{\delta+}\ldots H^{\delta-}$ interaction may be a crucial step in the desorption process. Furthermore, the ensuing lattice expansion and anion disorder of the sodium hydride hydroxide could play a significant role in the diffusion of hydrogen in the solid state either or both as protons and hydride. The mobility of one or both of these species is likely to be key to mediating and controlling the hydrogen uptake and release kinetics in this and similar systems. Strategies involving nanostructuring, additives and catalysts are likely to be crucial in the development of cheap, abundant materials systems such as hydroxides into potentially useful hydrogen stores.

3. Experimental Section

Na_2O (Alfa Aesar, anhydrous, 90%), NaH (Sigma Aldrich, dry, 95%), and (α-)NaOH (Sigma Aldrich, reagent grade, 97%) were used as received. Before the hydrogenation or dehydrogenation experiments, Na_2O and NaH-NaOH mixtures (employing different NaH:NaOH molar ratios as indicated elsewhere in the text), respectively, were milled using a 50 mL stainless steel milling jar under argon atmosphere in a Retsch PM100 planetary ball mill. Milling was performed at 400 rpm for 1 min in one direction, paused for 1 min and then reversed to give a total milling duration of 1 h.

A ball:powder mass ratio of approximately 50:1 was employed throughout. All manipulations were performed in an Ar-filled recirculating glovebox (Saffron Scientific, 1 ppm H$_2$O, 1 ppm O$_2$).

Room temperature *ex-situ* powder X-ray diffraction (PXD) experiments were conducted with a Bruker D8 powder diffractometer in transmission geometry with spinning sealed capillaries. Diffraction data for phase identification were typically collected over $5° \leq 2\theta \leq 85°$ with a $0.017°$ step size and scan times of 1 or 10 h.

Synchrotron PXD data were collected using incident radiation with $\lambda = 0.9533$ Å using a Mython-II detector at the powder diffraction beamline, Australian Synchrotron. Time-resolved *in situ* high temperature measurements were conducted with a flow cell under an atmosphere of 1 bar argon (99.99%) using a Cyberstar hot air blower to heat the quartz capillary from 50 °C to 260 °C at a constant heating rate of 10 °C min^{-1}. Data were collected with an exposure time of 150 s at every 10 °C step. Data analysis was performed using the TOPAS 4.2 software package [30].

Hydrogen absorption experiments were performed on an intelligent gravimetric analyzer (IGA, Hiden, Warrington, UK) with samples of ~50 mg contained in stainless steel sample holders. Hydrogen gas (BOC, 99.98%; Motherwell, UK) at 18 bar was introduced into the reaction chamber and absorption was performed between 20 and 400 °C with a fixed ramp rate of 3 °C min^{-1}. The variation in sample mass and temperature with time was recorded.

Thermal behavior of NaH-NaOH samples was analyzed via simultaneous thermogravimetric-differential thermal analysis with coupled mass spectrometry (TG-DTA-MS; Netzsch STA 409 coupled to a Hiden Analytical HPR20 mass spectrometer; Selb, Germany) in an Ar-filled recirculating glovebox (MBraun UniLab; Garching, Germany, 1 ppm H$_2$O, 1 ppm O$_2$). Samples of *ca.* 20 mg were loaded into alumina pans and heated to 673 K under a flow of Ar gas at a rate of 5 °C min^{-1} in order to achieve complete decomposition of the samples.

4. Conclusions

In summary, the Na-O-H reversible hydrogen storage system has been examined in detail. The (de)hydrogenation behavior of the Na-O-H system was investigated by means of IGA measurements, simultaneous TG-DTA-MS, XRD, and *in-situ* synchrotron PXD. Na$_2$O starts to absorb hydrogen at temperatures slightly above ambient (30–50 °C) under a relatively low hydrogen pressure (18 bar). The hydrogenated products, NaH and NaOH, release hydrogen at a lower temperature than the binary hydride NaH. Moreover, differential thermal analysis shows an endothermic event near 170 °C in the NaH-NaOH system that cannot be associated with the dehydrogenation reaction of the mixture or with phase transitions of either of the individual components. *In-situ* synchrotron PXD results show that NaOH forms a solid solution with NaH that persists until 240 °C. The NaH and Na(H,OH) cubic phases demonstrate a large volume expansion in the temperature range of 160–240 °C. The resulting predominantly *intraphase* interaction of protic hydrogen and hydridic hydrogen provides a likely driving force for the subsequent dehydrogenation in the hydride-hydroxide system.

Acknowledgments

The research post for JM has received funding from the European Union's Seventh Framework 610 Programme (FP7/2007-2013) for the Fuel Cells and Hydrogen Joint Technology Initiative under Grant

611 Agreement number 303447. Part of the experiment was conducted at the Powder diffraction beamline at the Australian Synchrotron.

Author Contributions

Jianfeng Mao performed the synthesis, PXD and hydrogen performance characterization; Qinfen Gu performed the synchrotron measurements; Duncan H. Gregory was responsible for planning and supervising the research. All authors discussed the results and co-wrote the manuscript.

Conflicts of Interest

The authors declare no conflict of interest.

References

1. Hadjipaschalis, I.; Poullikkas, A.; Efthimiou, V. Overview of current and future energy storage technologies for electric power applications. *Renew. Sust. Energy Rev.* **2009**, *13*, 1513–1522.
2. Schlapbach, L.; Züttel, A. Hydrogen-storage materials for mobile applications. *Nature* **2001**, *414*, 353–358.
3. Mandal, T.K.; Gregory, D.H. Hydrogen: Future energy vector for sustainable development. *Proc. Inst. Mech. Eng. Part C J. Mech. Eng. Sci.* **2010**, *224*, 539–558.
4. Schüth, F.; Bogdanović, B.; Felderhoff, M. Light metal hydrides and complex hydrides for hydrogen storage. *Chem. Commun.* **2004**, *20*, 2249–2258.
5. Orimo, S.; Nakamori, Y.; Eliseo, J.; Züttel, A.; Jensen, C.M. Complex hydrides for hydrogen storage. *Chem. Rev.* **2007**, *107*, 4111–4132.
6. Mandal, T.K.; Gregory, D.H. Hydrogen storage materials: Present scenarios and future directions. *Ann. Rep. Prog. Chem. Sect. A Inorg. Chem.* **2009**, *105*, 21–54.
7. Reardon, H.; Hanlon, J.; Hughes, R.W.; Godula-Jopek, A.; Mandal, T.K.; Gregory, D.H. Emerging concepts in solid-state hydrogen storage: The role of nanomaterials design. *Energy Env. Sci.* **2012**, *5*, 5951–5979.
8. Hamilton, C.W.; Baker, R.T.; Staubitzc, A.; Manners, I. B–N compounds for chemical hydrogen storage. *Chem. Soc. Rev.* **2009**, *38*, 279–293.
9. Targets for Onboard Hydrogen Storage Systems for Light-Duty Vehicles. Available online: http://energy. gov/sites/prod/files/2014/03/f12/targets_onboard_hydro_storage.pdf (accessed on 23 April 2015).
10. Ibrahim, H.; Ilinca, A.; Perron, J. Energy storage systems—Characteristics and comparisons. *Renew. Sust. Energy Rev.* **2008**, *12*, 1221–1250.
11. Mao, J.F.; Gregory, D.H. Recent advances in the use of sodium borohydride as a solid state hydrogen store. *Energies* **2015**, *8*, 430–453.
12. Mao, J.F.; Guo, Z.P.; Liu, H.K. Improved hydrogen sorption performance of NbF_5–catalysed $NaAlH_4$. *Int. J. Hydrogen Energy* **2011**, *36*, 14503–14511.
13. Mao, J.F.; Guo, Z.P.; Nevirkovets, I.P.; Liu, H.K.; Dou, S.X. Hydrogen De-/Absorption improvement of $NaBH_4$ catalyzed by Titanium-based additives. *J. Phys. Chem. C* **2012**, *116*, 1596–1604.

14. Mao, J.F.; Guo, Z.P.; Liu, H.K.; Dou, S.X. Reversible storage of hydrogen in NaF-MB$_2$ (M = Mg, Al) composites. *J. Mater. Chem. A* **2013**, *1*, 2806–2811.

15. Mao, J.F.; Yu, X.B.; Guo, Z.P.; Liu, H.K.; Wu, Z.; Ni, J. Enhanced hydrogen storage performances of NaBH$_4$–MgH$_2$ system. *J. Alloys Compd.* **2009**, *479*, 619–623.

16. Xu, Q.; Wang, R.T.; Kiyobayashi, T.; Kuriyama, N.; Kobayashi, T. Reaction of hydrogen with sodium oxide—A reversible hydrogenation/dehydrogenation system. *J. Power Sources* **2006**, *155*, 167–171.

17. Yu, P.; Chua, Y.S.; Cao, H.J.; Xiong, Z.T.; Wu, G.T.; Chen, P. Hydrogen storage over alkali metal hydride and alkali metal hydroxide composites. *J. Energ. Chem.* **2014**, *23*, 414–419.

18. Gnanasekaran, T. Thermochemistry of binary Na-NaH and ternary Na-O-H systems and the kinetics of reaction of hydrogen/water with liquid sodium—A review. *J. Nucl. Mater.* **1999**, *274*, 252–272.

19. Mikheeva, V.I.; Shkrabkina, M.M. Solid solutions in the NaOH-NaH and KOH-KH systems. *Russ. J. Inorg. Chem.* **1962**, *7*, 1251–1255.

20. Jansson, S.A. Thermochemistry and solution chemistry in the sodium-oxygen-hydrogen system. In *Corrosion by Liquid Metals*, 2nd ed.; Draley, J.E., Weeks, J.R., Eds.; Springer: New York, NY, USA, 1970; Session V, pp. 523–560.

21. Myles, K.M.; Cafasso, F.A. The reciprocal ternary system Na-NaOH-Na$_2$O-NaH. *J. Nucl. Mater.* **1977**, *67*, 249–253.

22. Veleckis, E.; Leibowitz, L. Phase relations for reactions of hydrogen with sodium oxide between 500 and 900 °C. *J. Nucl. Mater.* **1987**, *144*, 235–243.

23. Sorte, E.G.; Majzoub, E.H.; Ellis-Caleo, T.; Hammann, B.A.; Wang, G.; Zhao, D.X.; Bowman, R.C., Jr.; Conradi, M.S. Effects of NaOH in Solid NaH: Solution/Segregation Phase Transition and Diffusion Acceleration. *J. Phys. Chem. C* **2013**, *117*, 23575–23581.

24. David, W.I.F.; Jones, M.O.; Gregory, D.H.; Jewell, C.M.; Johnson, S.R.; Walton, A.; Edwards, P.P. A mechanism for non-stoichiometry in the Lithium Amide/Lithium imide hydrogen storage reaction. *J. Am. Chem. Soc.* **2007**, *129*, 1594–1601.

25. Bleif, H.-J.; Dachs, H. Crystalline modifications and structural phase transitions of NaOH and NaOD. *Acta. Cryst. A* **1982**, *38*, 470–476.

26. Douglas, T.B.; Dever, J.L. Anhydrous sodium hydroxide: The heat content from 0 to 700 °C, the transition temperature, and the melting point. *J. Res. Natl. Bur. Stand.* **1954**, *53*, 81–90.

27. Kiat, J.M.; Boemare, G.; Rieu, B.; Aymes, D. Structural evolution of LiOH: Evidence of a solid–solid transformation toward Li$_2$O close to the melting temperature. *Solid State Commun.* **1998**, *108*, 241–245.

28. Vajo, J.J.; Skeith, S.L.; Mertens, F.; Jorgensen, S.W. Hydrogen-generating solid-state hydride/hydroxide reactions. *J. Alloys Compd.* **2005**, *390*, 55–61.

29. Leardini, F.; Ares, J.R.; Bodega, J.; Fernández, J.F.; Ferrer, I.J.; Sánchez, C. Reaction pathways for hydrogen desorption from magnesium hydride/hydroxide composites: Bulk and interface effects. *Phys. Chem. Chem. Phys.* **2010**, *12*, 572–577.

30. *User Manual*; Bruker AXS: Karlsruhe, Germany, 2005.

Permissions

List of Contributors

Wai Hong Lee
Nanotechnology & Catalysis Research Centre (NANOCAT), Institute of Postgraduate Studies (IPS), University of Malaya, Kuala Lumpur 50603, Malaysia

Chin Wei Lai
Nanotechnology & Catalysis Research Centre (NANOCAT), Institute of Postgraduate Studies (IPS), University of Malaya, Kuala Lumpur 50603, Malaysia

Sharifah Bee Abd Hamid
Nanotechnology & Catalysis Research Centre (NANOCAT), Institute of Postgraduate Studies (IPS), University of Malaya, Kuala Lumpur 50603, Malaysia

Mubarak Y. A. Yagoub
Department of Physics, University of the Free State, PO Box 339, Bloemfontein, ZA 9300, South Africa
Department of Physics, Sudan University of Science and Technology, Khartoum 11113, Sudan

Hendrik C. Swart
Department of Physics, University of the Free State, PO Box 339, Bloemfontein, ZA 9300, South Africa

Luyanda L. Noto
Department of Physics, University of the Free State, PO Box 339, Bloemfontein, ZA 9300, South Africa

Peber Bergman
Department of Physics, Chemistry and Biology, Linköping University, Linköping S-581 83, Sweden

Elizabeth Coetsee
Department of Physics, University of the Free State, PO Box 339, Bloemfontein, ZA 9300, South Africa

Seongkyum Kim
Department of Civil Engineering, Kongju National University, Cheonan 330-717, Korea

Kwanho Lee
Department of Civil Engineering, Kongju National University, Cheonan 330-717, Korea

Piergiorgio Gentile
School of Clinical Dentistry, University of Sheffield, 19 Claremont Crescent, Sheffield S10 2TA, UK

Caroline J. Wilcock
School of Clinical Dentistry, University of Sheffield, 19 Claremont Crescent, Sheffield S10 2TA, UK

Cheryl A. Miller
School of Clinical Dentistry, University of Sheffield, 19 Claremont Crescent, Sheffield S10 2TA, UK

Robert Moorehead
School of Clinical Dentistry, University of Sheffield, 19 Claremont Crescent, Sheffield S10 2TA, UK

Paul V. Hatton
School of Clinical Dentistry, University of Sheffield, 19 Claremont Crescent, Sheffield S10 2TA, UK

Chao-Ming Hsu
Department of Mechanical Engineering, National Kaohsiung University of Applied Science, No. 415 Chien Kung Road, Kaohsiung 807, Taiwan

Wen-Cheng Tzou
Department of Electro-Optical Engineering, Southern Taiwan University, No. 1, Nan-Tai Street, Yungkang Dist., Tainan City 710, Taiwan

Cheng-Fu Yang
Department of Chemical and Materials Engineering, National University of Kaohsiung, No. 700 Kaohsiung University Road, Nan-Tzu District, Kaohsiung 811, Taiwan

Yu-Jhen Liou
Department of Chemical and Materials Engineering, National University of Kaohsiung, No. 700 Kaohsiung University Road, Nan-Tzu District, Kaohsiung 811, Taiwan

Jing Liu
School of Information Engineering, Jimei University, Xiamen 361021, China
China–Australia Joint Laboratory for Functional Nanomaterials, Xiamen University, Xiamen 361005, China

Yushan Chen
School of Information Engineering, Jimei University, Xiamen 361021, China

Haoyuan Cai
School of Information Engineering, Jimei University, Xiamen 361021, China

Xiaoyi Chen
Department of Physics, National University of Singapore, Singapore 117551, Singapore

Changwei Li
China–Australia Joint Laboratory for Functional Nanomaterials, Xiamen University, Xiamen 361005, China

Cheng-Fu Yang
Department of Chemical and Materials Engineering, National University of Kaohsiung, No. 700, Kaohsiung University Rd., Nan-Tzu District, Kaohsiung 811, Taiwan

Ming La
College of Chemistry and Chemical Engineering, Pingdingshan University, Pingdingshan 467000, Henan, China

Lin Liu
College of Chemistry and Chemical Engineering, Anyang Normal University, Anyang 455000, Henan, China

Bin-Bin Zhou
College of Chemistry and Chemical Engineering, Pingdingshan University, Pingdingshan 467000, Henan, China
College of Chemistry and Chemical Engineering, Anyang Normal University, Anyang 455000, Henan, China

Yuanfeng Yang
School of Materials, the University of Manchester, Manchester M13 9PL, UK

Robert Akid
School of Materials, the University of Manchester, Manchester M13 9PL, UK

Kevin E. Bennet
Division of Engineering, Mayo Clinic, Rochester, MN 55905, USA
Department of Neurologic Surgery, Mayo Clinic, Rochester, MN 55905, USA

Kendall H. Lee
Department of Neurologic Surgery, Mayo Clinic, Rochester, MN 55905, USA

Jonathan R. Tomshine
Department of Neurologic Surgery, Mayo Clinic, Rochester, MN 55905, USA

Emma M. Sundin
Department of Physics, University of Texas at El Paso, El Paso, TX 79968, USA

James N. Kruchowski
Division of Engineering, Mayo Clinic, Rochester, MN 55905, USA

William G. Durrer
Department of Physics, University of Texas at El Paso, El Paso, TX 79968, USA

Bianca M. Manciu
Department of Physics, University of Texas at El Paso, El Paso, TX 79968, USA

Abbas Kouzani
School of Engineering, Deakin University, Waurn Ponds, Victoria 3216, Australia

Felicia S. Manciu
Department of Physics, University of Texas at El Paso, El Paso, TX 79968, USA

Siqi Huan
College of Material Science and Engineering, Northeast Forestry University, Harbin 150040, China

Guoxiang Liu
College of Material Science and Engineering, Northeast Forestry University, Harbin 150040, China

Guangping Han
College of Material Science and Engineering, Northeast Forestry University, Harbin 150040, China

Wanli Cheng
College of Material Science and Engineering, Northeast Forestry University, Harbin 150040, China

Zongying Fu
College of Material Science and Engineering, Northeast Forestry University, Harbin 150040, China

Qinglin Wu
School of Renewable Natural Resources, Louisiana State University Agricultural Center, Baton Rouge, LA 70803, USA

Qingwen Wang
College of Material Science and Engineering, Northeast Forestry University, Harbin 150040, China

Sudeok Shon
School of Architectural Engineering, Korea University of Technology and Education, Cheonan 330-708, Korea

Seungjae Lee
School of Architectural Engineering, Korea University of Technology and Education, Cheonan 330-708, Korea

Junhong Ha
School of Liberal Arts, Korea University of Technology and Education, Cheonan 330-708, Korea

Changgeun Cho
School of Architecture, Chosun University, Gwangju 501-759, Korea

Paulina Dobrowolska
Institute of Optoelectronics, Military University of Technology, Warsaw 00-908, Poland

Aleksandra Krajewska
Institute of Optoelectronics, Military University of Technology, Warsaw 00-908, Poland

Magdalena Gajda-Rączka
Institute of Optoelectronics, Military University of Technology, Warsaw 00-908, Poland

Bartosz Bartosewicz
Institute of Optoelectronics, Military University of Technology, Warsaw 00-908, Poland

Piotr Nyga
Institute of Optoelectronics, Military University of Technology, Warsaw 00-908, Poland

Bartłomiej J. Jankiewicz
Institute of Optoelectronics, Military University of Technology, Warsaw 00-908, Poland

Godfrey Keru
School of Chemistry and Physics, University of KwaZulu-Natal, Private Bag X54001, Durban 4000, South Africa

Patrick G. Ndungu
School of Chemistry and Physics, University of KwaZulu-Natal, Private Bag X54001, Durban 4000, South Africa
Department of Applied Chemistry, University of Johannesburg, P.O. Box 17011, Doornfontein, Johannesburg 2028, South Africa

Genene T. Mola
School of Chemistry and Physics, University of KwaZulu-Natal, Private Bag X54001, Durban 4000, South Africa

Vincent O. Nyamori
School of Chemistry and Physics, University of KwaZulu-Natal, Private Bag X54001, Durban 4000, South Africa

Jianfeng Mao
WestCHEM, School of Chemistry, Joseph Black Building, University of Glasgow, Glasgow G12 8QQ, UK

Qinfen Gu
Australian Synchrotron, Clayton, Victoria 3168, Australia

Duncan H. Gregory
WestCHEM, School of Chemistry, Joseph Black Building, University of Glasgow, Glasgow G12 8QQ, UK